Bacterial Cell Surface Techniques

MODERN MICROBIOLOGICAL METHODS

Series Editor Michael Goodfellow, *Department of Microbiology, University of Newcastle-upon-Tyne*

Methods in Aquatic Bacteriology (1988)
Edited by Brian Austin

0 471 91651 X

Bacterial Cell Surface Techniques (1988)
Ian Hancock and Ian Poxton

0 471 91041 4

Bacterial Cell Surface Techniques

Ian Hancock
Department of Microbiology
University of Newcastle-upon-Tyne

and

Ian Poxton
Department of Bacteriology
University of Edinburgh

with contributions from
A. R. Archibald, D. R. Bundle, G. Dougan, B. Glauner, J. E. Heckels,
P. A. Lambert, P. Messner, D. E. Minnikin, D. C. Old, D. Parratt,
R. R. B. Russell, U. Schwartz, U. B. Sleytr, M. Virji

A Wiley–Interscience Publication

JOHN WILEY & SONS
Chichester · New York · Brisbane · Toronto · Singapore

Library of Congress Cataloging-in-Publication Data

Bacterial cell surface techniques.

(Modern microbiological methods)
'A Wiley–Interscience publication.'
1. Bacterial cell walls—Research—Technique.
I. Hancock, Ian. II. Poxton, Ian. III. Series.
QR77.3.B296 1988   589.9'0875   87-27932
ISBN 0 471 91041 4

British Library Cataloguing in Publication Data

Bacterial cell surface techniques.—(Modern
microbiological methods).
1. Bacterial cell walls 2. Bacteriology
—Technique
I. Hancock, Ian II. Poxton, Ian
III. Series
589.9'0874'028   QR77.3

ISBN 0 471 91041 4

Typeset by Acorn Bookwork, Salisbury, Wiltshire
Printed and bound in Great Britain by
Bath Press Ltd., Bath, Avon

Contents

Contributors

A. R. Archibald *Department of Microbiology, Medical School, University of Newcastle, Newcastle upon Tyne NE2 4HH, UK*

D. R. Bundle *Immunochemistry Section, Division of Biological Sciences, National Research Council of Canada, Ottawa, Ontario K1A OR6, Canada*

G. Dougan *Wellcome Biotech, Langley Court, Beckenham, Kent BR3 3BS, UK*

B. Glauner and **U. Schwartz** *Max-Planck Institüt fur Entwicklungsbiologie, Abteilung Biochemie, Spemannstrasse 35, D-7400 Tübingen, West Germany*

J. E. Heckels and **M. Virji** *Department of Microbiology, University of Southampton Medical School, Southampton General Hospital, Southampton SO9 4XY, UK*

P. A. Lambert *Microbiology Research Group, Department of Pharmaceutical Sciences, Aston University, Birmingham B4 7ET, UK*

P. Messner and **U. B. Sleytr** *Zentrum für Ultrastrukturforschung, Universität fur Bodenkultur, A-1180 Wien, Austria*

D. E. Minnikin *Department of Organic Chemistry, University of Newcastle, Newcastle upon Tyne NE1 7RU, UK*

D. C. Old *Microbiology Department, University of Dundee Medical School, Ninewells Hospital, Dundee DD1 9SY, Scotland*

D. Parratt *Microbiology Department, University of Dundee Medical School, Ninewells Hospital, Dundee DD1 9SY, Scotland*

R. R. B. Russell *Dental Research Unit, Royal College of Surgeons of England, Downe, Orpington, Kent BR6 7JJ, UK*

Series Preface

The science of microbiology owes its existence as well as its underlying principles to the talent and practical prowess of pioneers such as Leeuwenhoek, Pasteur, Koch and Beijerinck. Interest in microbiology has recently increased quite significantly given the exciting developments in genetics and molecular biology and the growth of microbial technology. There was a time when most microbiologists were acquainted with many of the techniques used in microbiology. It is, however, now becoming increasingly difficult for research workers to keep abreast of the bewildering range of techniques currently used in microbiological laboratories. This problem is compounded by the fact that scientists in any one field increasingly need to apply techniques developed in other scientific disciplines.

The series 'Modern Microbiological Method' aims to identify specialist areas in microbiology and provide up-to-date methodological handbooks to aid microbiologists at the laboratory bench. The books will be directed primarily towards active research workers but will be structured so as to serve as an introduction to the methods within a speciality for graduate students and scientists entering microbiology from related disciplines. Protocols will not only be described but difficulties and limitations of techniques and questions of interpretation fully discussed.

In summary, this series of books is designed to help stimulate further developments in microbiology by promoting the use of new and updated methods. Both authors and the editor-in-chief will be grateful to hear from satisfied or dissatisfied users so that future books in the series can benefit from the informed comment of practitioners in the field.

MICHAEL GOODFELLOW

Preface

Perhaps since 1884, when Gram developed his staining technique, the bacterial cell wall has had a major influence on the development of microbiology. Early in this century it became clear that certain surface components such as capsules and endotoxin had roles in bacterial virulence. Great advances in understanding the structure and chemistry of bacterial surfaces were made in the 1950s and 1960s due to the developments in electron microscopy and carbohydrate chemistry.

In the past decade advances in analytical chemistry and immunochemistry have revealed unexpected heterogeneity and complexity in cell surface macromolecules. This has coincided with the development of genetic techniques that offer the possibility of manipulating the expression of surface components.

It is not just in the areas of medical microbiology that there is an interest in cell surfaces. An understanding of surface properties is required for adequate control of adhesion, adsorption and cell disruption in biotechnological processes, and surface components are increasingly being recognized as valuable chemotaxonomic markers for the identification and classification of bacteria.

The aim of this book is to assemble our present day understanding of the chemistry, biochemistry and immunochemistry of bacterial cell surfaces and to describe the techniques by which the individual components can be isolated, purified and analysed. It is very much a laboratory manual which will, we hope, be of equal use to the novice and the expert. This is the first time that all of these techniques have been collected under one cover.

IAN C. HANCOCK
IAN R. POXTON
December 1987

Acknowledgements

We are extremely grateful to those colleagues who have contributed to this book in areas where we ourselves have not had the necessary expertise.

We also thank Dr Sheila Patrick and Marie Byrne for several of the electronmicrographs, Dr Gary Cousland for help in the development of some of the immunochemical methods, and Clive Mellstrom in Newcastle and Robert Brown in Edinburgh for excellent and extensive technical assistance.

Much of our expertise has been gained during the tenure of grants from the Medical Research Council, the Science and Engineering Research Council, and the Scottish Home and Health Department to whom we give our thanks.

Finally we thank Michael Dixon and Patricia Sharp at John Wiley and Sons, Chichester, for help and encouragement at all stages of the production of the book.

Bacterial Cell Surface Techniques
I. Hancock and I. Poxton
© 1988 John Wiley & Sons Ltd.

1

Structure of Bacteria and their Envelopes

With contributions from
U. B. Sleytr, P. Messner, D. E. Minnikin, J. E. Heckels
M. Virji, and R. B. B. Russell

A. INTRODUCTION—BACTERIAL CELL WALLS

With the exception of Mollicutes (Mycoplasmas, *Spiroplasma*) the cell envelopes of prokaryotic microorganisms are characterized by the presence of two distinct components: an inner **cytoplasmic membrane**, which controls the substrate and electron transport processes of the cell and which is the site of biosynthesis of extracellular macromolecules; and a strong, outer **cell wall**, which maintains the shape of the cell and protects the mechanically fragile cytoplasmic membrane from rupture owing to the high osmotic pressure exerted on it by the cell cytoplasm. These components can be separated *in vitro* and are seen as two distinct layers by electron microscopy, in most bacteria (Figure 1.1). In glutaraldehyde-fixed specimens the wall can be seen to make intimate contact with the cytoplasmic membrane, however, at sites of cell wall synthesis, where nascent wall polymers may form transient covalent bridges between synthetic complexes in the membrane and sites of incorporation into the wall. These localized interactions have proved strong enough to permit the isolation of specific contact regions—'Bayer junctions'—in some Gram-negative bacteria. Recently, Cook *et al.* (1986) presented evidence for localized adhesion zones on either side of the nascent division septum, that effectively compartmentalize the periplasm at the division site to form 'periseptal' compartments.

Although the walls of all bacteria share the roles of providing mechanical strength, maintaining cell shape and, possibly, molecular sieving, they exhibit a wide diversity of structures and compositions (Table 1.1).

B. WALLS BASED ON PEPTIDOGLYCAN

The most widely distributed types, and the most thoroughly studied, are the walls of the typical Gram-positive and Gram-negative eubacteria,

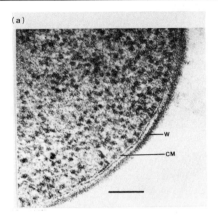
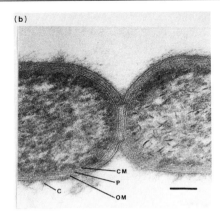

Figure 1.1 Electron micrographs of negatively stained thin sections of Gram-positive and Gram-negative bacteria. (a) A typical Gram-positive cell, *Clostridium perfringens*. CM: cytoplasmic membrane; W: wall. Bar marker: 0.1 μm. (b) A typical Gram-negative organism, *Bacteroides fragilis*, in the process of cell division. OM: outer membrane; CM: cytoplasmic membrane; P: peptidoglycan or 'rigid layer'; C: capsule. Bar marker: 0.1 μm. See Appendix 1 for staining details.

Table 1.1 Types of cell wall found in bacteria.

Eubacteria	
Mycoplasmas, 'Mollicutes'	No wall
Gram-positive, 'Firmicutes'	Thick, cross-linked peptidoglycan layer, with covalently linked anionic accessory polymers
Gram-negative, 'Gracilicutes'	Thin, cross-linked peptidoglycan layer (murein), with outer membrane
Pasteuria/Pirella	Aggregate of protein subunits
Achaebacteria	
Sulfolobus Halobacterium Methanococcus	Aggregate of protein/glycoprotein subunits
Methanobacterium	Covalently cross-linked glycopeptide
Methanosarcina Halococcus	Non-cross-linked, anionic heteropolysaccharide

where strength and shape are provided by **peptidoglycan** which is cross-linked by way of its peptide components to form a single, cell-shaped macromolecule. This material, sometimes described as 'murein', finds an analogue in **pseudomurein**, the principal wall component of the archaebacterial genus *Methanobacterium*. Both polymers contain a glycan backbone consisting of a disaccharide repeating unit of two different *N*-acetylated amino sugars, to one of which is attached a short peptide chain of between four and nine amino acid residues. Peptide bonds between the peptide chains of different glycan strands create a covalent network that has great mechanical strength as well as considerable elasticity. These wall types, and their components, are described below.

(i) Peptidoglycan

A generalized structure for peptidoglycan is shown in Figure 1.2. The structure is best considered in three parts—the glycan backbone, the linear muramyl tetrapeptide and the peptide cross-link (X)—each of which is subject to variation from strain to strain and may vary with growth conditions or during differentiation processes within a given strain.

The **glycan backbone** usually consists of the disaccharide repeating unit (*N*-acetylglucosamine β1–4 *N*-acetylmuramic acid) polymerized by β1–4 glycosidic linkages. Variations occur in which the muramic acid residues are partially *O*-acetylated at C6 (for example, in *Staphylococcus aureus* and *Neisseria gonorrhoeae*) or partially de-*N*-acetylated (for example, in *Bacillus cereus*). In *Mycobacterium* the amino group of the muramyl residues carries a glycolyl substituent instead of acetyl (Adam *et al.*, 1969), while in the spore cortex peptidoglycan of *Bacillus* the *O*-lactyl groups of some muramic acid residues lack peptide substituents and instead form internal lactams with the amino groups of the sugars (Warth & Stromiger, 1969, 1972). The *in vivo* chain length of the glycan strands has proved difficult to determine owing to the presence in cell walls of endogenous endoglycosidases (autolysins) that may catalyse chain degradation during the isolation of peptidoglycan. Taking precautions against autolysis, Ward (1973) measured chain lengths of approximately 45 disaccharide units in *Bacillus licheniformis* and of 79 units in a lytic-deficient strain of the same species. An average length of 35 units has been determined for *Escherichia coli*, though this may vary from 20 to 200 for individual chains. In some Gram-negative bacteria glycan chains may terminate in 1,6-anhydromuramic acid owing to the action of a murein transglycosylase (Holtje *et al.*, 1975).

The structure of the **tetrapeptide** attached to the lactyl group of the muramic acid residues is well conserved. In most types of peptidoglycan (types A) it consists of the sequence L-ala-D-glu-AA(3)-D-ala, in which the

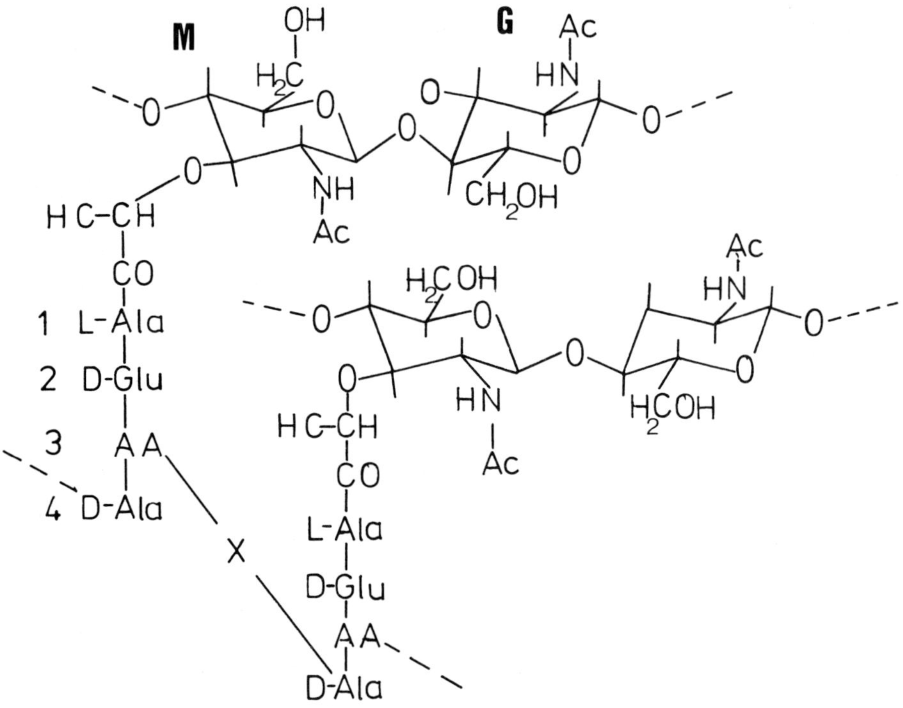

Figure 1.2 The primary structure of peptidoglycan. Two disaccharide tetrapeptide repeating units are shown cross-linked by the variable peptide unit X. Variations in the linkage of the glutamic acid residue (amino acid 2) and the diamino acid (amino acid 3) in the tetrapeptides, and in the nature of the cross-link, are described in the text. **M** is N-acetylmuramic acid; **G** is N-acetylglucosamine. The dotted lines indicate extensions of the polymeric structures.

glutamic acid residue is usually linked through its side-chain carboxyl to the amino acid at position 3. AA(3) is a species-specific diamino acid, most commonly L-lysine, an isomer of diaminopimelic acid (DAP), or ornithine, though several other variants occur (Schleifer & Kandler, 1972). Inter-species variety may also be introduced by the partial amidation of the side-chain carboxyl groups of D-glutamic acid and diaminopimelic acid and by the replacement of L-alanine at position 1 by glycine or serine in corynebacteria (type B peptidoglycans; Schleifer & Kandler, 1972). Important differences occur between the muramyl peptides of peptidoglycan from vegetative cells and from the cortex of endospores in *Bacillus*. In *B. subtilis* spores, about 20 per cent of muramyl residues carry only L-alanine (Warth & Strominger, 1972) instead of tetrapeptide, while in

B. sphaericus spores the diaminopimelic acid of vegetative cells is replaced by L-lysine (Tipper & Gauthier, 1972).

The nature of the **cross-linkage** (-X- in Figure 1.2) is the most variable feature of the peptidoglycan structure and is a valuable taxonomic character (Schleifer & Stackebrandt, 1983). A detailed description and classification of the many variants is beyond the scope of this book, but was thoroughly reviewed by Schleifer and Kandler (1972). In its simplest form the cross-link consists of a direct peptide bond between the side-chain amino group of the diamino acid at position 3 of one tetrapeptide and the carboxyl of the terminal D-alanine of a tetrapeptide on another peptidoglycan strand. This type occurs in *Bacillus subtilis* and in all Gram-negative eubacteria with the exception of spirochaetes and *Fusobacterium* (Kato *et al.*, 1979). Recent developments in the analysis of peptidoglycan structure (see Chapter 5B) have revealed an alternative cross-linkage in Gram-negative bacteria, between the diaminopimelic acid residues of two tetrapeptides. Gram-positive species exhibit a wide range of cross-bridges consisting of peptide chains of up to five amino acids. These may be chains of a single amino acid (for example, the pentaglycine bridge in *Staphylococcus aureus*) or more complex heteropeptides. An interesting type found in a number of strains of *Micrococcus* consists of a tetrapeptide identical to that linked to muramic acid, and there is good evidence that this cross-bridge originates as a muramic acid-linked tetrapeptide that is subsequently transferred to its new position in a transpeptidation reaction. The presence in these strains of muramic acid residues carrying no peptide chain is consistent with such a process. In type B peptidoglycans found in some plant-pathogenic corynebacteria, the cross-bridge is formed between the side-chain carboxyl of glutamic acid at position 2 of the tetrapeptide and the terminal D-alanine of another tetrapeptide.

The chemical nature and the extent of cross-linking in peptidoglycan are both subject to phenotypic variation. Schleiffer *et al.* (1976) reviewed the effects of growth conditions and described the interference with normal cross-link formation of high concentrations of certain amino acids, particularly alanine, serine and glycine, in the culture medium. For example, in *Staphylococcus aureus* a high concentration ratio of alanine to glycine leads to the attachment of an L-alanine residue to the side-chain amino group of the lysine at position 3 of the muramyl tetrapeptide, hence blocking cross-link formation. A high concentration ratio of serine to glycine in the medium results in the replacement of some of the glycine residues in the cross-bridge by serine. For a given strain of bacterium the extent of cross-linking of the peptidoglycan depends very much on growth conditions. Dobson & Archibald (1978) found that in *S. aureus* growing at a constant rate in a chemostat on fully defined medium, the proportion of muramyl tetrapeptide lysine residues whose side-chain amino groups were involved in peptide linkages varied between 62 and 90 per cent depending on the limiting nutrient in the medium.

The molecular architecture of peptidoglycan in the cell wall—the spacial distribution of cross-links, the degree of three-dimensional order in the polymer, its orientation in the wall and the possibility of heterogeneity through the depth of the wall—is a subject of considerable research interest but great uncertainty. Good discussions can be found in Shockman & Barrett (1983), Naumann *et al.* (1982) and Burman & Park (1983).

Although peptidoglycan determines the shape and strength of walls of Gram-positive and Gram-negative eubacteria, electron microscopy and chemical analysis reveal differences both in the architecture of the peptidoglycan layer and in accessory wall structures between the two groups. In Gram-positive bacteria the entire wall is seen as a single, amorphous layer with no internal features, that can be isolated, almost intact, as a covalently linked complex of peptidoglycan and accessory polymers. In its hydrated form it represents of the order of 10 per cent of the total cell volume and is between 20 and 50 nm thick. In Gram-negative bacteria the peptidoglycan forms only a thin layer; quantitative measurements of peptidoglycan and assumptions about its space-filling properties suggest that it may be only one molecule thick. Most of the Gram-negative wall consists of an outer membrane, seen in electron microscopy as a double-track membrane but chemically very different from the bacterial cytoplasmic membrane. Both groups of bacteria often display a regularly arrayed layer of protein units (S-layer) on the outer surface of the wall (see Chapter 4F), and the cells may be surrounded by a loosely-attached capsule of polysaccharide or protein (see Chapter 4H).

(ii) Walls of Gram-positive eubacteria

Chemical analysis of isolated walls from Gram-positive bacteria typically indicates that peptidoglycan represents between 50 and 80 per cent of their dry weight. The rest of the wall consists of 'accessory polymers'—teichoic acids, polysaccharides and proteins—covalently linked to the peptidoglycan throughout the thickness of the wall. A typical Gram-positive envelope is shown diagramatically in Figure 1.3. The walls of some coryneform and actinomycete bacteria, for example *Corynebacterium*, *Mycobacterium* and *Nocardia*, also possess covalently bound lipids. Immunochemical, enzymic and surface-labelling techniques reveal additional, non-covalently attached, surface components which are lost during wall isolation (see Chapter 3). While these may be minor wall components in quantitative terms, they may play major roles in pathogenicity, adhesion and wall turnover. The structures and properties of the wall components are summarized below.

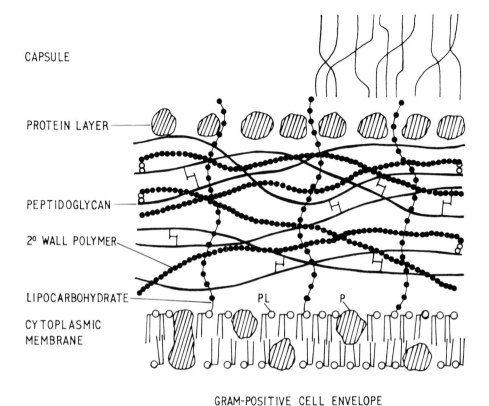

CAPSULE

PROTEIN LAYER

PEPTIDOGLYCAN

2° WALL POLYMER

LIPOCARBOHYDRATE

CYTOPLASMIC
MEMBRANE

GRAM-POSITIVE CELL ENVELOPE

Figure 1.3 The various components to be found in Gram-positive cell walls. Note that not all the components shown occur in every strain of Gram-positive bacterium. P: protein; PL: phospholipid.

1. Teichoic acids

These are short polymer chains, having the general structure shown in Figure 1.4. The repeating unit of a teichoic acid is shown, where n is 30–40. The polymers can be classified into four groups, depending on the structural type of the repeating unit:

Group 1: O–X–O is a simple alditol residue (glycerol, ribitol or mannitol) with its two primary alcohol groups involved in the phosphodiester linkages, often with a glycosyl substituent, R (commonly glucose, galactose, *N*-acetylglucosamine, *N*-acetylgalactosamine, which may be in the alpha or beta anomeric configuration). [e.g. Rol–P–Rol–P–. . .]

Group 2: O–X–O is a glycosyl alditol residue in which the glycosyl group

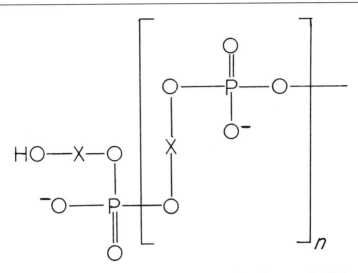

Figure 1.4 The generalized structure of teichoic acids. Rol: ribitol; –P–: phosphate in diester linkage; Glc: glucose; Gro: glycerol; GlcNAc: *N*-acetylglucosamine.

forms part of the main phosphodiester-linked polymer chain. [e.g. Glc–Gro–P–Glc–Gro–P–. . .]

Group 3: O–X–O is a glycosyl (or oligosaccharide)–1–phosphate–alditol residue in which the sugar–1–phosphate forms part of the main polymer chain. The sugar is commonly an *N*-acetylhexosamine. [e.g. GlcNAc-P-Gro-P-GlcNAc-P-Gro-P–. . .]

Group 4: O–X–O is a sugar (or oligosaccharide) residue to which one of the phosphodiester linkages is attached at C1. [e.g. GlcNAc–P–GlcNAc–P–. . .]. For abbreviations, see legend to Fig. 1.4.

Polymers of Groups 1,2 and 3 nearly always contain non-stoichiometric amounts of an amino acid ester-linked through its carboxyl group, usually to the alditol residues. In most cases the amino acid is D-alanine, but lysine has been reported in a species of *Streptomyces* (Skoblilova *et al.*, 1982) and ester-linked acetate or succinate replace alanine in a wide range of other actinomycetes (Naumova *et al.*, 1978).

Analysis of polymers of Group 1 frequently reveals non-stoichiometric amounts of glycosyl substituents on the repeating units. In some cases this may be due to substitution of only a proportion of the alditols within a single polymer chain. Immunological methods, however, have shown that in other cases such polymers are a mixture of fully glycosylated and entirely unglycosylated chains.

2. Polysaccharides and teichuronic acids

Polysaccharides occur as accessory polymers in a wide range of Gram-positive bacteria. They are invariably heteropolysaccharides, usually containing *N*-acetylamino sugars, and are commonly anionic in nature, owing either to the presence of uronic acid or sialic acid residues or to substitution with glycerolphosphate (Emdur *et al.*, 1974) or sugar phosphates (Pazur, 1982). Where the repeating unit contains a uronic acid residue, the polymer is often described as a 'teichuronic acid'. A few neutral polysaccharides have been detected, notably the group antigens of streptococci, but strains that possess these polymers usually contain a second, anionic, polysaccharide as well. Complete structures of neutral wall polysaccharides have been determined only rarely (Amano *et al.*, 1977). The teichuronic acids are the best-characterized wall polysaccharides and some representative structures are shown in Figure 1.5.

3. Proteins

For many years, studies of Gram-positive cell walls were founded on methods of preparation which included treatment with proteolytic enzymes. Hence it is no surprise that little protein was found associated with the peptidoglycan and that proteins were largely disregarded, while attention was focused on other wall polymers. If, however, wall preparations are cleaned of adherent membrane-derived and extracellular material by the use of detergents or other denaturing agents, the protein components can survive and be recognized. Studies of this kind (see Chapter 4F) have revealed wall-bound proteins in *Streptococcus* Groups A, C and G (Reis *et al.*, 1985), *Streptococcus mutans* (Russell, 1979), *Streptococcus*

[4 ManUANAc 1–6 Glc 1]$_n$ *Micrococcus luteus*

[4 GlcA 1–6 GalNAc 1]$_n$ *Bacillus subtilis*

[4 Glc 1–3 Rham 1–4 Rham 1]$_n$ *Bacillus megaterium*
 |
GlcA

[GlcA/glutamic acid]$_n$ alkalophilic *Bacillus* sp.

Figure 1.5 Structures of some teichuronic acids from Gram-positive bacteria. ManUANAc: *N*-acetylmannosaminuronic acid; GlcUA: glucuronic acid; Rham: rhamnose; Glc: glucose.

sanguis (Reusch, 1982), *Streptococcus salivarius* (Weerkamp & Jacob, 1982), *Staphylococcus aureus* (Sjoquist *et al.*, 1972; Esperson *et al.*, 1985), and *Bacillus subtilis* (Mobley *et al.*, 1983). In many cases these proteins have been shown to have specific receptor functions, such as the affinity for the Fc region of immunoglobulins found in staphylococcal protein A and several streptococcal proteins such as protein G, or the fibronectin-binding and fibrinogen-binding proteins of streptococci and staphylococci (Esperson & Clemmenson, 1982; Esperson *et al.*, 1985).

The amount of information available on the ways in which proteins may be bound to other wall components is as yet extremely sparse. There is the possibility of linkages to either the glycan or peptides of peptidoglycan, to teichoic or teichuronic acids or to other proteins. The surface location of proteins may also involve interaction with the cytoplasmic membrane. The amino acid sequences derived by translation of the DNA base-sequence in the genes for staphylococcal protein A and streptococcal M-protein have yielded some insight into the possible interactions which exist (Guss *et al.*, 1984; Hollingshead *et al.*, 1986). Both proteins are characterized by a carboxy-terminal stretch of hydrophobic amino acids, typical of a region which might be expected to be embedded in the cytoplasmic membrane. Next to this is a highly conserved region of 10–13 repeats of a proline-containing octapeptide—a length consistent with its spanning the peptidoglycan thickness of the cell wall. The C-terminus is blocked and it has been proposed that in *S. aureus* it is the site of covalent attachment of peptidoglycan, though it is difficult to reconcile this with the idea of the C-terminal sequence being membrane-bound. Both proteins then have a structure composed of a number of repeating units on the outside of the cell. While the generality of this arrangement is not yet known, it does suggest that disruption of bonds at a number of locations might be needed for release of intact protein.

Although it appears that there are at least some proteins in Gram-positive bacteria which are only found tightly linked to peptidoglycan, such as the Fc receptors of Group C streptococci (Reis *et al.*, 1984), most of those which have been characterized to date can also be found, in certain circumstances, loosely associated with the wall or free in the culture supernatant. The term 'wall associated proteins' has been introduced to refer to these. Examples include several proteins of *Streptococcus mutans* (Russell, 1979), staphylococcal protein A (Movitz, 1976), *Streptococcus salivarius* (Weerkamp & Jacob, 1982) and Group A streptococcal M-protein (Pinney & Widdowson, 1977). Recent electron microscopy studies of *Staphylococcus aureus* using freeze-substitution (Umeda *et al.*, 1987) have implicated teichoic acids in the non-covalent binding of proteins to the cell wall. Interpretation of the observation of soluble forms of wall proteins is complicated by the fact that wall constituents are released by wall turnover during growth of most Gram-positive bacteria. Identification of the origin of such proteins, therefore, may depend on detailed kinetic studies or chemical analysis of highly purified material.

4. Bound lipids

Mycobacterial cell envelopes are rich in unusual lipid types which are considered to be essential structural components of a characteristic outer membrane layer (Minnikin, 1982). This is probably based on a monolayer of high-molecular-weight mycolic acids (Figure 1.6) esterified to an arabinogalactan polysaccharide that in turn is covalently linked to peptidoglycan. It is possible that the membrane layer is completed by a range of free lipids designed to interact with the mycolic acid chains, though the evidence for the presence of individual free lipids specifically in the wall is often indirect—mycoside glycolipids, for example, are believed to be cell surface components because of the correlation between their presence in the cell and colony morphology (Goren & Brennan, 1979; Minnikin, 1982). In *Nocardia*, however, some free lipids have been shown to occur exclusively in the cell wall (Pommier & Michel, 1981). In many cases it is practically impossible to extract the plasma membrane and outer membrane free lipids separately.

5. Linkage of accessory polymers to peptidoglycan

Since the demonstration by Liu & Gottschlich (1967) that acid hydrolysis of the cell walls of a wide range of Gram-positive bacteria yields, *inter alia*, muramic acid 6-phosphate, and the recognition that teichoic acids and wall polysaccharides can be released from their wall attachment by mild acid hydrolysis, overwhelming evidence has accumulated that most, if not all, carbohydrate-related wall polymers are attached via terminal sugar–1–phosphate ester linkages to C6 of muramic acid residues in peptidoglycan. The linkage may be directly from the reducing-terminal sugar of a polysaccharide (Hughes, 1971), but in the case of teichoic acids and some teichuronic acids a specialized 'linkage unit' with a structure distinct from that of the polymer repeating unit intervenes between the terminal residue of the polymer and the muramic acid to which it is linked. In the case of teichoic acids (TA) most of the linkage units examined have the structure [TA–P]–(groP)$_{2-3}$ManNAc–GlcNAc–1–P–[muramyl] (Kaya *et al.*, 1984), but modifications have been reported in which 7 glycerolphosphate residues are present (*Bacillus pumilis*; Kojima *et al.*, 1985a), the glucosamine residue is not N-acetylated (*Lactobacillus plantarum*; Kojima *et al.*, 1985b), N-acetylmannosamine is replaced by glucose (*B. coagulans*; Kojima *et al.*, 1985c), or all but one of the glycerolphosphate residues are

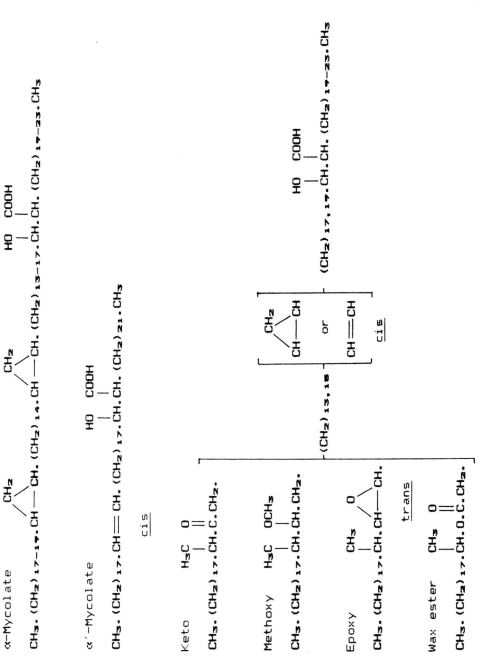

Figure 1.6 Representative mycolic acid structures. Reproduced by permission of D. E. Minnikin.

replaced by glucose (*Listeria monocytogenes*; Kaya *et al.*, 1985). The susceptibility of the N-acetylglucosamine–1–phosphate linkage to mild acid
hydrolysis, and of the glycerolphosphate–N–acetylmannosamine linkage
to alkali (Archibald *et al.*, 1971), account for the ease of extraction of
teichoic acids from walls in acid and alkali (Coley *et al.*, 1977).

The teichuronic acid in the cell wall of *Micrococcus luteus* has been
reported to be attached to a muramyl residue of peptidoglycan by an
N-acetylglucosamine–1–phosphate linkage unit (Hase & Matsushima,
1977) or by a direct phosphodiester link (Nasir-ud-Din *et al.*, 1985). The
former linkage is more consistent with the biosynthetic evidence (Stark *et
al.*, 1977). There is evidence for the attachment of other teichuronic acids
to peptidoglycan by direct phosphodiester linkage (*B. licheniformis*; Ward
& Curtis, 1982) or by a glycosidic linkage (*B. megaterium* M46; Ivatt &
Gilvarg, 1979).

Accessory polymers account for as much as 60 per cent of the weight of
the cell walls from many strains. In *B. subtilis*, for example, a glycerol
teichoic acid with an average chain length of about 40 repeating units
makes up about 50 per cent of the wall weight. This represents one
accessory polymer chain for every 12 disaccharide repeating units of
peptidoglycan, or about four chains per peptidoglycan strand, assuming a
glycan chain length of 45 disaccharide repeating units (see section B(i)
above). Nothing is known about the distribution of the linkage points for
accessory polymers along the peptidoglycan chain, although the results of
lysozyme digestion of cell walls suggest that the accessory polymers are
not clustered at a particular region. Ward & Curtis (1982) demonstrated
that when teichoic acid and teichuronic acid were synthesized concomit-
antly in *B. licheniformis* they always became attached to different strands
of peptidoglycan.

6. *Physicochemical properties of the walls*

Gram-positive walls are porous structures with well-defined molecular
size exclusion limits. Measurements with dextrans and polyglycols have
yielded exclusion limits of between 60 000 and 120 000 for the walls of
different strains (Scherrer & Gerhardt, 1971; Hughes *et al.*, 1975). There is
convincing evidence that the wall is an elastic structure in which three
physicochemical processes dominate: hydration of the wall polymers,
interactions between charged groups in the wall polymers, and the degree
of cross-linking in the peptidoglycan. The effects of hydration are most
clearly seen in electron microscopy where dehydrated thin sections of
walls appear markedly thinner than hydrated, freeze-etched specimens,
and it has been suggested that the material in contact with the cytoplasmic
membrane, at the inner surface of the wall, might be less hydrated and

therefore more tightly packed than that in the bulk of the wall (Tsien *et al.*, 1978). Torbet & Norton (1982) obtained neutron-scattering data from walls of *Staphylococcus aureus* consistent with a higher peptidoglycan packing density near the surfaces.

Buoyant densities of 1.44 and 1.05 g/ml have been measured for peptidoglycan from both Gram-positive and Gram-negative walls, in caesium chloride and 'Percol' gradients respectively, indicating that about 90 per cent of the bulk of the material is water. A measure of the average packing density in the wall can be obtained from the dextran-impermeable volume of the wall (Marquis, 1968) or from immersion refractometry (Marquis, 1973). Such measurements have shown that different packing densities result from changes in electromechanical interactions between charged groups within the wall. In typical Gram-positive walls having a high content of negatively charged accessory polymers distributed throughout the depth of the wall, charge repulsion tends to maintain an open structure. Shielding of the charged groups from one another in media of high ionic strength or low pH, or internal neutralization of some of the negative charges by positively charged amino groups in the wall, causes the wall to shrink. A low degree of peptidoglycan cross-linking would be expected to allow the wall more freedom to swell and shrink, and Marquis (1973) has demonstrated a correlation between these two parameters. Archibald (1985) has discussed the possibility of changes in packing density through the depth of the wall.

7. *Wall components that are not covalently linked to peptidoglycan*

The great sensitivity of modern analytical techniques, particularly immunochemical methods, has led to the detection of cell surface components on whole bacteria that were not observed during gross cell wall analysis. These substances, proteins and amphiphilic carbohydrates such as lipoteichoic acid, are often retained in crude, untreated cell wall preparations. They are lost in the course of purification techniques designed to remove contaminating membrane and cytoplasmic components, but their labelling at the surface of whole cells by macromolecular reagents that would not be expected to penetrate the cell wall confirms that they are genuine wall components. As has already been discussed in section B(ii)3, several proteins that are found tightly linked to the wall can also occur, under certain conditions, in a non-covalently associated form as well.

A characteristic of many of the non-covalently bound proteins in walls is their ability to be extracted in chaotropic reagents and non-ionic detergents. Thus, autolysins, an important group of wall-bound enzymes, have been extracted from a wide range of bacteria in lithium chloride solutions or in the detergent Triton X-100 (Rogers, 1979); lithium diiodosalicylate has been used to extract the sialic acid-binding lectins from oral streptococci (Murray *et al.*, 1986); and chaotropes have been widely used to isolate surface protein subunits (Koval *et al.*, 1983). The physiological roles of most wall-bound proteins have not been identified. Probably the best characterized in this respect are the autolysins. These wall-bound enzymes are usually recognized by their ability to catalyse hydrolysis of various linkages in peptidoglycan and hence cause bacterial lysis, but they are known to play an important part in normal growth of the cell wall. They are involved in cell wall turnover, which is particularly marked in rod-shaped bacteria, in separation of daughter cells after division, and in the remodelling of peptidoglycan during the ageing process (Burman & Park, 1983). Autolysins interact specifically with particular components of the cell wall, especially teichoic acids and teichuronic acids, and their proper functioning depends on these interactions (Tomasz *et al.*, 1971; Herbold & Glaser, 1975; Robson & Baddiley, 1976).

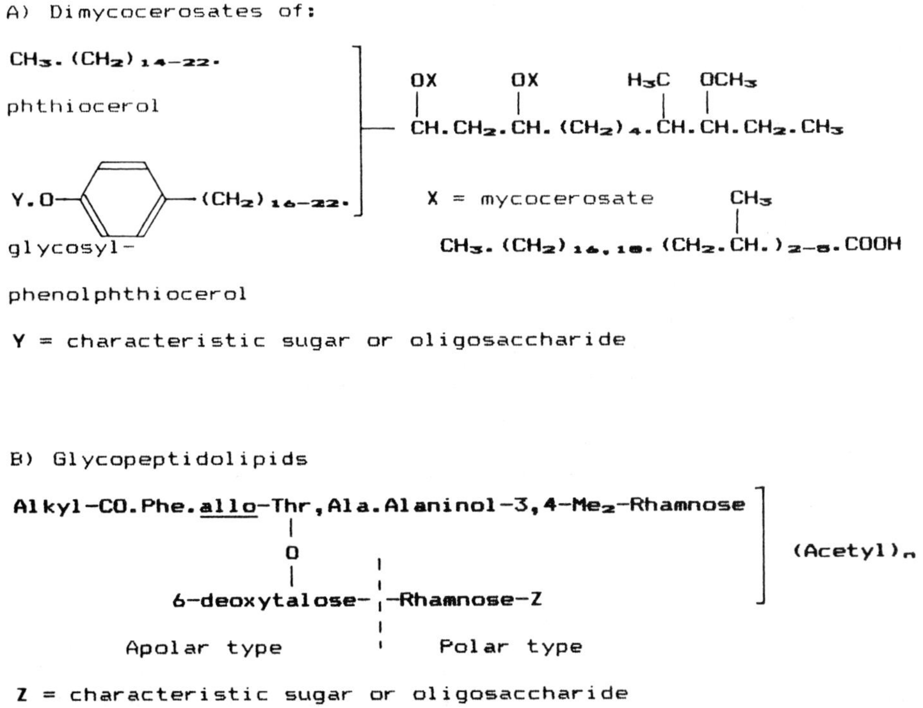

A) Dimycocerosates of:

$CH_3.(CH_2)_{14-22}.$

phthiocerol

$OX \quad OX \quad H_3C \quad OCH_3$

$CH.CH_2.CH.(CH_2)_4.CH.CH.CH_2.CH_3$

Y.O—⟨benzene ring⟩—$(CH_2)_{16-22}.$

glycosyl-

phenolphthiocerol

X = mycocerosate

$CH_3.(CH_2)_{14,16}.(CH_2.CH.)_{2-6}.COOH$ with CH_3 branch

Y = characteristic sugar or oligosaccharide

B) Glycopeptidolipids

Alkyl-CO.Phe.<u>allo</u>-Thr,Ala.Alaninol-3,4-Me$_2$-Rhamnose

|
O
|
6-deoxytalose- | -Rhamnose-Z

(Acetyl)$_n$

Apolar type | Polar type

Z = characteristic sugar or oligosaccharide

Figure 1.7 Some mycobacterial free lipids. Reproduced by permission of D. E. Minnikin.

Lipoteichoic acids (LTAs) form another group of polymers frequently detected as cell surface antigens and loosely associated with the cell wall (Chiu *et al.*, 1974). These are Group 1 glycerol teichoic acids (see section B(ii)1), but instead of a linkage to peptidoglycan they terminate in a diglyceride or glycolipid moiety, whose structure appears to be species-specific (Iwasaki *et al.*, 1986). Under appropriate conditions these poly-mers can be isolated in association with cytoplasmic membrane and for this reason they were originally described as 'membrane teichoic acids' (Hughes *et al.*, 1973; Lambert *et al.*, 1976). They are, however, also found associated with the cell wall and as extracellular products, sometimes in a deacylated form (Kessler & Shockman, 1979). Interactions have been demonstrated between LTAs and autolysins, of which they are effective inhibitors, and streptococcal M-proteins (Ofek *et al.*, 1982). Related amphiphilic polymers, particularly lipomannans, occur in some bacterial species (Hamada *et al.*, 1980). Owing to their amphiphilic properties the LTAs form micelles in aqueous solution and the critical micelle concentra-tions of some of them have been measured (Wicken *et al.*, 1986).

The presence of non-covalently bound lipids in the walls of some coryneform and actinomycete bacteria is mentioned above. These include long-chain wax esters such as dimycocerosates, glycolipids containing glyceric acid, and glycopeptidolipids in which sugars are found in glycosidic linkage to a peptide with an alkylated C-terminus (Figure 1.7) (Minnikin & O'Donnell, 1984). Some of these have been demonstrated to be cell surface antigens by immunochemical detection of components separated by thin-layer chromatography (Ridell *et al.*, 1986).

(iii) Walls of Gram-negative eubacteria

As described in section B(i), the walls of Gram-negative bacteria contain only a thin layer of peptidoglycan, which does not carry covalently-linked accessory polysaccharides or related compounds. A lipoprotein of low molecular mass (7.2 kilodaltons in *E. coli*), covalently linked to diaminopimelic acid in the peptide moiety of peptidoglycan through its C-terminal lysine residue, is, however, found in a wide range of Gram-negative bacteria. In *E. coli*, covalently linked lipoprotein makes up more than 40 per cent of the weight of detergent (sodium dodecylsulphate)-insoluble wall material, corresponding to one lipoprotein molecule for every ten disaccharide repeating units of the peptidoglycan, or about 2.5×10^5 molecules per cell.

The lipoprotein appears to link the peptidoglycan layer to the other part of the Gram-negative cell wall, the outer membrane, by an interaction between the N-terminal lipid of the lipoprotein and the lipophilic region of the outer membrane (Figure 1.8). In intact cells of *E. coli* the outer

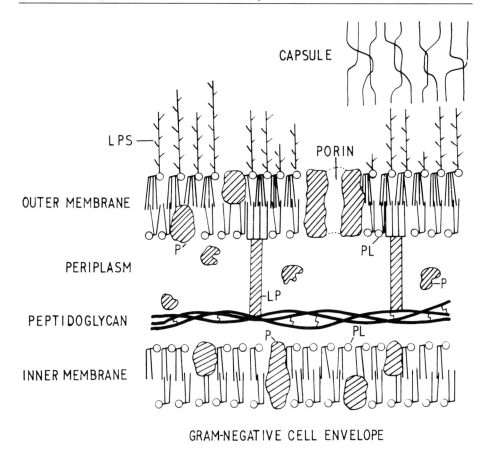

GRAM-NEGATIVE CELL ENVELOPE

Figure 1.8 The cell envelope of a Gram-negative bacterium. LP: lipoprotein; LPS: lipopolysaccharide; P: protein; PL: phospholipid.

membrane renders the lipoprotein inaccessible to anti-lipoprotein antibodies (Braun, 1973), but after exhaustive treatment of the cell wall with lysozyme to degrade peptidoglycan all the lipoprotein, still attached to fragments of peptidoglycan, is found in the outer membrane fraction (Braun & Bosch, 1973), where additional lipoprotein that has never been linked to peptidoglycan is also located (Inouye *et al.*, 1972).

The outer membrane of the walls of Gram-negative bacteria is seen as a trilaminar (i.e. bilayer) structure on electron micrographs of thin sections, and its composition and structure have been the subject of intense research that has been thoroughly reviewed (see, for example, Inouye, 1979; Lugtenberg & van Alphen, 1983; Nikaido & Vaara, 1985). Only the general features of its structure will be discussed here. Outer membrane proteins are described in Chapter 4G, lipopolysaccharides and enterobacterial common antigen in Chapter 4E. Immunochemical analysis of the *E. coli* outer membrane has revealed at least 24 distinct antigens in addition to the O-antigenic lipopolysaccharide (Smyth *et al.*, 1978).

Electron microscopy of thin sections and of freeze-fractured samples (Glauert & Thornley, 1969; Smit *et al.*, 1975; Cook *et al.*, 1986), and

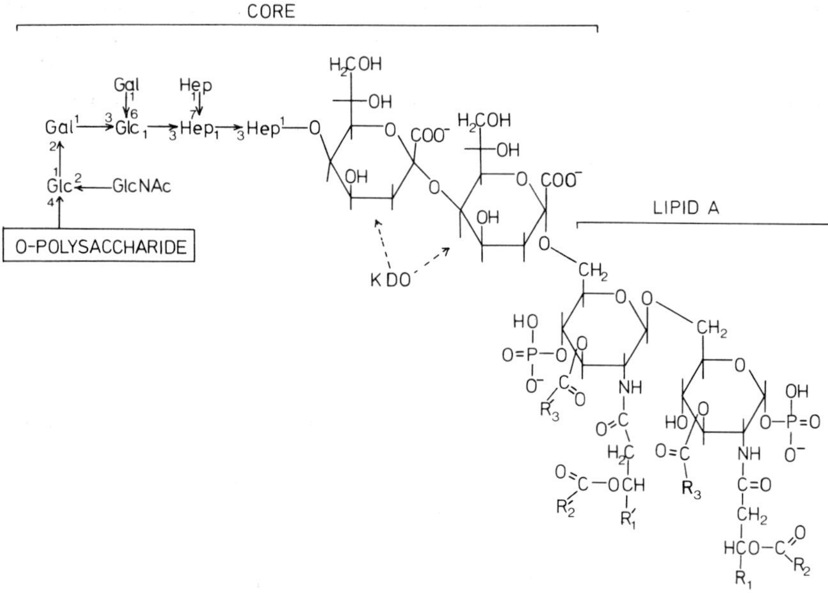

Figure 1.9 The structure of lipopolysaccharide. This is a composite structure showing features from several types of lipopolysaccharide. The lipid A portion is shown fully acylated, where R_1 and R_3 are usually the alkyl groups of long-chain alpha- or beta-hydroxy fatty acids and R_2 is derived from an ester-linked saturated straight-chain fatty acid. In some strains the glucosamine disaccharide is not fully acylated. 4-aminoarabinose may be attached through its anomeric carbon to a phosphate group in the lipid.

The core oligosaccharide shown here is that from *Salmonella typhimurium*; the structure of this region is strain-dependent. In addition to variations in sugar composition, modifications to this region may include phosphorylation of a heptose residue and substitution of heptose or KDO with phosphoethanolamine or pyrophosphoethanolamine. KDO monomer and dimer side-chains may occur in place of, or in addition to, one of the main-chain KDO residues.

The O-antigenic polysaccharide, which is a polymer of an oligosaccharide repeating unit, also has a strain-dependent structure that may vary in its degree of polymerization (see Chapter 5C). Rough (R) strains lack O-polysaccharide and may also have incomplete core oligosaccharides.

Key—KPO: 3–deoxy–D–*manno*–2–octulosonate; Hep: D– or L–glycero–D–*manno*–heptose; Gal: D–galactose; Glc: D–glucose; GlcNac: N–acetyl–D–glucosamine.

numerous physicochemical studies, indicate that the outer membrane has a bilayer structure. However, in *E. coli* the phospholipid content of the outer membrane is only about half that of the cytoplasmic membrane, and calculations by Smit *et al.* (1975) showed that this would only be sufficient to cover one leaflet of the bilayer; lipopolysaccharide (Figure 1.9) is present in sufficient quantity to make up the other leaflet of the bilayer. Subsequent surface labelling experiments by a number of workers (see Nikaido & Vaara, 1985 for a recent review) have clearly shown that the outer surface of the membrane in *E. coli*, *Salmonella* and *Pseudomonas aeruginosa* is largely occupied by lipopolysaccharide, while phospholipid makes up the inner leaflet. In enteric bacteria this arrangement may have evolved to provide the cell with protection against phospholipases and the detergent action of bile salts. There is much less evidence for asymmetry in the outer membranes of other species, and lipopolysaccharides of the enterobacterial type are absent from some Gram-negative bacteria (see, for example, *Rhodospirillum*, Evers *et al.*, 1986; *Deinococcus*, Thompson & Murray, 1981).

Proteins typically make up almost half the dry weight of an outer membrane and, with the exception of the peptidoglycan-bound and free forms of lipoprotein, are individually named, after the genes that encode them, as Omp proteins. *E. coli* outer membrane proteins have received most attention and a few of them have been described in great detail. The 'porin' proteins Omp C, Omp F and Pho E form relatively non-specific pores that allow the diffusion of small hydrophilic solutes, particularly negatively-charged ones in the case of Pho E, within a limited size range, across the membrane. Not surprisingly, the relative proportions of these proteins in the outer membrane, and the proportion of them that form active pores, vary with growth conditions. The amino acid sequences of these proteins show very pronounced homology (Mizuno *et al.*, 1983).

The purified proteins have been reconstituted in artificial lipid bilayer membranes (Nakae, 1976; Darveau *et al.*, 1983) to give functional pores. Nevertheless, controversy still exists about the exclusion limits of the pores (Yoneyama & Nakae, 1986) and about the nature and role of the interactions of porin proteins with lipopolysaccharide (Poxton *et al.*, 1985) and peptidoglycan, which in disrupted outer membranes are very strong and lead to the isolation of detergent-resistant complexes (see Lugtenberg & van Alphen, 1983 and Chapter 4G). Although the physiological significance of some of these interactions may be in doubt, they can be exploited very successfully for the purification of outer membrane proteins (see Chapter 4G).

The Omp A protein of *E. coli* and related proteins from other species are unusual in having decreased mobility on SDS-PAGE when they have been heated in SDS. They span the outer membrane, acting as external phage receptors while at the same time being accessible for chemical cross-linking to underlying peptidoglycan. There is evidence for interaction of Omp A with lipoprotein in the outer membrane (Palva, 1979) and mutants that lack it have unstable outer membranes, but no single role has been ascribed to it. Some outer membrane proteins, such as the Pho E porin, the Lam B phage receptor, which is a maltose-binding protein, and the proteins involved in the iron-scavenging mechanism, are inducible and become major components under certain conditions of growth (see Chapter 2).

(iv) The periplasm in Gram-negative eubacteria

Under the electron microscope, osmium-fixed thin sections of Gram-negative bacteria show a 'space' between the inner and outer membrane layers, except for a few adhesion zones where the membranes resist separation even after plasmolysis. Within this region, apparently held to the inner surface of the outer membrane by covalently linked lipoprotein, is the peptidoglycan layer. The region is known as the periplasm and has been shown to be iso-osmotic with the cell cytoplasm (Stock *et al.*, 1977). It contains a variety of proteins, many of which can be released selectively by osmotic shock (Nossal & Heppel, 1966) or by treatment with organic solvents (Ames *et al.*, 1984). They have been reviewed by Beacham (1979). Hobot *et al.* (1984) studied the periplasm by electron microscopy of thin sections embedded at low temperature and freeze-substituted, and by measurement of the buoyant density of isolated peptidoglycan from *E. coli*. They concluded that, in addition to the thin layer of highly cross-linked peptidoglycan visible in standard techniques of electron microscopy and which they believed to be in contact with the outer membrane, the periplasm was filled with a gel of highly hydrated peptidoglycan with a low degree of cross-linkage.

The periplasms of *E. coli* and several other species of Gram-negative bacteria contain 'membrane-derived oligosaccharides' (MDO) that consist of highly branched oligomers of nine or ten glucose residues, some of which bear glycerolphosphate and phosphoethanolamine groups in phosphodiester linkage at C6 (Schulman & Kennedy, 1979). In some species O-succinyl esters also occur. There is considerable evidence that these oligomers play a part in osmotic regulation in *E. coli* (Bohin & Kennedy, 1984).

C. ENVELOPES OF EUBACTERIA WITHOUT PEPTIDOGLYCAN

(i) Mollicutes

Bacteria of the genera *Mycoplasma*, *Acholeplasma*, *Spiroplasma* and *Ureaplasma* lack cell walls and contain only one type of membrane, the cytoplasmic membrane. They are osmotically extremely labile, growing only in isotonic or hypertonic media. These bacteria do, however, produce envelope-associated components in addition to the plasma membrane itself. In the spirally filamentous *Spiroplasma* a protein with a molecular weight of 55 000, 'spiralin', produces fibrils that constitute a shape-determining ribbon on the inner surface of the cytoplasmic membrane and also strands adhering to the outer surface (Townsend & Plaskitt, 1985). A number of species also produce capsules or glycocalixes of acidic polysaccharides, detectable in electron micrographs of ruthenium red-stained preparations (see the review by Razin, 1978).

(ii) Ancient budding eubacteria

Budding bacteria of the genera *Planctomyces*, *Pirella* (formerly *Pasteuria*) and *Isosphera* have walls that do not contain muramic acid or diaminopimelic acid, indicating the absence of typical peptidoglycan (Konig *et al.*, 1984; Giovannoni *et al.*, 1987). Instead, the walls appear to consist of a stable protein layer that retains the shape of the cell even in 10% sodium dodecylsulphate at 100°C. The protein layer is rich in the amino acids alanine, glycine and proline and exhibits a high degree of cross-linking through disulphide bonds between cysteine residues (Liesack *et al.*, 1986). Only traces of sugars are detectable. In these respects, the walls resemble those of some archaebacteria, but Stackebrandt *et al.* (1984) have demonstrated convincingly that the two groups are not related and that these genera of budding micro-organisms are eubacteria.

(iii) Bacteria with walls that contain insufficient peptidoglycan to form a complete sacculus

The walls of several genera of Gram-negative bacteria contain very small, but measurable, amounts of peptidoglycan, as assessed by measurement of muramic acid and peptidoglycan-specific amino acids. *Xanthobacter*, *Flexithrix* and some species of *Myxococcus* and *Rhodospirillum* contain less than 1 per cent of the peptidoglycan found in *E. coli* as a proportion of cell dry weight (see, for example, Evers *et al.*, 1986). Average degrees of cross-linking of about 30 per cent have been measured, making it extremely unlikely that the measured amount of peptidoglycan could form a continuous sacculus around the cell. These bacteria also have outer membranes that are quite different from those of *E. coli* in terms of the types of Omps and in the amounts of conventional lipopolysaccharide they contain (Orndorff & Dworkin, 1980; Evers *et al.*, 1986; Irschik & Reichenbach, 1978). The features responsible for the structural stability of the walls of these organisms are unknown.

D. WALLS OF ARCHAEBACTERIA

Members of the archaebacteria (methanogenic bacteria, extreme halophiles and thermo-acidophiles) have evolved an interesting range of different cell wall types, to some extent reflecting the variety found in eubacteria. Only the walls of *Methanobacterium* contain a covalently cross-linked glycopeptide component, termed **pseudomurein**. In other genera walls consist of protein aggregates or uncross-linked polysaccharides. While some of these components have been characterized in a few species, knowledge of the overall composition of the walls is far from complete and their physical properties have not been investigated. Current information was reviewed by Kandler & König (1985).

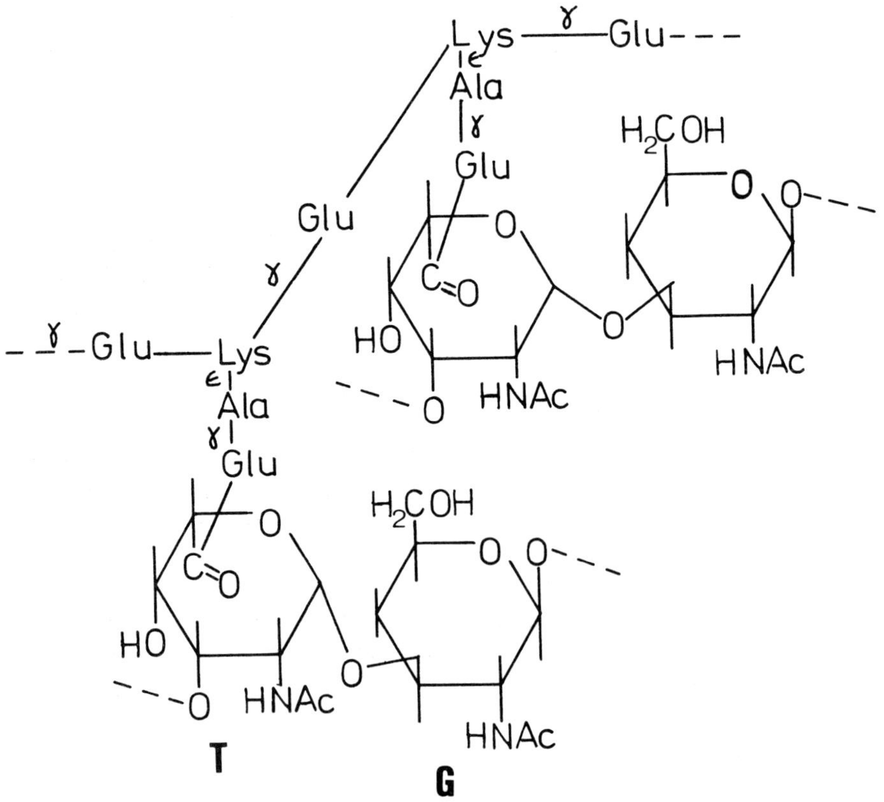

Figure 1.10 The primary structure of pseudomurein from *Methanobacterium*. Two disaccharide peptide repeating units of the glycan chains are shown with a peptide cross-link. Variations in the structure are discussed in the text. **T** is *N*-acetyl-D-talosaminuronic acid; **G** is *N*-acetyl-D-glucosamine.

(i) Walls containing pseudomurein

Treatment of *Methanobacterium* with hot detergent yields an insoluble residue retaining the shape of the cell, containing glycan chains cross-linked by peptides, but no muramic acid. König, Kandler and co-workers (König *et al.*, 1982) have characterized this material and named it pseudomurein. Its structure is shown in Figure 1.10. The glycan chains consist of β1–3 linked disaccharide repeating units containing N-acetylglucosamine and N-acetyltalosaminuronic acid (L-isomer). In most strains so far examined N-acetylglucosamine is partly replaced by N-acetylgalactosamine; it is not known whether the two sugars occur within the same glycan chain. Chain lengths of about 25 disaccharide units have been measured.

The talosaminuronic acid carboxyl groups carry amide-linked peptide chains through which adjacent glycan chains are cross-linked by a peptide bond between the side-chain amino group of a lysine residue in one chain and the side-chain carboxyl group of a glutamic acid residue in the other. Cross-linked disaccharide dimers were obtained from the polymer by treatment with 70% hydrofluoric acid under conditions that did not cleave peptide bonds (König *et al.*, 1982). Thus the polymer is very reminiscent of eubacterial peptidoglycan and like peptidoglycan its peptide component is subject to species and phenotypic variation (König *et al.*, 1982; König, 1985) and the action of an autolytic endopeptidase (König *et al.*, 1985; Kiener *et al.*, 1987). There is some evidence for the presence of covalently linked accessory polysaccharide containing neutral sugars, amino sugars and in some cases phosphate (König & Kandler, 1978).

(ii) Polysaccharide walls

The major component of the wall of *Methanosarcina* was found to be an uncross-linked polysaccharide fraction containing neutral sugars, N-acetylgalactosamine and a uronic acid. No covalently linked peptides were present (Kandler & Hippe, 1977). Nothing is known of the organization of these walls.

An acidic polysaccharide is also the principal component of the cell wall of *Halococcus morrhuae* (Steber & Schleiffer, 1975). In this case the negative charge is provided by substituent sulphate groups on the polysaccharide. The walls of this organism are quite strong, and resist osmotic lysis of the cells in hypotonic conditions.

(iii) Walls consisting of protein or glycoprotein subunits

The walls of *Halobacterium salinarium* (Mescher & Strominger, 1976a) and *H. halobium* (Wieland *et al.*, 1980) consist of regular arrays of glycoprotein subunits. The glycoproteins from the two species are very similar. They have a molecular weight of about 200 000 and contain both O-glycosidically linked di- or trisaccharides of neutral sugars and a single asparagine-linked oligosaccharide. Wieland *et al.* (1980) have shown that both types of saccharide are sulphated. The N-linked oligosaccharide appears to have a repeating structure containing galactose, N-acetylhexosamines, galacturonic acid and sulphate in the proportions 1 : 2 : 2 : 2. The glycoprotein units determine the cell shape; inhibition of N-glycosylation of the protein by bacitracin led to growth of the bacteria as cocci instead of rods (Mescher & Strominger, 1976b). The structure and biosynthesis of these walls has been reviewed by Sumper (1987).

Less well characterized protein subunit arrays form the basis of the walls of *Methanococcus* and *Sulfolobus*. In *Sulfolobus acidocaldarius* cytoplasmic membrane can be separated from the wall layer after cell breakage by extraction with the non-ionic detergent Triton X-100. The wall protein units remain insoluble and retain a sheet-like structure that reflects their arrangement in the whole bacterium (Weiss, 1974). In *Methanospirillum hungateii* chains of cells lacking a detectable sacculus are enclosed in a sheath consisting of protein subunits. This retains its structure when the cell contents are solubilized and extracted in boiling sodium dodecylsulphate solution, and is entirely resistant to proteases (Beveridge *et al.*, 1985). A sheath also encloses chains of *Methanothrix concilii* cells, which individually possess non-shape-forming walls; electron microscopy indicates that ingrowth of the sheath material is responsible for septation in this species (Beveridge *et al.*, 1986). The composition of walls and sheath is not known.

E. SUPERFICIAL WALL COMPONENTS

In all groups of bacteria examples are found of strains whose outer surfaces are covered by a more-or-less adherent layer of protein or polysaccharide (Figure 1.11). This superficial material may be in the form of a well-defined layer of hydrated gel of polysaccharide or polypeptide, forming a **capsule** whose thickness can be at least as great as the diameter of the cell. Capsular materials are firmly, though non-covalently, bound to the cell surface and form effective barriers to the penetration of macromolecules such as immunoglobulins. Thinner, less-well-defined layers of polysaccharide, often with a fibrillar structure, can be detected on the surfaces of a wide range of bacteria by electron microscopy of suitably stained preparations. These would not be identified as capsules by classical techniques of light microscopy. Costerton and co-workers have coined the term **glycocalyx** to describe all types of polysaccharide material adherent to the

Figure 1.11 An array of S-layer glycoprotein subunits on the surface of *Desulfotomaculum nigrificans* NCIB 8706: an electron micrograph of a freeze-etched replica of the whole cell. The S-layer lattice (p4-symmetry) is composed of identical glycoprotein subunits. Bar marker: 0.1 μm. Reproduced by permission of U. Sleytr.

cell surface but outside the integral elements of the cell envelope (Coster-ton *et al.*, 1981). In some cases these have been shown to be involved in cell adhesion. Electron microscopy is also revealing an increasing number of cases in which the bacterial surface is covered by a regularly packed array of protein subunits, termed an **S-layer**. In many cases these consist of glycoprotein and would be included in Costerton's definition of the glycocalyx.

(i) Protein S-layers

Investigations by electron microscopy have revealed that superficial crystalline surface layers of protein (S-layers) are present on a variety of cells of eubacteria and archaebacteria (Sleytr, 1976; Beveridge, 1981; Sleytr & Messner, 1983; Kandler & König, 1985; Sleytr *et al.*, 1986a). When viewed in thin sections, some archaebacteria that have no rigid envelope layers (Kandler & König, 1978) possess S-layers as their only shape-determining structure outside the cytoplasmic membrane. In the majority of Gram-positive eubacteria and archaebacteria the S-layers overlie the thick, apparently amorphous wall layers of peptidoglycan, pseudomurein or sulphated polysaccharide, though in several thermophilic bacilliaceae there is a relatively thin peptidoglycan layer that is comparable to those of Gram-negative organisms. The cell envelope architecture of Gram-negative bacteria is more complex but when present S-layers are clearly discernible as the outermost macromolecular structure, associated with the outer membrane.

(ii) Capsules and glycocalyces

Polysaccharide capsules are found in a wide range of bacteria: in mollicutes (Razin, 1973), in Gram-positive and Gram-negative eubacteria (Sutherland, 1982; Jann & Jann, 1977) and in cyanobacteria (Rippka *et al.*, 1979). Observation of bacterial microcolonies in nature suggests that most of them possess glycocalyces, though they are frequently lost during laboratory subculture. They are, moreover, often difficult to detect. Nega-tive staining may reveal capsules that are very thick relative to the cell diameter, such as those in pneumococci and *Klebsiella* but thinner ones, such as those in encapsulated strains of *Staphylococcus aureus*, cannot be observed reliably by light microscopy. Ruthenium red has been used widely as an electron microscopy stain for extracellular anionic polysac-charides, including those in capsules (Figure 1.12), but in such stained preparations dehydration artifacts are likely, so that the true morphology of the material may not be observed. Freeze-etched and critical-point dried specimens reveal the capsular material more successfully. Neverthe-

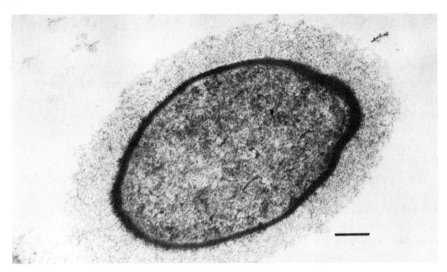

Figure 1.12 Electron micrograph of a thin section of a capsulate *Clostridium perfringens* cell stained with Ruthenium red. Bar marker: 0.2 μm. See appendix 1 for staining details.

less, it is still far from clear whether the network of fibrilar strands seen in most of these preparations accurately represents the *in vivo* architecture.

The capsules usually consist of heavily hydrated gels of heteropolysaccharides with repeating unit structures that contain negatively-charged groups. Capsular material from a strain of *E. coli* has been shown by X-ray diffraction studies to have a highly ordered secondary structure (Moorehouse *et al.*, 1977). Most commonly the anionic components are uronic acids, aminouronic acids or aldonic acids, but phosphate-containing polymers, resembling teichoic acids, have been found in the capsules of a variety of Gram-negative bacteria (Fischer *et al.*, 1982; Jennings *et al.*, 1979; Eagan *et al.*, 1982). A wide variety of structural types is produced by different strains of some species such as *Klebsiella*, *Streptococcus pneumoniae* and *E. coli* and can be valuable determinants in the serology of the species. In other species only one or two capsular types may be found.

The mode of attachment of the capsular polysaccharides (CPS) to the cell surface is unknown. Extraction of CPS usually requires conditions that cause some physical or chemical disruption of the cell wall. Treatment with phenol, for Gram-negative bacteria, and with dilute acid, for Gram-positive bacteria, have frequently been found necessary. Evidence is accumulating that the CPS of Gram-negative bacteria carry phospholipid substituents at their reducing termini (Gotschlich *et al.*, 1981; Schmidt & Jann, 1982; Kuo *et al.*, 1985). This suggests that they may be attached to the outer membrane by integration of the acyl chains of their terminal phospholipid into the apolar region of the outer membrane. Sutherland (1982) has discussed the possibility of tertiary structure formation between CPS and a different cell surface polysaccharide, as a mechanism for retaining the capsule at the cell surface, while Costerton *et al*. (1981) suggest that the participation of bivalent metal cations, as observed with some S-layer assemblies, may be a general feature.

Electron microscopy of bacteria growing under conditions resembling their natural habitats, particularly on solid surfaces, has shown that under these conditions the cells are usually associated with a fibrillar glycocalix (Costerton *et al*., 1981). This is usually involved in adhesion of the bacteria to surfaces, but similar material plays an important role in the adhesion of cells to one another. In *Sarcina ventriculi*, 50 or more cells are held together in a thick, fibrillar layer of cellulose (Holt & Canale-Parola, 1967), while enormous numbers of cells of *Zooglea ramigera* form flocculent masses in a matrix of fibrous exopolysaccharide at a late stage of growth (Farrah & Unz, 1976). An uncharacterized fine fibrous glycocalyx appears to be responsible for the association of square tablets of *Lampropedia* cells into extensive floating pellicles (Murray, 1984).

F. SURFACE APPENDAGES

(i) Pili and fimbriae

Pili (or fimbriae) are filamentous, non-flagellar, protein appendages which can be visualized by electron microscopy on the surfaces of a wide range of bacteria, especially Gram-negative species, and particularly when they have been freshly isolated from their natural environment. A precise definition of their appearance and general characteristics is difficult since large variations in morphology and other properties are seen even between types of pili produced by different strains of the same bacterial species. Following the initial pioneering studies in the laboratories of Duguid (Duguid *et al*., 1955) and Brinton (1965) there has been controversy over the name that should be used for these organelles. In this section the term pili will be used.

Studies on the properties of bacterial pili were stimulated by the observation that agglutination of erythrocytes by *E. coli* was associated with the presence of pili on the bacteria (Duguid *et al*., 1955). Subsequently pili have been shown to enhance the adhesion of a large number of bacterial species to a range of animal cells. The role of pili in the pathogenesis of a wide range of bacterial infections and their potential as vaccines is therefore a subject of continuing interest.

Pili that are purified from different strains of gonococci are assembled from pilin proteins of different subunit molecular masses. Even greater heterogeneity between strains is revealed by immunological techniques. Rabbit antisera raised against pili from one strain typically show 10 per cent or less cross-reactivity with other strains as measured by an ELISA inhibition assay (Buchanan, 1978). It has been suggested that there is almost unlimited potential antigenic diversity between strains (Brinton *et al.*, 1981) and modern genetic techniques are beginning to define the mechanisms by which such diversity can be generated (Haas & Meyer, 1986; Segal *et al.*, 1986). Structural studies show that the antigenically distinct pilins share regions of amino acid sequence homology. However, these conserved regions are immunorecessive, so that structural differences in the immunodominant variable regions of the molecule result in antigenically distinct pili (Rothbart *et al.*, 1984).

The pili of meningococci are structurally and immunologically related to those of gonococci (Stephens *et al.*, 1985). They, too, undergo antigenic variation both in laboratory subculture and during the course of infection (Olafson *et al.*, 1985; Tinsley & Heckels, 1986). Pili of *Pseudomonas* and *Moraxella* species also appear to be related to those of the *Neisseria*, sharing highly conserved N-terminal amino acid sequences (Saastry *et al.*, 1983; Froholm & Sletten, 1977). However, it is not yet clear whether they are capable of antigenic shift during infection.

The pili produced by the enterobacteriaceae, particularly *E. coli*, have been the subject of several recent reviews (Klemm, 1985; Smyth, 1986). Pilus expression in *E. coli* is frequently plasmid-encoded and this may provide an alternative mechanism to that of the *Neisseria* for switching pilus expression. This may be particularly relevant to a species that is able to colonize a number of sites in a variety of animal species. Thus K88 and K99 pili are associated with adhesion to porcine intestine, CFA pili with human intestine and PAP pili with the human urinary tract. In some cases the nature of the receptor on the epithelial cell has been identified with some certainty. S pili, for example, recognize sialyl galactoside present on human cells (Korhonen *et al.*, 1984), and it is now clear that the pili have lectin-like carbohydrate binding properties. However, the binding sites have not been identified in detail. The PAP pili recognize the digalactosyl oligosaccharide that forms part of the P blood group substance and is present on uroepithelial cells, but recent genetic studies suggest that the carbohydrate receptor may not be present on the main pilin subunit but rather on a specialized polypeptide which is in some way associated with the pilus structure (Normark *et al.*, 1986). Strains of uropathogenic *E. coli* produce several distinct PAP pilus types, differing in antigenic properties (Pere *et al.*, 1986).

(ii) Flagella

Flagella are whip-like appendages responsible for the motility and chemotaxis of many bacteria. They have a role as virulence factors in species such as *Vibrio cholerae* and *Campylobacter jejuni*, where they are thought to propel the bacterium through the mucus lining the small intestine. The 'H' or flagellar antigens of enterobacteria, especially *Salmonella* species, are used for serotyping.

The arrangement of flagella on the cell varies between species. They can be found singly at a pole (monotrichous), in tufts at a pole (lopotrichous), singly at both poles (amphitrichous) or over the whole surface of the bacterium (peritrichous). This arrangement can readily be seen in electron microscopy by negative staining or shadowing (Figure 1.13), or in the light microscope with special stains.

The flagellum has a complex basal body which permits the filament of the flagellum to rotate. This is located in the cell envelope and therefore differs in structure in Gram-positive and Gram-negative bacteria. The main part of the appendage, which is several micrometres in length, is made up of subunits of flagellins, small globular proteins with molecular masses in the region of 50 kDa. Chemoreceptors on the bacterial surface cause the bacterium to move towards or away from a chemical stimulus. The exact mechanism by which flagella convert biochemical energy into

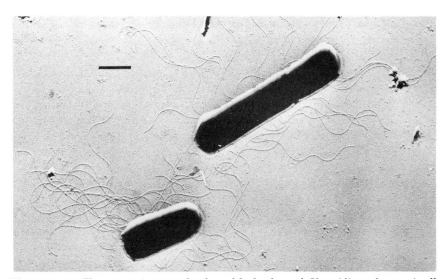

Figure 1.13   Electron micrograph of a gold-shadowed *Clostridium chauvoei* cell with peritrichous flagella. Bar marker: 1.0 μm. See Appendix 1 for staining details.

rotation is unknown, but the apparently random motion is controlled by the direction in which the flagellum is turned. One direction allows movement in a straight line while the other results in tumbling. The final direction of movement results from a combination of periods of tumbling and linear movement.

2

Culture of Bacteria

A. SELECTION OF CONDITIONS FOR BATCH CULTURE

As a result of empirical development over many years, the compositions of culture media and the choices of other growth conditions for particular bacterial species have become fairly well standardized. However, the principal criteria for their adoption have been the support of copious growth or the ability to promote selective growth of a particular species. Far less is known about the way growth conditions affect macromolecular synthesis. The physiology of cell wall synthesis is still too poorly understood to permit rational selection of culture conditions that will optimize the formation of a particular envelope component. Therefore, we can only summarize the ways in which growth conditions are known to influence envelope composition, in the hope that this will indicate some of the factors that must be taken into account in choosing conditions for a particular culture. The factors that can be controlled during growth, and their relationship to general cell physiology, were discussed in a recent review by Meyer *et al.* (1985).

(i) Growth rate

Until the advent of the chemostat as a practical laboratory culture technique, it was impossible to distinguish between direct effects of growth rate changes and other factors involved in the technique used to alter growth rate, on cellular physiology. Changes of carbon sources have been used widely to study the relationship between growth rate and particular biochemical activities, but the problems in interpreting the results of such experiments are obvious, particularly where complex processes such as cell wall synthesis are involved. The use of a chemostat permits the alteration of growth rate, by changing the dilution rate,

33

independently of many other factors (see section B below). Studies of this kind have revealed effects of growth rate on peptidoglycan cross-linking and on the balance of accessory wall polymers in Gram-positive bacteria (Ellwood & Tempest, 1972; Dobson & Archibald, 1978; Jacques *et al.*, 1979; Kruyssen *et al.*, 1980), and on the protein and lipopolysaccharide composition of the outer membranes of Gram-negative bacteria (Ellwood & Tempest, 1972; Driehuis & Wouters, 1987; Ombaka *et al.*, 1983; Williams *et al.*, 1984).

(ii) Medium composition

The influence of medium composition on peptidoglycan chemistry (Schleifer *et al.*, 1976), on synthesis of accessory anionic polymers in Gram-positive bacteria (Ellwood & Tempest, 1972; Hancock & Baddiley, 1985), on the production of capsular and extracellular polysaccharides (Sutherland, 1985; Sutherland, 1982), and on the synthesis of outer membrane proteins (Lugtenberg & van Alphen, 1983; Williams *et al.*, 1984; Ombaka *et al.*, 1983; Chart *et al.*, 1986) have been studied in some detail. The amino acid and peptide composition of the medium can have a profound effect on the peptide composition and degree of cross-linking of peptidoglycan. The main effect is on the cross-linking peptide unit in Gram-positive bacteria, and the muramyl tetrapeptide composition has in the past been considered to be well conserved. However, recent more detailed analysis of peptidoglycan by HPLC (see Chapter 5B) has revealed heterogeneity in this part of the molecule as well, and this might also be influenced by medium composition. The biochemical basis of some of the alterations in peptidoglycan composition is well understood and has been used to explain the effects of alterations in the make-up of complex broth media, such as differences between commercial yeast extracts (Schleifer *et al.*, 1976). Equally pronounced differences in peptidoglycan composition, involving the extent of cross-linking and degree of *O*-acetylation of the glycan, have been described, although their biochemical basis is unknown.

In batch culture, medium composition inevitably changes during growth with, in many cases, unpredictable effects on cell physiology. Exhaustion of a component is likely to produce particularly abrupt changes before growth ceases, but this period between the nutrient concentration becoming limiting and cellular biosynthetic activity ceasing may be quite short, so that heterogeneity of cell wall composition becomes likely. This effect is more readily studied during specific growth limitation in a chemostat, and the techniques and mathematical treatment of such experiments are described in section B. The effect of phosphate starvation in inhibiting teichoic acid synthesis and activating teichuronic acid synthesis in *Bacillus*, for example, is likely to result in walls containing teichoic acid in their outermost regions but teichuronic acid towards the interior, as can be seen very clearly in chemostat-grown cells that have been subject to a short pulse of phosphate-limited growth (Lang *et al.*, 1982). The induction of synthesis of specific outer membrane proteins in Gram-negative bacteria by limitation of phosphate or iron is also well known and may, for example in the case of iron limitation, dramatically alter the overall outer membrane composition. Low levels of inorganic nutrients such as sulphate, phosphate and nitrogen sources in media containing large amounts of carbon and energy sources often stimulate the production of extracellular polysaccharides (Duguid & Wilkinson, 1961).

Effects such as those described above should always be considered in the design of batch culture experiments, particularly when the bacteria are likely to be harvested after the end of exponential growth.

(iii) Other culture conditions

Other culture conditions can have profound effects on cell composition and may fluctuate during batch culture. At high biomass concentrations depletion of dissolved oxygen and changes in pH and dissolved carbon dioxide concentration are all likely to occur, and all are known to affect wall composition.

The pH of a culture of a Gram-positive bacterium affects the extent of substitution with alanyl ester groups, and hence the net charge, on both wall teichoic acids and lipoteichoic acids (MacArthur & Archibald, 1984). It also influences the balance between teichoic acid and teichuronic acid in the cell wall (Ellwood & Tempest, 1972), flagellum formation in *Proteus*, and encapsulation of *Streptococcus pneumoniae* (Lacey, 1961). The accumulation of fermentation products by *Clostridium* and *Zymomonas* is pH-dependent and in turn affects membrane composition.

Dissolved oxygen and carbon dioxide concentrations affect cell surface composition in a variety of ways, though the biochemical basis is in most cases unknown and the general applicability of the information is unclear. Oxygen limitation was reported to lead to low levels of surface antigens such as M-protein in Group A streptococci and Vi antigen in *Salmonella typhimurium*, and to affect fimbrial phase variation in *Shigella* (Duguid & Wilkinson, 1961), while a high CO_2 concentration favoured capsular protein production in *Bacillus anthracis* and capsular polysaccharide formation in *Staphylococcus aureus*. Even when such effects are recognized, it may be difficult to control them. Fermentation equipment in which the dissolved oxygen level is continuously monitored by an oxygen electrode and automatically controlled by stirrer speed is available but expensive. Independent control of dissolved CO_2 is possible, but rare in practice, except during small-scale culture on solid media. In the absence of monitoring, the effects of foaming, or alternatively of the addition of antifoam with its potential affects on surface tension and bubble size, are unquantifiable.

Although a particular temperature is optimum for growth of a given species, it may not be ideal for the expression of a particular surface component. The temperature sensitivity of flagellum formation in *Salmonella* and *Bordetella* has been known for many years, and similar dependence on temperature is shown by Vi antigen synthesis in *Salmonella typhi* and *Escherichia coli* and by capsule formation in *Pasteurella* and *Klebsiella* (Lacey, 1961). Lipopolysaccharide and outer membrane protein production in *Pseudomonas aeruginosa* are influenced by temperature (Kropinski *et al.*, 1987).

(iv) Defined media

The use of a chemically defined growth medium may in particular cases offer advantages such as reproducibility, cheapness and ease of subsequent treatment of the biomass and its subcellular components. While manufacturers of media make every effort to ensure consistency in their products, components such as peptones, yeast extracts and nutrient broths are by their nature likely to vary from batch to batch. If large-scale cultures in media containing such components are contemplated it is always worthwhile checking the suitability of the particular batch on a small scale first. Some batches of nutrient broth, for example, have been found to support yields of stationary-phase staphylococci and micrococci only 20 per cent of the size obtained from other batches from the same manufacturers.

The production of extracellular polysaccharide by *Xanthomonas campestris* (Davidson, 1978) is a case in which all the benefits of a chemically defined medium were obtained without any sacrifice in yield. On the other hand, undefined complex media often permit higher growth rates or yields, or give different wall compositions, and even if the growth factors involved can be identified their addition to a defined medium may be expensive or complicated to carry out. *Staphylococcus aureus* normally has a complex growth requirement which can only be met in a defined medium by the addition of 18 amino acids, thiamine, nicotinic acid, biotin and trace metal salts. The bacteria can be 'trained' to grow on a medium containing only one amino acid (see section B below), but this leads to a reduced maximum growth rate and substantial alterations in cell wall composition, including loss of protein A (A. R. Archibald, unpublished). The trace element balance and the buffering properties (or lack of them) in a complex medium may be impossible to mimic in a defined medium, in practice. The decision as to whether it is worthwhile substituting a defined medium for a well-established complex broth medium must therefore be based on empirical knowledge of the system.

B. CONTINUOUS CULTURE—GROWTH KINETICS AND WALL TURNOVER

A. R. Archibald

(i) Introduction

As described in section A, bacteria grown under different conditions can have widely differing surface properties and cell envelope chemistry. Such differences can be observed in bacteria grown in batch culture, but the use of chemostat culture enables individual growth parameters to be varied in a defined and controlled way so that the factors responsible can be identified. It is thus possible to grow bacteria so that they have varied amounts of particular wall components; this can be of value in assessing quantitative aspects of the relation between wall composition and particular wall or surface properties. Although better known as a way of achieving 'steady state' or 'balanced' growth under defined conditions, chemostat culture can also be used to study transitions between steady state conditions. This can give valuable information on the regulation of wall polymer synthesis and the dynamics of envelope assembly and metabolism.

(ii) Media for continuous culture

In principle any liquid medium that supports growth in batch culture can be used in the chemostat. Generally, however, pH is actively controlled in the chemostat so that buffering may be unnecessary. Usually, too, the nutrient concentration is diminished for chemostat culture, at least for the limiting nutrient. This is necessary to obtain a specified limitation and to ensure that aerobes can be sufficiently oxygenated. Even in baffled chemostats stirred at high speed it can be difficult to maintain adequate aeration in cultures that contain more than 2–3 mg dry weight of bacteria per millilitre, especially at higher growth rates.

The experimental criterion for growth limitation by a particular nutrient is that the bacterial density increases when the concentration of that nutrient is increased. A most important consideration in the design of media to be used in changeover or pulse experiments is that the transition should not give rise to greatly increased concentrations of the originally limiting nutrient before the next limitation has taken effect. Most medium formulations ensure specific nutrient limitation by holding other nutrients in substantial excess. Changeover between two such media gives a stage at which increased concentrations of both nutrients are present. This results in an increase in the growth rate under essentially unlimited conditions and can lead to a substantial increase in culture density. While any disturbance to the steady state can lead to uncertainties in relation to chemostat kinetics, such uncertainties are less where gross disturbances in growth rate are avoided and the culture density remains constant.

1. Complex media with limiting glucose or phosphate

Complex media can be used quite satisfactorily for chemostat studies provided that a defined limitation can be achieved. For example, glucose limitation is possible with a number of complex broths. Phosphate limitation is also possible provided that excess phosphates present in the complex media are removed. A method for this follows (D. C. Ellwood, personal communication):

(a) Tryptone (90 g) is dissolved in 4.5 litres of water and adjusted to pH 9 with aqueous ammonia.
(b) The inorganic phosphate present is then precipitated by the addition of a solution (45 ml) of 20% w/v $MgCl_2$ $6H_2O$ and is removed by centrifugation at 23 000g for 30 minutes. The medium is then adjusted to pH 7 by the addition of 5M H_2SO_4.
(c) Amino acids and vitamins are added:
L-glutamic acid, 9 g; L-alanine, 4.5 g; L-aspartic acid, 4.5 g; cysteine, 0.9 g; thiamine, 9 mg; nicotinic acid, 45 mg; biotin, 1 mg; 3,4-dihydroxybenzoic acid, 135 mg; K_2SO_4, 1.8 g.
The volume is made up to 17.5 litres with water, and autoclaved.
(d) A sterile solution of glucose (270 g) in 500 ml of water is added.

The medium should contain about 23 μg/ml of phosphorus, 98 per cent of it present as organic phosphate. The medium should support growth of *Staphylococcus* or *Bacillus* species to a concentration of 1–2 mg dry wt bacteria/ml before the bound phosphate is fully depleted and it can be used for phosphate-starved batch culture or for continuous culture.

2. Defined media with various limitations

Growth of *Staphylococcus aureus* has been achieved (Dobson & Archibald, 1978) in a defined medium of the following composition:

A: $(NH_4)_2SO_4$, 16 mM; K_2SO_4, 12 mM; NaH_2PO_4, 10 mM; $MgCl_2$, 6 mM; NaCl, 24 mM

B: citric acid, 1 mM; $CaCl_2$, 100 μM; $FeCl_3$, 100 μM; $MnCl_2$, 25 μM; $ZnCl_2$, 25 μM; $CuCl_2$, 5 μM

These components can conveniently be made up at ten times the stated concentration (10 × salts), by fully dissolving each component before adding the next, in the sequence given.

C: D-glucose, 140 mM

D: L-cysteine, 400 μM, and other amino acids as required

E: thiamine, 1 mg/l; nicotinic acid 1 mg/l; other vitamins as required

The medium is prepared by dissolving components A in about three-quarters of the final volume of deionized water and then adding one-tenth of the final volume of 10× salts B. The volume is adjusted appropriately and the mixture sterilized at 121°C. Glucose is sterilized separately, as a 30% w/v (1.67 M) solution, at 115°C for 15 minutes then added aseptically, together with filter-sterilized solutions D and E, to the reservoir.

Limitation is effected by reducing the concentration of certain of these nutrients individually as shown below, all other nutrients being present at the concentration given above:

glucose	30 mM
cysteine	80 mM
phosphorus	0.65 mM
nitrogen	3.2 mM
potassium	0.5 mM

The organism grown in this medium was a 'trained' derivative of *Staphylococcus aureus*, but other strains can be grown in similar media supplemented with various amino acids and biotin or other vitamins. In later work (R. N. Nwabueze and A. R. Archibald, unpublished) a modification of the above medium has been used in which the concentration of $MgCl_2$ is reduced to 2 mM; this concentration is not limiting but has the advantage of reducing the formation of insoluble salt precipitates, such as ammonium magnesium phosphate, which can otherwise be troublesome, particularly when ammonia is used to neutralize the culture.

Several media have been described for growth of *Bacillus* strains under different limitations (Ellwood & Tempest, 1969). Archibald & Coapes (1976) used the following media for changeover experiments involving *Bacillus subtilis* W23:

Phosphate-limiting medium

A: NaH_2PO_4, 1.44 mM; $(NH_4)_2SO_4$, 25 mM; K_2SO_4, 4.3 mM; $MgCl_2$, 1.0 mM

B: citric acid, 1.0 mM; $CaCl_2$, 0.1 mM; $FeCl_3$, 0.1 mM; $MnCl_2$, 25 μM; $ZnCl_2$, 25 μM; $CuCl_2$, 5.0 μM; $CoCl_2$, 5.0 μM; Na_2MoO_4, 5.0 μM

These components are made up at ten times the stated concentration as a 10× salts solution prepared by fully dissolving each component before adding the next, in the order given.

C: D-glucose, 170 mM

The medium is prepared by dissolving components A in about 75 per cent of the final volume of deionized water and then adding 10 per cent of the final volume of 10× salts solution. The volume is adjusted appropriately and the mixture is sterilized at 121°C. Glucose, sterilized separately as a 1.67 M solution at 115°C for 15 minutes, is added aseptically to the reservoir to the final concentration given above. The pH of this medium can be adjusted to 7.0 by the addition of sterile 2 M aqueous ammonia, though the medium can be used without neutralization provided the chemostat is not inoculated until the pH in the vessel has been brought to 7.0 by automatic addition of base.

Potassium-limiting medium

The composition is as above except that the concentration of NaH_2PO_4 is 1.44 mM and the concentration of K_2SO_4 is 3 mM. This medium provides the minimum amount of phosphate to sustain teichoic acid synthesis, and so changeover between potassium and phosphate limitation proceeds with little or no change in culture density.

Media providing mixed limitations

(a) Potassium-limiting medium, prepared as above but containing varied concentrations of phosphate from 5 mM, where phosphate is in considerable excess, to 0.5 mM, where phosphate is limiting. At intermediate concentrations of phosphate a mixed limitation is achieved in which growth rate is potassium-limited but teichoic acid and teichuronic acid synthesis are limited by the phosphate concentration, so that proportions of the two wall polymers are produced intermediate between those present when phosphate is limiting and when it is in excess (Anderson *et al.*, 1978).

(b) Magnesium-limiting medium contained K_2SO_4, 30 mM; $(NH_4)_2SO_4$, 50 mM; $MgCl_2$, 75 μM; glutamic aid, 5.0 mM; and the same 10× salts solution as above. The concentration of NaH_2PO_4 was varied between 5.0 mM, when magnesium was limiting and phosphate in excess, and 0.25 mM, when phosphate was limiting. In experiments with *Bacillus subtilis* 168 (Lang *et al.*, 1982) such media were used with glycerol, 30 mM, as the principal carbon and energy source and tryptophan, 20 mg/l, was added to satisfy the auxotrophic requirement.

(iii) Techniques

Information on setting up and maintaining chemostat equipment is given by individual manufacturers, and a helpful series of leaflets, *Hints and Tips—Continuous Culture Fermentation*, produced by L. H. Fermentation, will be of interest to those planning to use chemostat culture. Basically the operation of the chemostat for studies on cell envelopes does not present any special technical problems and the advantages of chemostat cultivation are essentially the same—control and definition—whatever the technique is used for. The fundamental reason for using the chemostat is that it can give information that cannot be obtained easily, or at all, from batch culture.

Although the chemostat provides 'steady state' conditions, and although cultures can be maintained for several months, problems can be caused by adhesion to the sides of the growth chamber and other surfaces, and by selection of variants better adapted to the particular growth conditions. Because of this we usually set up a large stock of ampoules of the freeze-dried organism under study and then prepare inocula from these for individual experiments, which are carried out on cultures that have been allowed to reach steady state by growth for between six and ten generation times beforehand.

Timing during pulse or changeover experiments needs to be given particular attention. Transients effected by direct injection of compounds into the growth chamber can be timed quite accurately, but it is necessary in changeover experiments to allow for the time it takes the new medium to pass through the delivery tube to reach the growth chamber. A small air space is usually present in the line after connecting new reservoirs, and the progress of this will show when new medium starts to flow into the chemostat. If the air space is large it will result in a cessation of medium addition and this must be avoided.

Medium flow rate and culture volume must also be known accurately. We monitor outflow continuously during experiments and then make an accurate determination of the culture volume at the end of the experiment. A very accurate measure of dilution rate is obtainable in radiolabel chase experiments since the fall in total activity in the chemostat is determined simply by the dilution rate (Clarke-Sturman & Archibald, 1978). Although commercially available chemostats are equipped with sampling devices, we have found it better to sample and collect directly from the overflow line rather than to alter the culture volume by withdrawing samples. In preparing media for chemostat cultures most workers find it convenient to use reservoirs of 20 litre capacity, which we use to contain 18 litre batches. In sterilizing such batches it is most important to allow for the time it takes large volumes of static liquid to reach the target temperature on autoclaving. The use of autoclaves fitted with temperature probes that can detect when the appropriate temperature is reached in a dummy reservoir is desirable. In our experience it takes about an hour for a 20 litre reservoir to reach 121°C, so that in the absence of appropriate control, sterilization times of about 90 minutes should ensure that the medium is held at full temperature for at least 15 minutes. Autoclaving causes a loss of liquid from the reservoir by evaporation; this can be as much as 500 ml and must be allowed for or corrected.

(iv) Effects of nutrient limitation

A variety of physiological responses to nutrient limitation has been observed in both Gram-positive and Gram-negative bacteria (see Harder & Dijkhuisen, 1983). These include changes in growth efficiency, in energy coupling, in the composition and structure of the outer and inner membrane components of the cell envelope, and in transport systems. One of the most striking phenotypic changes in cell envelope composition is the complete replacement of teichoic acid, which normally constitutes about half of the wall, by teichuronic acid in bacilli grown under conditions of phosphate limitation (Tempest *et al.*, 1968; Ellwood & Tempest, 1972). This provides a good example of the experimental value of the chemostat.

Phosphate-limited *B. subtilis* contains about one-third as much total cell phosphate as do the bacteria grown in the presence of excess phosphate (Lang *et al.*, 1982; Anderson *et al.*, 1978). This increased efficiency in the utilization of phosphate is achieved by the bacteria switching off synthesis of wall teichoic acid and by diminished synthesis of phospholipids (Minnikin *et al.*, 1972): the lipoteichoic acid content of phosphate-limited bacteria has also been reported to be reduced to one-tenth of that present in bacteria grown under other conditions (Button *et al.*, 1975).

As well as simple phosphate limitation, bacilli respond to mixed limitation in media containing a growth-limiting amount of one nutrient and phosphate in amounts intermediate between limiting and excess. On growth in such media *B. subtilis* strains W23 (Anderson *et al.*, 1978) and 168 (Lang *et al.*, 1982) make amounts of wall teichoic acid intermediate between the trace amounts present under phosphate limitation and the major amount present when phosphate is in excess. By controlling the composition of the medium it is possible to grow the bacteria so that they contain any desired amount of teichoic acid between the maximum and minimum values. This can be of interest in examining quantitative relationships between wall composition and surface properties such as phage binding (Givan *et al.*, 1982) or lectin binding (Lang & Archibald, 1983).

The nature of substrate limitation also has effects on peptidoglycan structure. In fully defined and simplified media the number of penta-glycine cross-bridges and the extent of cross-linking in *Staphylococcus aureus* varies substantially (Dobson & Archibald, 1978). Changes in envelope proteins, depending upon substrate limitation, are also seen in several Gram-positive and Gram-negative bacteria.

In studying the effects of specific limitation it is desirable to use dilution rates that are not much greater than $\mu_{max}/2$, so that the steady-state concentration of limiting nutrient is not much greater than the K_s value. As the dilution rate is increased beyond this, so the steady-state substrate concentration increases (see section B(viii)). This can substantially affect the physiological response to substrate limitation so that, for example, teichoic acid synthesis is not fully suppressed in bacilli growing in phosphate-limiting media at high growth rates (Lang *et al.*, 1982; Kruyssen *et al.*, 1980).

(v) Effects of growth rate

Variations in growth rate can be effected in batch culture by varying the growth temperature, pH or the composition of the medium. Consequent changes in bacterial properties could be due either to the difference in growth rate alone, or to some other effect caused by nutrition, pH or temperature. This ambiguity is avoided in chemostat culture which has been used to show that it is specifically because of the altered growth rate that, for example, wall content is increased in bacteria growing at lower rates (Ellwood & Tempest, 1972) and that the rate of killing effected by β-lactam antibiotics is less in slow-growing bacteria (Tuomanen *et al.*, 1986). Pulse or changeover experiments can also be carried out at different dilution rates so as to determine the effect of growth rate on wall assembly and metabolism.

While less ambiguous than comparable batch culture experiments, chemostat cultures run at different dilution rates do differ from each other in that the steady-state concentration of limiting nutrient is greater in cultures grown at higher rates. As indicated above, this can influence the nature of the physiological response to nutrient limitation.

(vi) Use of the chemostat to study the kinetics of synthesis and turnover of cell wall

Kinetics of wall synthesis and metabolism have been studied both in cultures growing in steady-state conditions and in cultures undergoing changeover or pulsed release of limitation.

1. Steady-state conditions

Kinetic studies using 'steady-state' growth conditions have involved label and chase experiments using radiolabelled components. For example, we have used ^{3}H-labelled N-acetylglucosamine to study wall turnover in *B. subtilis* growing under steady conditions under phosphate and potassium limitation (Clarke-Sturman & Archibald, 1982). Information was obtained by following wall labelling both during uptake of radiolabel and after chasing by changeover to medium lacking the radioisotope but containing excess of cold N-acetylglucosamine. The likelihood that addition of a cold chase might affect the steady-state growth conditions seems small provided that the culture is not carbon- or nitrogen-limited (in which case the N-acetylglucosamine could be used as a growth substrate) but an alternative would be to grow an auxotroph in limiting amounts of the required nutrient. Pulses of radioactive nutrient would be rapidly and completely taken up and could be provided in amounts too small chemically to affect balanced growth.

In addition to its obvious value in permitting the study of the effect on wall turnover of parameters such as growth rate, cell density, pH and nutrient limitation, chemostat culture makes it easy to follow labelling for extended periods of chase. This is particularly valuable for studies designed to determine the kinetics of turnover. Although the kinetic order is important in relation to mechanism, it can be difficult to distinguish zero-order and first-order kinetics in experiments that are terminated before the majority of the labelled material has been removed. For example, turnover that results in the loss of 55 per cent of the original wall label in a given period would at all prior times give results for the proportion of label retained that would differ from each other by no more than 6 per cent, whether loss was zero or first order; such differences can be within experimental error. In chemostat studies it is quite easy to extend the period of observation, even where turnover proceeds relatively slowly, and this can clarify the kinetic analysis. Thus by following turnover for several generation times in chemostat cultures of *B. subtilis*, Clarke-Sturman & Archibald (1982) were able to show that substantial proportions of labelled wall (in fact the polar caps) undergo much slower turnover than the cylindrical wall and, further, that kinetics derived from batch experiments conducted before this phenomenon was known considerably understated the rate at which the turnover-susceptible portion of the wall is removed.

2. Transitional growth conditions

It is important to note that since balanced growth is inevitably disturbed by pulse and changeover experiments some reservation must be exercised in applying strict kinetic analysis to the results, though it seems valid to do so provided growth rate is unaffected. As explained in section B(ii), a change in growth rate resulting from altered medium composition would lead to a change in bacterial density in the chemostat. Experiments in which the culture density remains constant must therefore have proceeded with no gross effect on growth rate, and so these may be analysed with little ambiguity. The rate at which wall chemistry changes during transitions depends on turnover kinetics as well as growth rate, as is discussed further below.

(vii) Use of chemostat culture to study wall assembly and the regulation of wall polymer synthesis

The location of newly synthesized material in the wall, and its accessibility to surface probes such as phages and lectins, has been studied in *B. subtilis* undergoing transitions to and from phosphate limitation. These transitions have been effected both by medium changeover (Archibald & Coapes, 1976; Archibald, 1976) and by injection of small amounts of inorganic phosphate solution directly into the growth chamber (Anderson *et al.*, 1978). The latter technique leads to transient release of phosphate limitation and can be used to study very brief pulses of teichoic acid synthesis. The duration of such pulses has been determined by observing whether whole bacteria, taken at intervals from the chemostat, can synthesize teichoic acid when incubated in medium containing chloramphenicol and excess phosphate (Anderson & Archibald, 1981).

Chemostat culture has been particularly valuable in studies on the biochemistry of the regulation of teichoic acid and teichuronic acid synthesis. Anionic polymer synthesis has been examined in bacteria growing under controlled nutritional conditions in the steady state (Cheah *et al.*, 1982) or in transition (Hancock, 1983). Phosphate-starved batch cultures of bacilli lose the ability to synthesize teichoic acid and acquire the enzymes necessary for synthesis of teichuronic acid, but the bacteria soon stop growing; this precludes several kinds of observations that can, however, be carried out in chemostat culture. For example, Hussey *et al.* (1978) showed that after changeover to phosphate limitation, *B. licheniformis* lost CDP-glycerol pyrophosphorylase and poly(glucosylglycerolphosphate) polymerase activities at a rate that showed that not only had enzyme synthesis stopped but that the enzymes were subject to specific inactivation during phosphate-limited growth. Similar studies with *B. subtilis* W23 showed that the bacteria lost CDP-ribitol pyrophosphorylase at a rate that was consistent with washout but not with specific inactivation, whereas they lost CDP-glycerol pyrophosphorylase activity at a rapid rate consistent with specific inactivation (Chea *et al.*, 1981). This kind of information can clearly not be obtained from batch culture since the bacteria stop growing soon after they reach phosphate starvation.

(viii) Kinetics of growth in continuous culture

Monod (1950) proposed that the relation between growth rate and substrate concentration is similar to that between reaction rate and substrate concentration for enzyme catalysed reactions. That is

$$\mu = \mu_{max} \, S/(K_S + S) \qquad (1)$$

where μ is the specific growth rate constant, μ_{max} is the maximum growth rate, S is the substrate concentration in the growth chamber, and K_S, the substrate constant, is the substrate concentration at which the culture grows at half the maximum rate.

Although entirely empirical in its application to growth (for discussion, see Button, 1985) this relation fits the available data quite well. Growth rate in chemostat culture is therefore determined by the substrate concentration. Changes in growth rate, effected by changing the rate at which medium is pumped into the chemostat, must ultimately depend on changes in the steady-state substrate concentration. The relation between growth rate, dilution rate and substrate concentration can be derived (Herbert, 1958) from the equation

$$dx/dt = \mu x - Dx \qquad (2)$$

where x is the bacterial concentration in the chemostat, μ the specific growth rate constant and D the dilution rate.

If conditions are such that the bacteria can grow at a rate greater than D (that is $\mu > D$) then dx/dt is positive and the bacterial concentration increases. As a consequence the utilization of substrates increases, so that the concentration of substrates in the growth chamber falls. Since (other things being equal) growth rate is determined by the substrate concentration, as the bacterial density increases and the residual substrate concentration falls the growth rate will decrease until it becomes equal to D. At this point growth μx is exactly balanced by washout, Dx, so that the bacterial concentration becomes constant and a steady state is reached; that is $dx/dt = 0$ and so $\mu = D$.

Thus the growth rate is the same as, and is fixed by, the dilution rate. This can be varied up to a maximum value above which the culture is washed out (i.e. $D > \mu_{max}$ so that Dx/dt is negative). Because K_S values are low (of the order of 10^{-6} M or less for many substrates) the steady-state concentration of the limiting substrate is very low. Correspondingly low concentrations of substrate are reached in batch culture only at the very last stages of the exponential growth phase, but since substrate is not being added continuously growth stops almost immediately since residual substrate is quickly depleted. In the chemostat substrate is continuously added so that the bacteria continue to grow exponentially. The mean generation time or doubling time, T_D, in the chemostat is related to D in just the same way as it is to the specific growth rate constant in batch culture:

$$T_D = \ln2/D \tag{3}$$

(ix) Kinetics of synthesis of wall polymer

In bacteria growing under balanced or 'steady-state' conditions the rate constant for synthesis of all cellular components is, by definition, the same, and is given as the specific growth rate constant μ as shown in equation (2). Thus

$$dW/dt = \mu W \tag{4}$$

where W is the amount of cell wall present.

In many bacteria, wall material is shed by turnover so that the total amount of cell wall material present is the sum of that present in the cells and in shed 'turnover product'. During balanced growth both the cell wall and the turnover product will increase at the growth rate, so that the ratio a of the turnover product T to intact cell wall W will be constant (Pooley, 1976). Provided that there is no direct secretion of wall polymers from the bacteria the total amount of wall material present (WM) will be the sum of that present in the bacteria (W) and that present in the culture fluid $(Tg = aW)$:

$$WM = W + aW = (1 + a)W \tag{5}$$

It follows that although the rate of increase of cell wall is that given by equation (4), the rate of synthesis of cell wall material (peptidoglycan plus associated polymers) is given by:

$$dWM/dt = \mu WM = \mu\,(1 + a)W \tag{6}$$

Thus, because of wall turnover greater amounts of cell wall material are made than would be expected simply from the rate of growth of the intact cell wall. It might therefore be expected that the rate of wall polymer synthesis will be determined by both the bacterial growth rate and the extent of wall turnover, so that the greater the turnover (i.e. the greater the value of a in equation (6)) the greater will be the amount of wall material synthesized. Alternatively, however, it is also possible that at least in some cases, altered turnover does not affect the rate of wall polymer synthesis. For example, changes in the turnover ratio a, caused by mutation or by changes in growth conditions, would not necessarily affect wall polymer synthesis if they were accompanied by compensatory changes in the proportion of the total wall polymer that remains in the intact cell wall. If cessation of turnover were accompanied by an increase in the wall content of the bacteria, so that at the new steady state it was $(1 + a)$ times as much as previously, the rate of production of wall material would be unaffected. The cellular wall content would thus represent a balance between synthesis and turnover of wall material.

(x) Rate of incorporation of new material into the wall

The rate of change in wall chemistry following pulse or changeover experiments has been studied in *B. subtilis* during transitions to and from phosphate limitation. Ellwood & Tempest (1969) pointed out that material is constantly removed from the chemostat by washout so that any bacterial component that ceased to be synthesized would be removed at the washout (dilution) rate; that is

$$X_t/X_0 = e^{-Dt} \qquad (7)$$

where X_0 and X_t are the concentrations of that component present at the time its synthesis ceased and at some later time t.

Similar considerations would apply to any cellular component that ceased to be synthesized even if growth continued at the original rate. Therefore, if at the moment of changeover to phosphate limitation synthesis of teichoic acid stopped totally, but the teichoic acid present in the wall was not excised or degraded, then the cell wall teichoic acid content would diminish at the rate shown in equation (7).

Similarly, if at the moment of changeover synthesis of teichuronic acid started and continued at the overall rate of biomass synthesis (that is, at the growth rate), its concentration in the walls would increase at a rate given by

$$Z_t/Z_s = 1 - e^{-Dt} \qquad (8)$$

where Z_t is the concentration of the new polymer at time t after changeover, and Z_s is the final steady-state concentration of teichuronic acid in the wall.

Equations (7) and (8) can be made general by replacing X_t and Z_t by WX and WZ to represent respectively the 'old' material made before changeover and the 'new' material synthesized after changeover. Obviously these equations are applicable to the composition of the walls only if changeover does not affect the bacterial growth rate. This condition was met in changeover experiments described by Ellwood & Tempest (1969) since the bacterial density was very nearly constant throughout. However, the changes in wall composition observed in these experiments proceeded much more rapidly than predicted by equations (7) and (8), showing that the newer polymer was incorporated at a greater rate than expected and that the older polymer was more rapidly removed. Correspondingly rapid changes were seen following changeover in the opposite direction. These observations showed that the change in wall chemistry was a phenotypic change and was not due to the selection of variants having different wall chemistry; they also showed that the wall was undergoing a rapid turn-over so that changes in wall chemistry resulting from altered growth conditions occur more rapidly than would be expected from the overall rate of biomass synthesis (i.e. the growth rate). Subsequent work has shown that this turnover of wall material is not special to changes in composition but that it proceeds even during growth under steady-state conditions.

In wild type bacilli, turnover product can constitute more than a third of the total wall material present in the chemostat, so that, as described in section B(ix), substantially more wall is synthesized than would be neces-sary simply to provide material for new intact wall. New wall material is, however, not susceptible to removal by turnover until a considerable time, which may be designated t^*, after it is incorporated. The value of t^* can be of the order of one to two generation times (Pooley, 1976; Archibald & Coapes, 1976). Since turnover involves removal of only older wall material, all of the new wall material will remain in the wall and this will contain a greater proportion of 'new' material than forecast by equations (7) and (8). The amount of new material in the wall can be derived as follows.

Since, during balanced growth, $W = W_0e^{\mu t}$ (see equation 4), the rate of synthesis of wall material shown in equation (6) can be written

$$dWM/dt = \mu\,(1 + a)\,W_0e^{\mu t} \tag{9}$$

from which the amount of wall material synthesized between zero time and time t is

$$WM_t - WM_0 = W_0\,(1 + a)\,(e^{\mu t} - 1) \tag{10}$$

When $t < t^*$ all of this new material will be present in the intact cell wall, which thus contains $(1 + a)$ times as much new wall as would otherwise be expected.

If new wall is resistant to turnover until a time t^* after it is incorporated, then for $t < t^*$ the amount of intact wall present will be the sum of the wall material synthesized from time zero to time t and the amount (WX) of wall material present initially that has not been removed by turnover up to time t:

$$W_t = W_0e^{\mu t} = WX + W_0\,(1 + a)\,(e^{\mu t} - 1) \tag{11}$$

Therefore

$$WX/W_0 = 1 - a(e^{\mu t} - 1) \tag{12}$$

and

$$WX/W_t = e^{-\mu t} (1 + a) - a \tag{13}$$

The intact wall, W_t, consists of WX, the remaining proportion of the wall that was present at time zero (the time of changeover), and WZ, the wall material made after time zero. Therefore

$$WZ/W_t = (1 + a) (1 - e^{-\mu t}) \tag{14}$$

Equations (13) and (14) thus describe the fall in concentration of older polymer and the rise in concentration of the newer polymer that would be expected following changeover. They can more generally be applied to wall of different age as well as of different composition. With appropriate values of a they give a reasonably good fit to the experimental data reported by Ellwood & Tempest (1969) and Archibald & Coapes (1976).

The above treatment considers only the earlier stages of changeover up to the time when the newer wall material becomes susceptible to removal by turnover. Thereafter the kinetics can become rather complex, particularly where different regions (pole and cylinder) undergo turnover at different rates.

Perhaps the simplest case reported is that where after time t^* turnover proceeds at a first-order rate (Pooley, 1976). The subsequent loss of wall in this case is given by

$$- dW^*/dt = kW \tag{15}$$

where W^* is the material in the wall that is older than t^* and k is the specific turnover rate of the turnover-susceptible fraction of the wall. Integration and substitution in equations (12) and (13) gives

$$WX/W_0 = [1 - a(e^{\mu t^*} - 1)] e^{k(t^*-t)} \tag{16}$$

and hence

$$WX/W_t = [1 - a(e^{\mu t^*} - 1)] e^{k(t^*-t)-\mu t} \tag{17}$$

Provided that t^* is replaced by t when $t^* > t$, equation (16) gives a complete description of the time course for the loss of wall material present initially during growth and turnover for time t.

Equation (17) gives the proportion of material in the wall that is older than t. This equation thus describes the change in concentration of a wall component, for example teichoic acid, whose synthesis stopped at $t = 0$, for example, by changeover to phosphate limitation. The rise in concentration of new material, for example teichuronic acid following changeover to phosphate limitation, can similarly be calculated on this basis. Equations (16) and (17) show a good fit with the experimental data for *B. subtilis* provided that allowance is made for the presence of a wall fraction that undergoes slower turnover.

The above equations describing changes in wall composition during growth and turnover of the wall are, in essence, similar to those derived for turnover by de Boer *et al.* (1981) on the basis of a particular wall model in which turnover involves removal of material from an outer compartment that contains all of the old wall. The de Boer model has been extended by Koch & Doyle (1985) who have carried out computer simulations. These mechanistic models have been used to show that the mechanism of wall growth in bacilli results in certain relationships between growth and turnover; for example, that the turnover rate constant and the growth rate constant will normally have the same value.

The kinetic analysis described here depends only on the postulate (Pooley, 1976), supported by experimental observations on *B. subtilis* by several authors, that new wall is initially resistant to turnover but after a delay becomes susceptible to removal by a process that follows first-order kinetics. This analysis does not depend on any particular model of wall assembly and metabolism, and it does not impose restrictions on possible turnover rates. It can be used to describe wall metabolism and composition changes. While this treatment gives a satisfactory description of what is observed, particularly if allowance is made for the fraction of the wall (concentrated at the poles) that is resistant to turnover, it will be noted that the kinetics are complex and may only approximate to the 'initial complete resistance and then first-order loss' profile described: several mechanistically different models give results which approximate closely to these kinetics.

C. *IN VIVO* RADIOLABELLING OF CELL WALL COMPONENTS

It is often valuable, in studies of cell walls, to have information on the kinetics of synthesis and the sites of insertion of wall components into the intact envelope structure. This is best gained by specific radiolabelling of the component of interest during growth. The isotopic label also aids subsequent purification of the compound in question. The ideal label would be one that identified a single molecular species in the cell wall. In practice, labelling of a particular class of wall polymer is the best that can be achieved, though in a few cases, for example where a wall contains only a single type of polysaccharide or teichoic acid, the result is the same. In most cases it is impossible to label a specific component of a mixture of polysaccharides or proteins *in vivo*. Even if a polysaccharide contains a unique sugar it is unlikely that addition of the radioisotopically labelled sugar to the medium would result in its uptake and incorporation into polysaccharide in an unchanged form—most rare sugars are formed from the common monosaccharides at the sugar phosphate or sugar nucleotide level and bacteria are therefore unlikely to contain the appropriate kinase or, more importantly, sugar nucleotide pyrophosphorylase for activation of the added sugar. For proteins, the rare case in which the molecule of interest lacks a particular amino acid species—the Braun lipoprotein of the *E. coli* outer membrane is an example—can be exploited by introducing a general protein metabolic label in a medium that lacks the missing amino acid, but normal growth is inhibited in such a situation and continued synthesis of the protein of interest will depend on the stability of its mRNA. Even attempts to label a particular class of compound can be complicated. The choice of an amino acid label for protein must avoid amino acids that are found in the peptidoglycan of the species in question, or that could be metabolized to form such a component. Sugars, such as glucose or *N*-acetylglucosamine, that are likely to be taken up and incorporated into polysaccharides or lipopolysaccharides are also likely to be metabolized to a wide range of other sugars, labelling other components such as peptidoglycan and glycolipids, or may even be catabolized and cause general labelling of celullar constituents. [^{32}P]-phosphate can be used to label teichoic acids or lipopolysaccharides, but labelled nucleic acids and phospholipids may cause problems during isolation and measurement of the wall components. Peptidoglycan is frequently labelled by addition of radioactive *N*-acetylglucosamine to the growth medium, but the sugar is also incorporated into lipopolysaccharides, teichoic acids and polysaccharides that contain amino sugars.

The use of appropriate mutants can overcome some of the problems of general metabolism of the added label and may also prevent reduction of the specific radioactivity of the label by dilution with endogenously synthesized unlabelled material. An elegant example of the use of a mutant strain was provided by the studies of Osborn and Leive's groups on the incorporation of lipopolysaccharide (LPS) into the outer membrane in *Salmonella* and *E. coli* (Osborn *et al.*, 1962; Levy & Leive, 1968). They used mutants that lacked UDP–galactose–4–epimerase and therefore were unable to inter-convert UDP–galactose and UDP–glucose. In a wild type strain UDP–galactose would be synthesized either from UDP–glucose or from galactose by way of galactokinase and UDP–galactose pyrophosphorylase, but in the mutants only the second pathway was available and in the absence of an exogenous supply of galactose no UDP–galactose synthesis was possible. Synthesis of the galactose-containing repeating unit of the *O*-antigenic side-chains of the LPS in the mutants was therefore dependent on addition of galactose to the growth medium.

Addition of exogenous radiolabelled galactose in the presence of excess glucose 'switched on' LPS synthesis and the labelled sugar was incorporated undiluted into LPS without any redistribution of the label into other sugars. All the radiolabel in the cell wall was found in LPS, as galactose. A similar technique has permitted the use of radioactive glycerol as a label for teichoic acid and phospholipids in *Bacillus subtilis* without general metabolism of the compound, in a glycerol auxotroph lacking glycerol-phosphate dehydrogenase (Mindich, 1970).

Because metabolic labelling techniques are not entirely specific in practice, the use of *in vivo* labelling for quantitative experimentation requires a method for the preparation of cell walls completely free of contamination by other cellular constituents, and an efficient and well-understood technique for fractionating the wall into its component macromolecules. Table 2.1 lists some published applications of *in vivo* radiolabelling to cell wall studies. Labelling of proteins is discussed in Chapter 4F and a method for specifically labelling peptidoglycan in LysA mutants of *E. coli* is described in Chapter 5B. A more general method for labelling peptidoglycan for studies of wall growth kinetics, turnover and autolysis is given below.

Table 2.1 Radiolabelling wall polymers in growing cultures.

Wall component	Label added to culture	Distribution of label	References
Peptidoglycan	[³H]- or [¹⁴C]-N-acetylglucosamine in presence of excess glucose	95% in wall, of which 57% in peptidoglycan (S. aureus) or 80% in peptidoglycan (B. licheniformis)	Wong et al. (1974) Elliott et al. (1975)
	[¹⁴C]-glucosamine	In wall as peptidoglycan and lipopolysaccharide, in E. coli and N. gonorrhoese	Greenway & Perkins (1985)
	[³H]-diaminopimelic acid	In wall as peptidoglycan, in LysA mutants of E. coli	See Chapter 5B
Teichoic acids	[³H]-glycerol in presence of excess glucose	All label in wall was in teichoic acid in B. licheniformis, B. subtilis	Elliott et al. (1975) Glaser & Loewy (1979)
	[¹⁴C]-glucose	Label in all wall components	Rosenberger (1976)

(i) Labelling of peptidoglycan by growth in the presence of N-acetyl[^{14}C]-glucosamine

Bates and Pasternak (1965) showed that, within the concentration range 0.01 to 1.5 mM, radioactive N-acetylglucosamine was taken up by *Bacillus subtilis* and incorporated into peptidoglycan at rates proportional to the weight of biomass and independent of the N-acetylglucosamine concentration. The label appeared in both amino sugar components of the peptidoglycan, but also in the galactosamine-containing teichoic acid of the cell wall. Based on this information, Pooley (1976) developed a procedure for *Bacillus subtilis* 168 which is applicable to a wide range of bacteria. Bacteria are grown in a liquid medium containing 50 μM of N-acetyl[^{14}C]glucosamine (40–50 μCi/μmol) for at least five generations, to give steady-state labelling. Media containing yeast hydrolysates and extracts, or digested meat products, will contain variable amounts of N-acetylamino sugars that will affect the specific activity of the added radioactive sugar. Whenever possible, therefore, such constituents should be avoided. If necessary, acid-hydrolysed casein can be used as a source of amino acids and peptides. Incorporation of label into peptidoglycan is measured as described by Rogers & Forsberg (1971) as follows:

(a) A 1 ml sample of culture is mixed with an equal volume of 10% w/v aqueous trichloroacetic acid (TCA) and filtered on to a 25 mm diameter glass-fibre filter (Whatman, GF-C).
(b) The filter is washed at room temperature with 5% TCA containing unlabelled N-acetylglucosamine (10 mM).
(c) The filter is transferred to a glass vial or beaker and heated at 90°C for 10 minutes in 5% TCA.
(d) The filter is washed with phosphate buffer (0.1 M in phosphate, pH 7.5) until the washings are neutral, then incubated in 2 ml of the same buffer containing trypsin (1 mg/ml) for 7 h at room temperature.
(e) The filter is washed with absolute ethanol, then diethyl ether, and dried for 1 h at 105°C.
(f) Radioactivity on the filter is measured in a toluene-based scintillation fluid without a solubilizer. The filter must be completely dry. Under these conditions counting efficiency for [^{14}C] should be about 90 per cent. The efficiency can be measured by drying and counting filters on which measured volumes of radioactive N-acetylglucosamine solution of known activity have been absorbed.

3

Isolation and Purification of Cell Walls

A. INTRODUCTION

It is often impractical to isolate cell wall components other than surface appendages (see Chapter 4A) and surface protein arrays (Chapter 4F) from intact bacteria. Even when this is possible, extraction is unlikely to be completely selective and subsequent purification can be extremely difficult. For most purposes, therefore, preliminary isolation of a clean cell wall fraction provides a valuable purification step and permits quantitative studies of wall composition. It must be recognized, however, that in all procedures non-covalently bound wall components, including pili, fimbriae and flagella, will be lost.

On a large scale, efficient separation of walls from cell contents and cytoplasmic membrane is best achieved following mechanical breakage of the cells. Subsequently, repeated extraction with boiling anionic surfactant solutions efficiently isolates the insoluble peptidoglycan sacculus and its covalently attached accessory polymers but also solubilizes all non-covalently bound wall components, including the outer membrane in Gram-negative bacteria. On a smaller scale, the insoluble, covalently linked components of the cell wall can often be obtained in an acceptably clean form by treatment of unbroken bacteria with boiling anionic surfactant.

All effective enzymic methods of cell disruption degrade peptidoglycan and therefore solubilize some wall components, but they can be useful for the isolation of specific subfractions of the wall.

(i) Mechanical cell disruption

Most mechanical disruption processes for bacteria (Table 3.1) rely on the effects of local cavitation produced in a liquid suspension. This can be induced by ultrasonic radiation, by explosive decompression (French Press, Manton–Gaulin Homogenizer) or by rapid agitation with small, rigid beads (Braun Homogenizer, Bead Beater, Dynomill, Mini-Mill).

Table 3.1 Mechanical disruption techniques.

Technique and equipment*	Typical sample size	Time	Breakage (%)	References
Shaking with beads	30–50% wet w/v	1–3 min	75–100	Anderson & Archibald (1975)
Braun Homogenizer	Up to 60 ml			
Bead Beater	Up to 200 ml			
Mini-Mill		25 min		Wickus & Strominger (1972)
French Press	Up to 80% wet w/v 10–50 ml	1–5 min/pass	75–100	Hughes *et al.* (1973) Burnell *et al.* (1980)
Manton–Gaulin Homogenizer	Continuous	Adjustable	Up to 50/pass	
X Press	30% wet w/v Up to 25 ml	<1 min/pass after freezing	20–30/pass	
Ultrasonics	25% wet w/v 20 ml	At least 1.5 min	10–90	Taku & Fan (1979)
Grinding	25% wet cells/ 75% alumina Up to 50 g wet cells	About 5 min	Unpredictable	Bordet & Perkins (1970)

*For suppliers of equipment, see Appendix 2.

Grinding with sand or alumina has also been used, although the mechanism of disruption is unclear and the results are unpredictable. The X-Press and the Hughes Press employ a different principle: compression of a frozen bacterial suspension, at a temperature low enough to prevent melting, causes a phase change in the ice that shears the cell walls. This technique avoids the problems of overheating that may be encountered with the other procedures, but can only be carried out on a small scale and may require several applications to produce complete disruption of the sample, with intervening periods of freezing. On a large scale, shaking with glass beads and the French Press and related techniques are the most useful.

(ii) Purification of walls and outer membrane

Following cell disruption, walls may be recovered by differential centrifugation or by density gradient centrifugation. The latter procedure gives better separation from unbroken cells and from membrane fragments and ribosomes; but however the walls are isolated they often contain cytoplasmic membrane fragments bound to the inner surface of the peptidoglycan layer. Washing with buffers and salt solutions may not remove this material. The choice of the subsequent purification procedure will therefore depend on the purpose of the preparation, and it is best considered for Gram-positive and Gram-negative bacteria separately.

1. Gram-positive bacteria

Because the major wall components are covalently linked, treatment with strongly surface-active agents is generally used, but loss of non-covalently bound wall components such as proteins or lipoteichoic acids is inevitable. Boiling 4% sodium dodecylsulphate (SDS) and aqueous 80% w/v phenol at 0°C both effectively purify the insoluble wall material. Alternative procedures use proteases and nucleases to free the insoluble wall of membrane and cytoplasmic contaminants, but proteases are likely to remove both covalently and non-covalently bound wall proteins as well. Whatever purification process is employed, thorough subsequent washing with saline, then water, is essential. The washed material is best stored lyophilized, and should be white.

2. Gram-negative bacteria

The wall or envelope fraction isolated by centrifugation following cell disruption usually contains some of the cytoplasmic membrane (inner membrane, IM) attached to the peptidoglycan layer. Washing does not remove all the IM and may damage the outer membrane. It is not always possible, therefore, to isolate intact cell wall (that is, peptidoglycan plus outer membrane, OM) completely free of IM, although this has been done very successfully with *E. coli* broken in the French Press (Burnell *et al.*, 1980). OM may be purified but with at least partial degradation of the peptidoglycan layer; the peptidoglycan component, with covalently linked lipoprotein, may be obtained only with destruction of the outer membrane. The latter preparation involves boiling with SDS, as employed for Gram-positive bacteria. The detergents sodium lauryl sarcosinate ('Sarkosyl') and Triton X-100 selectively solubilize IM from the crude envelopes of some strains and this can provide a method for examining OM proteins without degradation of the peptidoglycan layer. The method will not, however, provide intact outer membrane as it results in considerable loss of lipid and lipopolysaccharide; nor is it generally applicable. For a critical discussion of this, see Lugtenberg & van Alphen (1983).

Isolation of pure outer membrane requires treatment with lysozyme to degrade the peptidoglycan layer sufficiently to separate the inner and outer membranes. In some procedures the bacteria are treated with lysozyme before disruption, but this requires permeabilization of the outer membrane with EDTA or by osmotic shock to permit access of lysozyme to the underlying peptidoglycan, and this can result in the loss of OM components or reorganization of its structure (Lugtenberg & van Alphen, 1983). After treatment of the crude wall with lysozyme, the OM and IM can be separated by isopycnic density gradient centrifugation. The OM usually has the higher density, but this cannot be taken for granted— the OM of *Myxococcus xanthus* has the lower density (Orndorff & Dworkin, 1980). During the preparation of OM, hybridization of the IM and OM, or separation of different functional regions of the envelope, may give rise to bands of intermediate density. Density gradient flotation has been used to improve the resolution of these minor components (Ishidate *et al.*, 1986). Because the stabilities of membranes from different strains of bacteria vary widely, the results of the application of any technique to a new strain, even of a previously investigated species, need critical evaluation.

(iii) General precautions for wall isolation

Degradation of peptidoglycan by autolytic enzymes is a serious problem during wall preparation. Autolysin activity can be induced in whole bacteria by anaerobiosis, by osmotic shock and by reagents that disorganize wall or membrane structure, such as EDTA and surfactants. The autolytic enzymes remain associated with the wall on cell disruption and rapidly degrade the peptidoglycan under appropriate conditions. Precautions against autolysis must be taken at all stages of wall preparation, from harvesting of bacteria onwards. All procedures with material in which autolysins may still be active must be carried out quickly and at as low a temperature as possible. Boiling SDS inactivates autolysins, but at lower temperatures detergents may activate autolysis and proteolytic enzymes, so in all techniques employing hot detergents the bacteria or walls should be added to already boiling detergent solution.

Endogenous phospholipase A in the outer membrane of Gram-negative bacteria is activated on cell disruption and by non-ionic detergents and can cause substantial degradation of both the inner and outer membranes (Burnell *et al.*, 1980; Hancock & Williams, 1986). The *E. coli* enzyme requires bivalent metal ions and has a pH optimum of 8–8.5 (Albright *et al.*, 1973), so the presence of a chelating agent such as EDTA together with buffering at a lower pH and maintenance of a low temperature all help to reduce phospholipase activity.

The chemical stability of the wall components and of their attachment to one another must be taken into account. In Gram-positive bacteria the linkage of accessory polymers to peptidoglycan is susceptible to dilute acid hydrolysis owing to the presence of a sugar–1–phosphate group, and is also sensitive to alkaline hydrolysis of a phosphodiester linkage between glycerol and an N-acetylamino sugar. Some teichoic acids have similar acid- or alkali-labile linkages in the main polymer chain. A particularly labile group is the D-alanine that occurs esterified to some of the alditol residues in most wall teichoic acids and lipoteichoic acids. Hydrolysis of this ester linkage is possible at any pH value higher than 6 (Archibald *et al.*, 1973).

B. METHODS

(i) Wall purification without mechanical disruption

This method has been used successfully for the analysis of wall composition in Gram-positive bacteria such as *Streptococcus sanguis* (Reusch, 1981), *Bacillus megaterium* (Gilvarg, 1977), *Bacillus subtilis* and *Streptomyces* spp. It is also effective in isolating the intact murein sacculus of Gram-negative bacteria (see Chapter 5B; Rosenbusch, 1974; Hoyle & Beveridge, 1984).

(a) Wet bacterial paste or lyophilized bacteria (autoclaved bacteria are not suitable) are suspended in ice-cold water (2–4 mg dry wt/ml) and immediately added, dropwise, to an equal volume of boiling 8% sodium dodecyl sulphate (SDS) solution. The mixture is brought to boiling point again as quickly as possible and boiled for 30 minutes with continuous stirring.

(b) The sample is cooled and left at room temperature overnight, then the insoluble wall material is recovered by centrifugation at 30 000g for 15 minutes (for Gram-positive bacteria) or at 130 000g for 60 minutes (for Gram-negative bacteria). Refrigeration must be avoided, to prevent precipitation of SDS.

(c) The extraction with boiling 4% SDS is carried out on the insoluble residue twice more, or until monitoring of the soluble extracts for the presence (see Appendix 1) of protein and organic phosphorus indicates that no more membrane or intracellular material is being removed.

The efficiency of the extraction depends on the sensitivity of the cell membranes to detergent and on the porosity of the cell wall, and thus varies from one strain to another. Reusch (1982) employed five cycles for *S. sanguis*, but we have found three cycles adequate for *Bacillus* and *Streptomyces* species. For the isolation of the murein sacculi of Gram-negative bacteria one extraction is usually sufficient.

(d) The extracted cells are subsequently washed, by resuspension and centrifugation at room temperature, four times with water, twice with 2 M sodium chloride and four more times with water. The use of potassium phosphate buffers must be avoided as it can cause effectively irreversible precipitation of dodecylsulphate in the wall pellet. The final residue may be frozen or lyophilized for storage. During washing, extraction of SDS can be monitored by the method described by Hayashi (1975).

With many species of Gram-positive bacteria the above procedure yields clean walls, as indicated by electron microscopy (see Figure 3.1a) and chemical analysis. However, Reusch (1982) found that walls of streptococci prepared in this way, but at 85°C instead of at 100°C, retained some intracellular material, while White & Gilvarg (1977) reported the retention of poly-β-hydroxybutyrate granules in cells of *B. megaterium* treated with

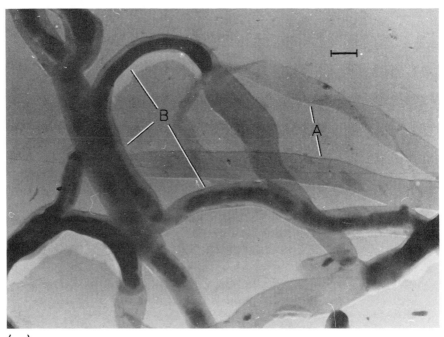

(a)

Figure 3.1 (a) Electron micrograph of *Streptomyces* sp. filaments after the first treatment with boiling SDS (detergent) solution by the method described in section B(i), showing clean, empty cell walls (A) and cells that still contain protein and nucleic acid (B). At the end of the procedure only cell walls, as at (A), are observed. Shadowed preparation. Bar marker: 1 μm.

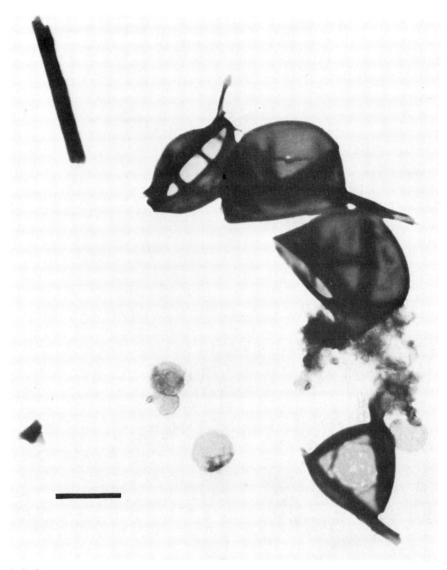

(b)

(b) Electron micrograph of cell walls of *Clostridium perfringens* produced by breakage in a French Press, as described in section B(ii). Note how the ends of the cells retain their shape while the cylindrical parts of the walls roll up. Negatively stained with phosphotungstate. Bar marker: 1.0 μm. See Appendix 1 for staining details.

10% SDS at room temperature. Glycogen may be retained even after prolonged boiling in SDS, but can be removed by digestion with alpha-amylase (from *Bacillus subtilis*; 100 μg/ml in 10 mM Tris-HCl, pH 7.0).

(ii) Procedures for cell disruption

1. *Disruption by shaking with glass beads*

This technique has been used extensively for the large-scale preparation of the cell walls of Gram-positive bacteria, including the genera *Bacillus*, *Staphylococcus*, *Streptococcus*, *Lactobacillus* and *Micrococcus*. It has also proved successful for the preparation of the cell envelope of the Gram-negative bacterium *Methylobacterium organophilum* (Hancock & Williams, 1986) and of the murein sacculi of a variety of Gram-negative bacteria (see, for example, Rosenbusch, 1974).

(a) Freshly harvested bacteria at 0°C are suspended (30% wet w/v) in cold 0.9% NaCl or an appropriate buffer solution, and up to 50 ml of the suspension is shaken with 30 ml of no. 11 glass Ballotini beads for up to 3 minutes in a Braun MSK homogenizer at maximum speed, with cooling by evaporation of liquid CO_2. Unless the homogenizer pot is completely full, the addition of one drop of polypropylene glycol as antifoam is advisable. The length of time taken for disruption of all the cells depends on the strain of bacterium. Cocci are usually more resistant than bacilli.

(b) The glass beads are rapidly removed by filtration through a no. 1 glass sinter, under suction, and the walls are recovered from the filtrate by centrifugation at 0°C (35 000*g* for 15 minutes). When unbroken bacteria are present they form a distinctly visible layer below the cell wall layer in the centrifuge tube. The walls can be scraped off, or gently resuspended in water or buffer, with minimum disturbance to the whole cell pellet. Further unbroken cells may be removed in the same way in subsequent centrifugation steps. Inevitably some loss of wall by entrapment in the whole cell pellet will occur. Density gradient centrifugation may be used to minimize this problem (Bracha *et al.*, 1978).

2. *Disruption in the French Press*

The French Press has found most use in the preparation of the envelope fractions from Gram-negative bacteria, although it has been used to obtain Gram-positive cell walls that retained functional fragments of cytoplasmic membrane and lipoteichoic acid (Hughes *et al.*, 1973). Burnell *et al.* (1980)

have described a procedure for preparing cell wall of *E. coli* (murein plus outer membrane) essentially free of cytoplasmic membrane (inner membrane), from cells broken in the French Press. Since this disruption process does not shear nucleic acids as effectively as shaking with glass beads, it is essential to treat the broken cell suspension with nucleases before purification of the walls. An example of an electron micrograph is shown in Figure 3.1(b).

(a) Harvested bacteria are washed twice in 0.9% NaCl at 0°C and frozen overnight at −20°C.
(b) The thawed bacteria are resuspended in 0.05 M-Tris-HCl, pH 8.0, containing 1 mM MgCl$_2$, 0.2 mM dithiothreitol and 50 μg each of RNAse and DNAse per ml, to a final concentration of 25–50% wet w/v.
(c) The cell suspension is passed through the French Press at about 10^7 Pa. Two passages may be necessary to provide adequate disruption.
(d) Unbroken bacteria are removed by centrifugation at 17 000g for 1 h at 0°C and washed twice with the disruption buffer, with the addition of KCl to 0.2 M.

(iii) Purification of the wall fraction

Subsequent treatment of the crude wall fraction obtained by either of the above disruption procedures depends on the purpose of the preparation. Unless non-covalently attached membrane, protein, polysaccharide or lipoteichoic acid must be retained, treatment with boiling SDS provides the most satisfactory method for simultaneous inactivation of autolysins and removal of contaminating membrane and cytoplasmic material.

1. Detergent treatment

Where it is intended to analyse the amino acid content of the wall peptidoglycan, protein may be removed from the wall either following, or instead of, SDS treatment, by incubation with trypsin at about pH 8. However, at this pH esterified D-alanine is lost from teichoic acids. If this must be avoided, treatment at pH 5 with pepsin or papain is preferable.

(a) The crude wall is suspended in the minimum volume of cold water and poured into boiling 4% SDS to a final concentration of 30 mg dry wt/ml, with stirring. The suspension is kept in a boiling water bath for 15 minutes before recovery of walls by centrifugation (45 000g for 15 minutes at 20°C).
(b) If the pellet remains pigmented or inhomogeneous, the SDS treatment is repeated.
(c) The wall pellet is washed repeatedly, by resuspension and centrifuga-

tion, first with 0.9% NaCl, then with water at 20°C, to remove traces of SDS.

(d) SDS-treated wall is suspended in 0.01 M-Tris-HCl, pH 8.2, to 30 mg wet wt/ml, for trypsin digestion.

(e) The wall suspension is mixed with trypsin to about 40 U/ml (assayed with BAEE as substrate) and incubated with gentle mixing at 30°C overnight, with the addition of a drop of toluene to prevent microbial contamination.

(f) Walls are recovered by centrifugation (45 000*g* for 15 minutes) and washed once with boiling 1% SDS to remove insoluble peptides. The residue is washed repeatedly with water.

2. Treatment with phenol*

A most effective final cleaning of walls, to remove traces of protein, nucleic acids and polysaccharides, is provided by treatment with aqueous phenol, though this may lead to some loss of material owing to the difficulty of recovering the walls.

(a) Walls are suspended in water (5% w/v) and cooled to 0°C, then mixed with an equal volume of 80% w/w aqueous phenol. The mixture is stirred for 30 minutes at 0°C.

(b) The suspension is centrifuged at 15 000*g* for 20 minutes. This should yield an upper aqueous layer, a lower phenol-rich layer and insoluble wall material at the interface. The liquid layers are removed with a Pasteur pipette, leaving the wall layer.

(c) The walls are washed exhaustively with water and recovered by centrifugation.

(iv) Purification of walls with attached outer membrane from Gram-negative bacteria

Density gradient centrifugation provides the most satisfactory separation of the murein–outer membrane complex from inner membrane and cytoplasmic material although, as described in the Introduction, it is not always possible to remove all traces of inner membrane without partial degradation of the peptidoglycan layer with lysozyme. However, the procedure of Burnell *et al.* (1980), based on the work of Schnaitman (1970) and Smit *et al.* (1975), is satisfactory for many strains of *E. coli* and *Salmonella*.

---

*CAUTION—see Chapter 4E for precautions when working with phenol.

(a) The crude wall fraction is washed twice with buffer consisting of 0.05 M Tris-HCl, pH 8.0, 1 mM $MgCl_2$, 0.2 mM dithiothreitol, 0.2 M KCl at 0°C, recovered by centrifugation at 200 000g for 30 minutes at 0°C, and finally resuspended in the same buffer at 0.25 g wet wt/ml.

(b) Stepwise gradients are prepared at 0°C, consisting of equal volumes of sucrose solutions covering the range 40–65% w/w, in 5% w/w increments.

(c) The wall suspension (not more than 15 per cent of the volume of the sucrose gradient) is layered on top of the gradient and centrifuged at 65 000g for 16 h in a swinging-bucket rotor.

(d) The wall–outer membrane complex bands with a buoyant density of 1.24–1.25 g/ml (residual inner membrane bands at about 1.19 g/ml). Material from the denser band is diluted with Tris buffer and recovered by centrifugation at 200 000g for 1 h.

4

Separation and Purification of Surface Components

A. DETECTION AND PREPARATION OF SURFACE APPENDAGES

(i) Pili (fimbriae)

J. E. Heckels and M. Virji

A detailed description of the preparation, characterization, biological properties and immunology of pili from a wide range of species is beyond the possible scope of a book such as this. This section will therefore present general principles which should be applicable to all pili but will focus detailed descriptions largely on studies of pili from gonococci and the experimental procedures which have proved successful in the authors' laboratory.

Early studies used a variety of methods to shear pili from gonococci, followed by precipitation and differential centrifugation to remove other contaminating materials (Buchanan *et al.*, 1973; Punsalang & Sawyer, 1973). Brinton *et al.* (1978) suggested that such preparations are likely to contain other contaminating proteins and they formulated a strategy for purification based on observations of pilus solubility. Pili form crystals under conditions of neutral pH, high ionic strength and in the presence of bivalent cations, whereas they are soluble at high pH, high sucrose concentrations and low ionic strength (Brinto *et al.*, 1978). Pili can therefore be purified by suspending gonococci in conditions favouring pilus solubility and removing bacteria and insoluble debris by centrifugation. The conditions are then adjusted to favour crystallization and pili are separated from soluble impurities by centrifugation. The cycle of disaggregation and crystallization is repeated several times to yield a pure pilus preparation.

SDS-polyacrylamide gel electrophoresis of purified pili shows that they are assembled from a single polypeptide subunit, pilin, with a molecular mass in the region of 20 kDa (Buchanan, 1977; Robertson *et al.*, 1977). The way in which pilin subunits are arranged to form the pilus structure is not yet known and must await a detailed X-ray crystallographic analysis, although one model for pilus assembly suggests analogies with tobacco mosaic virus coat protein (Deal *et al.*, 1985).

1. *Phase variation in pilus expression*

Gonococcal pili were first discovered because loss of their expression on laboratory subculture resulted in a change in colonial morphology. This resulted from the phenomenon of phase variation in which expression of a given phenotype is metastable (Brinton, 1965). Gonococci can exist in a piliated (Pil$^+$) or non-piliated (Pil$^-$) phase and may switch from one to the other spontaneously. In general loss of piliation occurs at a higher frequency than recovery of piliation, so that repeated non-selective subculture in the laboratory ultimately leads to a population that is predominantly Pil$^-$. Fortunately, since Pil$^+$ and Pil$^-$ colonies differ it is possible to select colonies of the correct morphology to maintain clones of either phase for laboratory experiments. Gonococci isolated from the genital tract of patients with gonorrhoea are invariably piliated, demonstrating a selection pressure *in vivo* that confirms the importance of pili in the pathogenesis of the natural infection.

2. *Pilus antigenic variation*

More detailed investigation of gonococcal colonial morphology changes led to the discovery that pili are also subject to antigenic variation, where one strain may express a particular phenotype from a number of different alternatives. A single gonococcal strain growing on laboratory media will produce colonies which vary in their opacity. Colonial variants which produce opaque colonies express one or two molecular species of an outer-membrane protein (PII) from a total repertoire of about six for each strain (Swanson, 1978; Lambden & Heckels, 1979). Colonies that differ in opacity differ in the nature of the PII species expressed and those with a transparent phenotype produce no PII. When pili were purified from colonial opacity variants of strain P9 those from the transparent variant were found to be assembled from a single pilin polypeptide of 19.5 kDa while those from the opaque variant contained pilin of 20.5 kDa (Lambden *et al.*, 1980). Subsequent studies (Lambden *et al.*, 1981) showed that the association with opacity variation was fortuitous and that pilus variation is independently regulated. Four different pilus types could be identified and purified from variants of strain P9. A more systematic method of obtaining pilus variants involves selecting Pil$^-$ colonies, subculturing these, selecting Pil$^+$ revertants and repeating the procedure (Hagblom *et al.*, 1985). By this means successive generations are obtained which frequently express pili with different subunit molecular weights. The total repertoire of pilus expression by any single strain remains unknown, but at least 12 different forms have been reported (Swanson & Barrera, 1983) and recent genetic studies suggest that the potential diversity may be even greater (Haas & Meyer, 1986).

Four pilus types, alpha, beta, gamma and delta, have been isolated from variants of strain P9 (Lambden *et al.*, 1981) and used to immunize rabbits. Each of the resulting antisera showed only 5–15 per cent cross-reactivity with the three heterologous pilus types (Virji *et al.*, 1982), confirming that variations in subunit molecular weight are associated with antigenic differences between the pili that are expressed by variants of a strain.

3. Detection of pilus expression

Electron microscopy. The only generally applicable method for detection of pilus expression is direct observation of pili by means of electron microscopy. This can be most easily achieved by negative staining of freshly grown cultures. In our laboratory gonococci are suspended in phosphate-buffered saline (PBS) to a concentration of about 10^8 cfu/ml and a drop of the suspension is placed on a carbon/formvar-coated copper grid. After 1 minute excess liquid is removed and replaced with 2% uranyl acetate which, after a further minute, is also removed and the grids are allowed to dry. Grids are then examined by transmission electron microscopy (Lambden *et al.*, 1980).

Pili appear as filamentous appendages surrounding the bacterial cell. The number, distribution and size of the pili observed varies widely between different species. With some species, such as *Escherichia coli*, the exact morphological appearance of the pili produced by different strains may be one of the important criteria in identifying their type (Brinton, 1965). The pili expressed by different variants of a single strain of gonococcus may also differ markedly in morphology, particularly with respect to the degree of pilus–pilus aggregation that is observed (Lambden *et al.*, 1980; Perera *et al.*, 1982).

Haemagglutination. Since the pili of many bacteria bind to erythrocytes of a variety of animals, this property can be used to detect and quantify pilus expression in appropriate cases. Haemagglutination has proved particularly successful in purification of pili since it can be used to monitor each step of the procedure. A typical method is the one Salit and Gotschlich (1977) used in their study of Type 1 pili from *E. coli*. Serial two-fold dilutions of the bacteria or partially purified pili in PBS (25 μl) were made in the wells of a round-bottomed microtitre plate. An equal volume of an erythrocyte suspension (0.25%) was added and after mixing the plate was incubated for 2 to 3 hours. The haemagglutination titre was recorded as the minimum concentration of bacteria or pili that gave visible haemagglutination. Less quantitative procedures examine haemagglutination by mixing the bacterial suspension with erythrocytes on a glass slide.

Immunological detection. The availability of high-titre antipilus antisera provides convenient methods for detection of pilus expression by the immunochemical techniques that are described in Chapter 6.

4. *Purification of gonococcal pili*

The standard method for the isolation of gonococcal pili is a modification of that of Brinton *et al.* (1978) which uses cycles of alternate treatment with ethanolamine buffer at high pH to disaggregate pili, followed by precipitation of pilus crystals in the presence of ammonium sulphate (Duckworth *et al.*, 1983).

(a) Gonococci are grown overnight on an appropriate solid medium to maintain the piliated phase. They are harvested (typically 2 g wet weight) directly into 0.15 M ethanolamine–HCl buffer of pH 10.2 (25 ml).
(b) The cell pellet is homogenized for 2 minutes with a motor-driven pestle or in the small chamber of a Sorvall Omnimixer.
(c) The suspension is centrifuged at 20 000g for 30 minutes to remove bacteria and cell debris.
(d) Saturated ammonium sulphate solution (2.8 ml) is slowly added to the supernate containing the disaggregated pili to achieve 10 per cent saturation and the mixture is stirred gently for 1 h.
(e) The precipitated crude pili are collected by centrifugation at 15 000 g for 30 minutes and stirred in ethanolamine buffer (5 ml) until dissolved. This usually takes about 10 minutes, but much longer may be required for some pilus types.
(f) Contaminating cell debris is removed by centrifugation and the pili are precipitated again with ammonium sulphate (0.6 ml) as in (*d*).
(g) The precipitate is recovered by centrifugation, resuspended in 1 M NaCl containing 0.05% sodium azide and stored at 4°C.

Since pili are normally composed of a repeating array of a single pilin subunit the purity of the final product is conveniently assessed by SDS-PAGE (see below). If necessary, the pili can be subjected to further cycles of disaggregation and precipitation until a product of the required purity is obtained.

5. SDS-polyacrylamide gel electrophoresis

SDS-polyacrylamide gel electrophoresis (SDS-PAGE) provides an important method for analysis of pilus purity and is an essential tool for immunological analysis of pilus expression by Western blotting or RIP (see below). Pili can usually be depolymerized to pilin by the usual reaction with SDS and 2-mercaptoethanol, but in some cases—for example *E. coli* Type 1 pili—more vigorous dissociation conditions are required (Eshdat *et al.*, 1981). Because of their variable and relatively low molecular masses we have found that optimum resolution of gonococcal pilins is achieved by separation on gradient polyacrylamide gels. Whole bacterial lysates (containing 10–20 μg protein) or purified pili (1–5 μg) are subjected to SDS-PAGE on linear gradients of 10–25% w/v polyacrylamide at 200 V for 16 h (Zak *et al.*, 1984). Purified pili can be visualized by staining with Coomassie blue. Staining of gels with silver to detect small amounts of other proteins and LPS (Hitchcock & Brown, 1983) will reveal the presence of outer membrane components, which are the most usual contaminants. Pili are most readily identified in whole cell extracts separated on SDS-PAGE by use of specific antisera and Western blotting (see Chapter 6D(iii)).

6. Radiolabelling of pili

The ability to obtain radiolabelled pili is essential for a variety of structural and immunological studies. Depending on the experiments to be performed, pili may be labelled on the intact bacteria or after purification.

Surface labelling of piliated bacteria. The standard method of surface labelling with ^{125}I and lactoperoxidase has proved to be unsuccessful with gonococcal pili. Methods using the chloramine-T reaction are more useful. Bacteria are harvested and suspended in PBS containing 2×10^{-6} M potassium iodide to give an A_{550} of about 6. The suspension is added to a glass tube containing one Iodobead (Pierce Chemical Co.), followed by 200 μCi Na^{125}I. The tube is incubated at room temperature for 10 minutes with occasional shaking, then the contents are diluted with 0.75 ml of PBS. The suspension is centrifuged to pellet the bacteria, which are then washed with PBS and resuspended in PBS containing freshly added PMSF (1 mM) and sodium azide (0.05%). This procedure labels both pili and outer membrane proteins (Tinsley & Heckels, 1986). Surface antigens labelled by this method are particularly useful in antibody detection by RIP (see Chapter 6D, vii).

Labelling of purified pili. Purified pili can be labelled readily using ^{125}I and chloramine-T (Virji & Heckels, 1984). Pili (1 mg) are suspended in 0.05 M potassium phosphate buffer of pH 7.5 (0.5 ml) and reacted with 500 μCi Na[125]I and 1% w/v chloramine-T solution (0.05 ml). After 5 minutes at room temperature the reaction is terminated by the addition of 0.3 M KI (0.5 ml) and 0.13 M $Na_2S_2O_3$ (0.5 ml). Saturated ammonium sulphate solution (1 ml) is added to precipitate the pili, which are recovered by centrifugation at 20 000g for 1 h and washed three times with 50% saturated ammonium sulphate solution. The pili are finally resuspended in 1 M NaCl containing 0.05% sodium azide and stored at 4°C.

In a comparative study of pilus structure it was found that the chloramine-T reaction labels too few tyrosine residues to be useful for peptide mapping. For such studies pili are first carboxymethylated with iodoacetic acid, then free amino groups are labelled with iodohydroxy-benzimidate (Lambden, 1982) or Bolton–Hunter reagent.

7. Production of antipilus antisera

Polyclonal antisera. Since pili may be obtained in high purity the production of specific antipilus antibodies is relatively straightforward. The method described in Chapter 6A(ii) is suitable, although quite different schedules may be equally effective. Typically the antisera produced by this method give a titre of $1-10 \times 10^6$ when measured by ELISA with purified pili as antigen.

Monoclonal antibodies. These are powerful tools for probing both pilus immunochemistry and the biological effect of antibodies directed against specific epitopes. The procedures that have been used to produce monoclonal antibodies directed against gonococcal pili have been described in detail elsewhere (Virji et al., 1983; Heckels & Virji, 1985). In general three types of antipilus monoclonal antibodies are likely to be obtained from the methods described: those that are highly specific and react only with a single pilus type, those that react with a limited range of pili, and those which cross-react with all pili of a particular class. Antibodies with limited antigenic cross-reactivity are most likely to be useful in serotyping, while those with wider reactivity are more suitable for diagnostic purposes and for detection of epitopes that may contain important functionally conserved domains. The reactivity of monoclonal antibodies can be monitored by ELISA using purified pili from the immunizing strain as antigen. However, for detecting the reactivity of an antibody with large numbers of strains a method that does not require pilus purification, such as 'dot blotting' (see Chapter 6) is required.

(ii) Flagella

Flagella are responsible for the motility of bacteria. They are longer and thicker than pili, but their preparation is in many respects similar. The number of flagella per cell and the phase of growth of the bacteria will influence their yield. For most purposes an actively growing exponential-phase liquid culture is the best source for isolation of flagella. As with many other cell surface components growth conditions such as temperature, aeration and medium composition affect formation of flagella. For each species the optimum growth conditions should be established, either by monitoring motility by phase contrast microscopy or by detecting the presence of flagella by electron microscopy of negatively stained preparations (see Appendix 1). The methods for isolation of flagella are all based on the preliminary removal of the appendages from the cell surface by shearing forces. Their subsequent purification is based on a range of conventional biochemical techniques.

1. *Preparation of flagella*

(a) Motile bacteria are harvested by centrifugation from a liquid exponentially growing culture (1 litre) at $10\,000g$ for 15 minutes. The bacteria are resuspended *gently* in 1 litre of PBS, pH 7.4, and recovered by centrifugation again.

(b) The bacteria are resuspended in 20 ml of PBS and homogenized in a rotating-blade (Waring-type) blender at maximum speed for 1 or 2 minutes.

(c) Bacteria are removed by at least two cycles of centrifugation ($10\,000g$ for 15 minutes), monitored by microscopy.

(d) The bacteria-free supernate is centrifuged at $100\,000g$ for 1.5 h to pellet crude flagella. This can be confirmed by preparing a negatively stained sample with 2% phosphotungstic acid or, preferably, by shadowing and examination by electron microscopy (EM; Figure 4.1).

At this stage the degree of purity required should be assessed. If intact flagella are required, the following stages (e) to (g) should be sufficient. If highly purified flagellin protein is required it will be necessary to continue to stage (j) and then follow standard protein purification techniques.

(e) 5 ml of caesium chloride solution (1.3 g cm^{-3} density) is prepared by dissolving solid CsCl (optical grade) in PBS, pH 7.4, to give a refractive index of 1.3630 (approx 2.125 g of CsCl is required). The solution is added to the pellet of flagella from (d) and the flagella are thoroughly resuspended. The suspension is centrifuged at $180\,000g$ for 20 h in a swinging bucket rotor (Chandler & Gulasekharam, 1974).

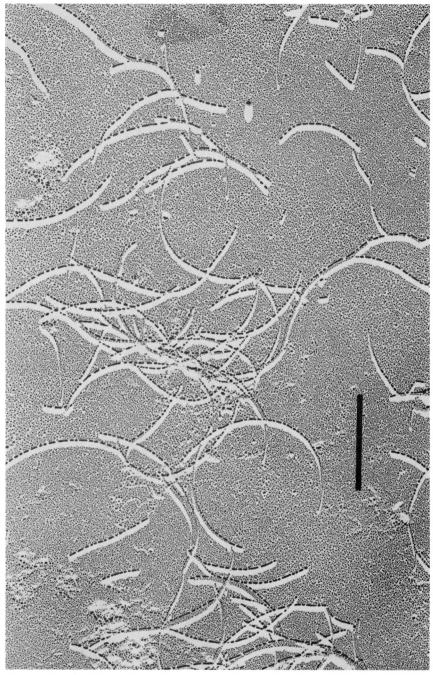

Figure 4.1 Gold-shadowed electron micrograph of purified flagella from *Clostridium sordelli*. Bar marker: 1 μm. For shadowing technique, see Appendix 1.

(f) Flagella should form a band near the centre of the gradient, and are collected by puncturing the tube with a 26-gauge syringe needle at the position of the band and aspirating into the syringe. The presence of flagella is again monitored by EM.

(g) The CsCl solution is diluted in about 10 ml of water and the flagella recovered by centrifugation at 100 000g for 1 h. The pellet might not be visible by eye, but should be resuspended in water and lyophilized.

Further purification may be achieved by the method developed for *Campylobacter jejuni* by Logan and Trust (1983). This involves dissociation of the flagella into flagellin subunits, removal of insoluble contaminants by centrifugation and reaggregation of the flagellin.

(h) The flagella are suspended in distilled water and the pH adjusted to 2.0 with 1 M HCl. The suspension is allowed to stand for 15 minutes at 0°C to ensure complete dissociation of the flagella.

(i) Insoluble material is removed by centrifugation at 100 000g for 1 h.

(j) The pH of the supernate is adjusted to 7.0 with 1 M NaOH, and after 30 minutes the reassociated flagella are recovered by centrifugation and lyophilized.

The final stages can be monitored by EM and SDS-PAGE.

B. PEPTIDOGLYCAN

Cross-linked peptidoglycan reflecting the structure of the polymer in the cell wall is usually isolated as an insoluble residue following treatment of whole bacteria (see Chapter 5B) or isolated walls (see Chapter 3) with hot detergent, acid, or both. Since both the glycan backbone and the cross-linking peptides are very susceptible to degradation by autolytic enzymes, rigorous precautions need to be taken during harvesting and wall preparation (see Chapter 3) if the native structure of the peptidoglycan is to be retained. Treatment with detergent or acid alone does not provide peptidoglycan entirely free of other wall components. The accessory wall polymers of Gram-positive bacteria (polysaccharides and teichoic acids) remain covalently attached to peptidoglycan during detergent treatment of walls. Subsequent mild acid hydrolysis cleaves these covalent linkages but leaves the linking phosphate groups attached to C6 of as many as 10 per cent of the muramic acid residues of the peptidoglycan. Neither detergent nor acid removes covalently linked proteins found in the walls of both Gram-positive and Gram-negative bacteria, nor the lipids from mycobacterial walls (see section I). Subsequent treatment with phosphomonoesterases and proteases is necessary to remove residual phosphate and protein from peptidoglycan. Removal of extractable lipids from the walls is essential during preparation of peptidoglycan from mycobacteria.

Soluble, uncross-linked peptidoglycan that is essentially nascent polymer never incorporated into the wall is excreted by a variety of Gram-positive bacteria, including *Bacillus* and *Streptococcus* species (Waxman *et al.*, 1980, and references therein) when they are incubated in a suitable medium in the presence of penicillin. Peptidoglycan obtained in this way is free of accessory polymers. If it is prepared from lytic-deficient (lyt⁻) strains it is of very high molecular weight and carries pentapeptide side-chains. It has been used, in radioisotopically labelled forms, for examining the enzymic activities of autolysins, D-carboxypeptidases and penicillin-binding proteins.

Establishment of the purity of a peptidoglycan preparation is best done by quantitative measurement of the amino sugars and amino acids in the material by the methods described in Appendix 1.

(i) Isolation of peptidoglycan from Gram-negative bacteria

(a) 'Sacculi' are prepared by treatment of whole bacteria with boiling SDS as described in Chapter 3B(i).

(b) Covalently bound lipoprotein is removed by treatment with pronase (from *Streptomyces griseus*). Any contaminating peptidoglycan hydrolases in the pronase should first be inactivated by heating a solution of the enzyme (10 mg/ml) in 10 mM Tris-HCl, pH 7.0 for 2 h at 60°C.

(c) The sacculi (material from 1 litre of mid-exponential culture in 2.5 ml of 10 mM Tris-HCl, pH 7.0) are incubated with treated pronase (200 μg/ml) at 60°C for 1 h.

(d) An equal volume of 8% SDS is added and the mixture is incubated at 100°C for 15 minutes.

(e) The mixture is diluted with three volumes of water and insoluble material is recovered by centrifugation at 130 000g for 60 minutes at room temperature, then washed three times with water.

This provides pure peptidoglycan from Gram-negative bacteria. If trypsin treatment is used to remove covalently linked lipoprotein it leaves a residual lysine from the C-terminus of the lipoprotein attached to about 10 per cent (in *E. coli* K12) of the muramic acid residues of the peptidoglycan.

(ii) Isolation of cross-linked, insoluble peptidoglycan from Gram-positive bacteria

(a) The method described for the preparation of protein-free walls by treatment with boiling sodium dodecylsulphate solution and trypsin after mechanical disruption of the bacteria (see Chapter 3B) is applied.

(b) The insoluble peptidoglycan is recovered by centrifugation at 130 000*g* for 60 minutes at room temperature.

(c) The water-washed insoluble material obtained from (b) is suspended in 0.1 M HCl (10 mg/ml) and incubated at 100°C for 1 h, or until no more phosphorus-containing material is released into solution (see Appendix 1 for analysis of phosphorus).

(d) Insoluble material is again recovered by centrifugation and washed twice with water.

(e) The insoluble residue is suspended in 0.1 M ammonium carbonate (5 mg/ml) and incubated with calf intestinal alkaline phosphatase (200 U/ml) at 37°C under toluene for 20 h.

(f) The insoluble peptidoglycan is recovered by centrifugation at 75 000*g* for 30 minutes, washed twice with water and lyophilized.

(iii) Isolation of cross-linked peptidoglycan from mycobacteria (based on Petit *et al.*, 1969)

(a) Bacteria are broken by shaking with glass beads, by ultrasonic treatment, or in the French Press and crude walls are isolated by centrifugation. The crude walls are heated at 100°C to inactivate autolytic enzymes, treated with trypsin (Chapter 3B) and lyophilized.

(b) Free lipids are extracted with chloroform/methanol (Hunter *et al.*, 1983) as follows. Lyophilized cells are extracted twice with chloroform-methanol (2 : 1 by volume; 40 ml/g dry wt wall) at 50°C, then once with chloroform-methanol (1 : 1 by volume; 20 ml/g dry wt wall).

(c) Bound mycolic acids are removed by treatment with 0.5% w/v KOH in ethanol for 48 h at 37°C, and the residue is washed with diethylether.

(d) Covalently-linked arabinogalactan is removed by hydrolysis in 0.05 M H_2SO_4 for 48 h at 60°C. The sulphuric acid is replaced every hour for the first 10 h to prevent accumulation of insoluble lipooligosaccharides.

(e) The peptidoglycan is washed repeatedly with water and lyophilized. It retains phosphomonoester groups originating in the attachment sites of arabinogalactan to peptidoglycan. These could be removed enzymically as described in section (ii)(e) above.

Draper *et al.* (1987) have described a procedure for small amounts of mycobacterial biomass in which walls are prepared essentially as described for mechanically broken cells in Chapter 3, then treated with organic solvents to remove free lipids before extraction of mycolic acids in methanolic KOH and arabinogalactan in acid.

(iv) Preparation of soluble, uncross-linked peptidoglycan

This method has only been demonstrated to be applicable to Gram-positive bacteria. Where a lyt⁻ strain is not available, the product is likely to be contaminated with wall autolysis products. The method described here is modified from that of Waxman *et al.* (1980) and is applicable to several species. Where modifications might be necessary, they are indicated in the text.

(a) Bacteria in the exponential phase of growth (about 2×10^8 cells/ml) are harvested quickly, by centrifugation at 10 000*g* for 5 minutes or, on a small scale, by membrane filtration at room temperature.
(b) The bacteria are washed once by rapid resuspension in 20 per cent of the original culture volume of potassium phosphate buffer, 80 mM in phosphate, pH 6.8, and centrifugation or filtration as above.
(c) The washed bacteria are resuspended in 20 per cent of the original culture volume of prewarmed (37°C) incubation medium. This consists of:
      80 mM potassium phosphate, pH 6.8
      0.2 mM D-alanine (plus, if required, ^{14}C-alanine, 3 μCi/ml, about 150 μCi/μM)
      1 mM L-glutamic acid
      1 mM diaminopimelic acid (mixed isomers) or L-lysine, depending on the amino acid composition of the wall peptidoglycan
      27.5 mM D-glucose
      3 μM thiamine
      8 μM nicotinamide
      1 mM MgCl$_2$ and 1 mM MnCl$_2$
      Penicillin G at 10 times the MIC for the bacterial strain being used.
(d) The suspension is incubated, with shaking at 200 r.p.m. in an orbital incubator, at 37°C for 30 minutes.
(e) Bacteria are removed by centrifugation and the supernatant incubation fluid is passed through a membrane filter (0.45 μm pore size) to remove residual cells. The filtrate is heated at 100°C for 5 minutes to inactive autolysins.
(f) The filtrate is concentrated by rotary evaporation, lyophilization or ultrafiltration to 20 per cent of its original volume and applied (up to 10 ml) to a column (1.2 $\times$ 100 cm) of Biogel A (0.5 m) equilibrated in water at room temperature. Other suitable gel-filtration media are Biogel P-300 (Biorad Labs), Sephacryl-300 and Sepharose CL-6B (Pharmacia Chemicals). The column is eluted with water and the eluant monitored for hexosamine (see Appendix 1), radioactivity or absorption of light at 220 nm.
(g) The peak of material containing peptidoglycan, eluting nearest the

void volume of the column, is bulked and lyophilized. Yields of about 10 mg/l starting culture can be obtained.

C. TEICHOIC ACIDS AND OTHER ACCESSORY CARBOHYDRATES FROM GRAM-POSITIVE BACTERIA

The walls of Gram-positive bacteria usually contain accessory polymers covalently linked to the peptidoglycan that constitutes the main structural element of the walls. Although proteins do occur (see Chapter 4F), the principal accessory polymers are usually polysaccharides or the related, phosphate-containing, teichoic acids. They can represent as much as 60 per cent of the weight of the wall and are frequently the dominant antigenic determinants at the cell surface. Acid hydrolysis of walls that contain accessory polymers yields muramic acid 6-phosphate that is believed to originate in phosphodiester linkages attaching the accessory polymers to peptidoglycan. As described in Chapter 1, the attachment of teichoic acids, the acidic polysaccharides known as teichuronic acids and of several polysaccharides to peptidoglycan, has been shown to involve phosphodiester linkages between C1 of a terminal sugar in the accessory polymer and C6 of a muramic acid residue in the peptidoglycan; and the lability of the sugar 1-phosphate group to dilute acid is the basis for the most generally applicable extraction procedure for accessory polymers. In addition, the linkage unit attaching teichoic acids to peptidoglycan contains a glycerophosphoryl-N-acetylhexosamine group that is exceptionally labile to alkali and this provides an alternative isolation procedure for this particular group of polymers. Both these techniques, however, may lead to more or less extensive degradation of the accessory polymer, depending on its precise structure, and corresponding loss of antigenic activity. Extraction into dilute aqueous dimethylhydrazine or phenylhydrazine, by a mechanism that is thought to involve a free radical attack, has been used successfully for retention of maximum serological activity (Anderson *et al.*, 1969). Where more than one type of accessory polymer occurs in a wall preparation, judicious choice of extraction conditions with acid or alkali may permit selective extraction (Duckworth *et al.*, 1972; Hughes, 1970a).

Because of the risk of polymer degradation during chemical extraction processes, it may sometimes be preferable to solubilize them by selective enzymic degradation of the cell wall, using muramidases, amidases, peptidases or autolysins. Such procedures give rise to polymers still attached to fragments of peptidoglycan and present more complex problems of purification than the selective chemical extractions, but provide undegraded accessory polymers.

In general, the isolated accessory polymers have molecular weights below 15 000 and may be purified by gel filtration, ion-exchange chromatography and affinity chromatography.

(i) Extraction with acid

The rate of extraction of accessory polymers into acid varies with different walls, but complete extraction cannot always be achieved without some degradation, particularly of teichoic acids. Some teichoic acids contain sugar 1-phosphate linkages that are as acid-labile as the attachment to the cell wall. The course of extraction may be followed by measuring (see Appendix 1) teichoic acid (as P), teichuronic acid (as uronic acid) or polysaccharide (as total hexose, N-acetylhexosamine or pentose) in the insoluble wall residue, at time intervals. Trichloroacetic acid has been used most frequently, but it leads to some degree of chain fragmentation of teichoic acids. Given walls of high purity, treatment of walls with buffers (pH 2–4), at 100°C may be used (Kojima *et al.*, 1983; Pavlik & Rogers, 1973) and can provide a degree of selectivity between polymers. For example, Pavlik & Rogers (1973) reported selective extraction of the polymer containing galactosamine and glucose phosphate from walls of *Bacillus subtilis* at pH 4, and of teichuronic acid from walls of *Bacillus licheniformis* at pH 3, with retention of teichoic acid on the wall in both cases.

1. *Extraction with trichloroacetic acid*

(a) Lyophilized cell walls (1 g) are stirred in 20 ml of 5% w/v trichloroacetic acid (TCA) for 24 h at 4°C (for teichoic acids) or at 37°C (for teichuronic acids and polysaccharides) to extract the polymer. All subsequent operations are carried out at 4°C.

(b) Insoluble material is removed by centrifugation at 20 000*g* for 30 minutes. If necessary, extraction may be repeated.

(c) Polymer is precipitated from the TCA extract by mixing with five volumes of ethanol for 16 h, and recovered by centrifugation at 30 000*g* for 15 minutes. The precipitate is redissolved in 5 ml of 5% TCA and any insoluble material removed by centrifugation. 25 ml of cold ethanol is added to precipitate the polymer, which is recovered by centrifugation, washed with ethanol and then diethyl ether and dried in a stream of air to give a white powder.

(d) As an alternative to (c), where the extracted polymer is to be immediately purified by column chromatography, most of the TCA can be removed from the extract by repeated (×4) extraction, by shaking in a separating funnel with equal volumes of diethyl ether. The aqueous solution is then neutralized with 1 M ammonia. Where retention of esterified alanine on teichoic acid is required, the pH should not be allowed to rise above 6.

2. *Extraction with dilute hydrochloric acid or glycine–HCL buffer*

Faster and more controlled extraction of the polymers may be achieved with dilute hydrochloric acid (Hughes & Tanner, 1968) or with hot, acidic buffers (Kojima et al., 1983). The rate of extraction may vary depending on the structure of the polymer, and the possibility of chemical degradation of the polymer.

Incubation of cell walls of a number of strains of *Bacillus* in 0.1 M HCl at 35°C gave almost complete extraction of teichoic acid and teichuronic acid within 1 h (Hughes & Tanner, 1968). Incubation of walls of *Bacillus subtilis*, *Staphylococcus aureus* and *Listeria monocytogenes* in 0.025 M glycine–HCl, pH 2.5, at 100°C for 10 minutes gave up to 80 per cent extraction of the teichoic acids (Kojima et al., 1983; Kaya et al., 1985), but this procedure is ineffective if the linkage region contains an unacetylated glucosamine residue, as found in *Lactobacillus plantarum* (Kojima et al., 1985).

In both procedures the residual insoluble material is removed by centrifugation and the supernatant is neutralized with 1 M aqueous ammonia, then desalted by dialysis.

(ii) Extraction of teichoic acids with alkali

Alkali can often be used to extract selectively teichoic acid from walls that also contain teichuronic acid or polysaccharide (Hughes, 1970a). It also gives more complete extraction of some teichoic acids, with less degradation, than does acid extraction (Archibald et al., 1969), but alanyl ester residues are lost. It should be noted that the peptidoglycan is also subject to solubilization by alkali in *Staphylococcus aureus* owing to the particular structure of its peptide cross-linking group (Archibald et al., 1969).

(a) Lyophilized cell walls (1 g) are stirred with 40 ml of 0.5 M NaOH for 1 h at 25°C, and insoluble material is removed by centrifugation at 20 000g for 30 minutes.

(b) The extract is passed through a column (2 × 30 cm) of Dowex 50 (NH_4^+ form) ion-exchange resin, washed off with 40 ml of water, and taken almost to dryness by rotary evaporation at 40°C and redissolved in water.

(c) As an alternative to (b), the alkali extract may be neutralized with 0.5 M HCl and dialysed against two changes of distilled water (5 litres), for 18 h each time, at 4°C. Precautions should be taken to prevent the material becoming hot during neutralization.

(iii) Extraction with N,N'-dimethylhydrazine (DMH) and phenylhydrazine (PH)*

This procedure has been used for the quantitative extraction of teichoic acid and polysaccharides from a range of bacteria (Anderson *et al.*, 1969) and for the preparation of serologically active teichoic acid from *Staphylococcus aureus* (Grov & Rude, 1967; Wheat *et al.*, 1984). The process requires oxygen.

(a) DMH is redistilled immediately before use and the fraction of boiling point 61–63°C is collected, dissolved in water to 2% v/v and neutralized (pH 7) with dilute formic acid.
(b) 125 mg lyophilized walls are suspended in water (100 ml) at 80°C and 25 ml of 2% DMF added. The suspension is stirred vigorously at 80°C, in a 1 litre conical flask to provide sufficient aeration, for 2 h.
(c) The suspension is cooled and insoluble material (often as a clear gel) is removed by centrifugation at 30 000*g* for 45 minutes.
(d) The extract is dialysed exhaustively against running tapwater, then distilled water, and freeze-dried, or may be used directly for chromatography.

(iv) Enzymic wall degradation

Isolated walls of Gram-positive bacteria that have not been heated or treated with detergent retain their autolytic activity, even after lyophilization. This activity is due to endogenous autolysin enzymes, which vary in character from species to species. All autolysins catalyse the hydrolytic degradation of peptidoglycan, but they fall into two main categories depending on their site of action: glycosidases, which cleave either the N-acetylmuramyl bonds (muramidases) or the N-acetylglucosaminyl bonds (N-acetylglucosaminidases), and peptidases, which attack various sites in the inter-chain peptide cross-bridges; the commonest peptidase is the muramyl L-alanine amidase that cleaves the peptide chain from the O-lactyl group of muramyl residues in the peptidoglycan (Rogers, 1979). Endogenous autolysins have been used to degrade walls in a number of structural studies (see, for example, Hughes, 1970b) and for the preparation of wall antigens (Grov *et al.*, 1978); but from the point of view of isolating pure accessory polymers the process suffers the disadvantage that the polymer remains attached to a fragment of peptidoglycan that may vary in size depending on the particular autolysins involved. This problem is partly overcome by the treatment of inactivated walls with exogenous lytic enzymes of known specificity, particularly with muramidases,

*CAUTION—DMH and PH are highly toxic and are suspected carcinogens.

which liberate accessory polymers with the minimum of peptidoglycan attached (Ghuysen, 1968). Unfortunately, few of these enzymes are commercially available and of those that are, the cheapest and most readily obtained, hen egg white lysozyme, a muramidase, has two major disadvantages: it is inactive against glycosidic linkages close to the point of attachment of the accessory polymer to the peptidoglycan and against glycosidic linkages to muramic acid residues that carry an O-acetyl substituent at C6 (as in *Staphylococcus*, for example) or lack an N-acetyl group (as in some *Bacillus* species). The *Chalaropsis* lysozyme suffers from neither of these problems, but appears not to be commercially available in pure form. Both the *Chalaropsis* enzyme and mutanolysin, a muramidase from *Streptomyces globisporus*, will attack the walls of *Streptococcus* species that are resistant to egg white lysozyme. Walls resistant to lysozyme digestion because of the presence of unacetylated amino sugars can be rendered sensitive by N-acetylation with acetic anhydride under mild conditions (Araki *et al.*, 1972). Table 4.1 lists the lytic enzymes that have been used most frequently for the isolation of wall components, and Table 4.2 gives references to published work exemplifying their use.

(v) Purification of accessory polymers isolated chemically or enzymically from cell walls

The acidic nature of many accessory wall polymers from Gram-positive bacteria makes anion-exchange chromatography the purification method of choice for most purposes. Gel filtration can also be valuable, particularly

Table 4.1 Lytic enzymes used for isolation of wall components.

Enzyme	Origin	References
Muramidases		
Lysozyme*	Hen egg white	
Mutanolysin—M-1 fraction*	*Streptomyces globisporus*	Siegel *et al.* (1981) Yokogawa *et al.* (1975)
Lysozyme CH	*Chalaropsis* sp.	Hash & Rothlauf (1967)
Streptomyces muramidase	*Streptomyces albus* G	Aksnes & Grov (1974)
Peptidases		
Lysostaphin*	*Staphylococcus staphylolyticus*	Browder *et al.* (1965)
Streptomyces amidase	*Streptomyces* sp.	Ghuysen *et al.* (1969)
L-11 peptidase	*Flavobacter* sp.	Kato & Strominger (1968)
AL-1 peptidase	*Myxobacter* sp.	Ensign & Wolf (1969)

*Commercially available in the UK, 1987.

Table 4.2 References to techniques for wall polymer isolation using lytic enzymes.

Example	Reference
Isolation of protein A from walls of *Staphyloccoccus aureus* digested with lysostaphin	Sjoquist *et al.* (1972)
Isolation of type and group antigenic determinants from walls of Group B, type III *Streptococcus* digested with mutanolysin	DeCueninck *et al.* (1982)
Characterization of wall components released from walls of *Bacillus subtilis* and *B. licheniformis* by autolysis	Hughes (1970b)
Isolation of the wall teichoic acid from walls of *Listeria monocytogenes* by digestion with lysozyme following chemical N-acetylation of the peptidoglycan	Kaya *et al.* (1985)

for enzymically extracted polymers that may be contaminated with low-molecular-weight wall degradation products, and for polymers contaminated with nucleic acid. Affinity chromatography on immunoabsorbents has also been used successfully, and when sufficient is known about the chemical structure of the polymer the use of immobilized lectins for affinity chromatography can be very effective.

1. Gel filtration chromatography

Nucleic acids are frequent contaminants of cell wall preparations and hence of solubilized wall polymers. A 'stacked' column of two grades of Sephadex (Slabyj & Panos, 1973) provides a satisfactory separation of nucleic acids from acidic wall polymers.

(a) A column, 15 mm diameter, is packed at a flow rate of 40 ml/h with an 18 cm bed of Sephadex G25, followed by a 70 cm bed of Sephadex G75, in 0.2 M LiCl.

(b) The sample (up to 5 ml) is applied, and eluted with 0.2 M LiCl at 40 ml/h. 2 ml samples are collected and monitored for UV absorption (A_{260}) and organic phosphorus, uronic acids or sugars (see Appendix 1) depending on the wall polymer being purified. Nucleic acid emerges first, followed by the wall polymer. Up to 200 ml of eluent may be required. LiCl may be removed by dialysis or by repeated extraction with diethyl ether.

2. Ion-exchange chromatography (see Table 4.3)

Anion-exchange chromatography on DEAE-derived supports such as DEAE-cellulose DE52 (Whatman), DEAE-Sephacel or DEAE-Sepharose CL (Pharmacia) is widely used for fractionation of acidic wall polymers and for separation of anionic and neutral polymers. The carbonate/bicarbonate form of the exchanger has been found most satisfactory for purification of phosphate-containing polymers (see, for example, Anderson & Archibald, 1975) and for the separation of these from carboxylate-containing components (for example, Ward & Curtis, 1982).

(a) The ease of conversion of the resin to the bicarbonate form depends on the initial counter ion on the resin. DEAE-celluloses are usually supplied in the hydroxide form, and are readily converted to the HCO_3^- form by passing through a packed column of the resin in water five bed volumes of 1 M sodium bicarbonate solution. At the end of the process the pH of the effluent should be the same as that of the bicarbonate solution. Conversion of chromatography materials supplied in the chloride form, such as DEAE-Sephacel, may require a much larger volume of the bicarbonate solution—up to 20 bed volumes. In this case, complete displacement of Cl^- must be checked by testing for chloride in the effluent (for example, using the mercuric thiocyanate reagent, available from BDH).

(b) Up to 500 mg of extracted, desalted wall polymer in up to 100 ml of water (adjusted to pH 8 with dilute aqueous ammonia) is applied to a column (2.6 × 50 cm) of the ion-exchanger in the bicarbonate form at 50–100 ml/h.

(c) Neutral polysaccharides are removed by elution with water, then a

Table 4.3 References for ion-exchange separation of cell wall accessory polymers.

Method	Reference
Separation of type and group antigenic determinants from a Group B, type III *Streptococcus*	DeCueninck *et al.* (1982)
Isolation of peptidoglycan-bound protein A from walls of *Staphylococcus aureus* digested with lysostaphin	Sjoquist *et al.* (1972)
Separation of a phosphate-containing polysaccharide and teichoic acid from the walls of *B. subtilis* 168	Duckworth *et al.* (1972)
Separation of teichoic acid and teichuronic acid isolated from the cell wall of *Bacillus licheniformis* by enzymic degradation of peptidoglycan	Ward & Curtis (1982)

linear gradient of 0–0.5 M ammonium carbonate (total volume 700 ml) is applied and the collected fractions are monitored for phosphorus, sugars, amino acids and antigenic activity as appropriate.

(d) Bulked fractions may be concentrated by rotary evaporation or vacuum dialysis and desalted either by repeated rotary evaporation to remove the volatile ammonium bicarbonate, by dialysis or by gel filtration on Sephadex G25.

Depending on their chemical composition, it may not be possible to separate teichoic acids from highly anionic polysaccharides by this method, particularly when the polymers have been obtained attached to peptidoglycan components by the use of wall-lytic enzymes. Ward & Curtis (1982) overcame this problem by using DEAE-cellulose equilibrated in 0.015 M triethylamine carbonate, pH 8.5, and eluting with a linear gradient of 0.015–0.5 M triethylamine carbonate. Under these conditions a teichuronic acid–glycan complex eluted before teichoic acid–glycan.

3. Affinity purification using lectins

This technique has been used mainly for the isolation of teichoic acids, but is widely applicable to polysaccharides and related polymers, given some knowledge of the qualitative sugar composition of the material. Because of the specificity of many lectins for sugars in a glycosidic linkage of a single anomeric configuration, they can be used for the separation of polymers indistinguishable by other chromatographic techniques. Thus, concanavalin A (Con A) coupled to Sepharose 4B (Pharmacia) can be used for the affinity chromatographic separation of the wall ribitol teichoic acid of *Staphylococcus aureus* into two fractions differing in the anomeric configuration of their N-acetylglucosaminyl substituents (Ndule & Flandrois, 1983) because of the specificity of Con A for glucosyl, mannosyl and N-acetylglucosaminyl groups in the alpha configuration. Similarly, Con A will distinguish between glucosylated and unglucosylated chains of the glycerol teichoic acid of *Bacillus subtilis* 168, between the alpha-glucosyl glycerol teichoic acid of *B. subtilis* 168 and the beta-glucosyl ribitol teichoic acid of *B. subtilis* W23 and between two glycerol teichoic acids from the walls of *Lactobacillus plantarum* that differ in the location of alpha-glucosyl groups within the chain or as substituents on the 2-hydroxyl group of the glycerols (Archibald & Coapes, 1971). Advantage has been taken of the affinity of wheatgerm lectin for oligomers of the disaccharide repeating unit of peptidoglycan to isolate teichoic acid chains linked to peptidoglycan following autolysis of *B. subtilis* (Hancock, 1981). This lectin also has sufficient affinity for beta-N-acetylglucosaminyl residues to be used for affinity chromatography of the N-acetylglucosaminyl ribitol teichoic acid of *Staphylococcus aureus* (Ndule *et al.*, 1981). Ricinus lectin was used for the purification of the alpha-galactosyl lipoteichoic acid of a Group N streptococcus (Wicken & Knox, 1975). Lectins are commercially available immobilized for affinity chromatography and in a variety of labelled forms suitable for detection of ligands separated on agarose or polyacrylamide gels or blotted on nitrocellulose, by ELISA, fluorescence or radioactivity. Unlabelled concanavalin A can also be used for detection by taking advantage of its affinity for the glycoprotein enzyme, peroxidase (Clegg, 1982; Faye & Chrispeels, 1985).

Affinity chromatography of wall polymers on Concanavalin A–Sepharose.

(a) A column (1 × 20 cm) of Con A–Sepharose 4B (Pharmacia) is equilibrated at 4°C in a buffer containing 0.1 M sodium acetate, 1 M NaCl, and 1 mM $MgCl_2$, $MnCl_2$ and $CaCl_2$, adjusted to pH 6 with acetic acid (buffer A), with 0.02% w/v sodium azide as preservative.*

(b) Sample is applied in up to 5 ml of buffer A and unabsorbed components are eluted from the column in two bed volumes of the same buffer.

(c) Specifically absorbed material is eluted with 0.1 M methyl alpha-mannoside in buffer A: one bed volume of eluant is applied and then flow is halted for 30 minutes to ensure complete dissociation. Elution with two bed volumes of the mannoside-containing buffer should then give quantitative release of absorbed material. The mannoside may be removed by dialysis or by gel-filtration.

Concentrations of alpha-mannoside up to 0.5 M may be necessary for exceptionally tightly bound components. Amphiphilic molecules such as lipoteichoic acid and lipomannans can only be eluted in the presence of non-ionic surfactants such as Triton X-100 (1% w/v) (Powell *et al.*, 1975).

Affinity chromatography of peptidoglycan-linked accessory polymers on immobilized wheatgerm lectin. Lotan *et al.* (1975) showed that uncross-linked peptidoglycan exhibited a specific interaction with wheatgerm lectin that could be inhibited by glycosides of N-acetylglucosamine and by high concentrations of the free sugar. This interaction permits the isolation of wall accessory polymers that have no affinity for the lectin, when they are covalently linked to soluble peptidoglycan, or glycan fragments. Such complexes may be obtained by autolysis of cell walls, by treatment of walls with peptidases or by partial digestion of walls with muramidases (see above).

---

*CAUTION—Azides should not be present during freeze-drying, because of the risk of explosion.

(a) A column (1 × 20 cm) of wheatgerm lectin–Sepharose 6MB (Pharmacia) (or equivalent) is equilibrated with buffer (buffer A) containing sodium phosphate (0.05 M in phosphate) and 0.2 M sodium chloride, pH 7.0, with 0.01% sodium azide as preservative, at 4°C. (See the foregoing warning about sodium azide.)

(b) The sample is applied in up to 5 ml of buffer A, and unabsorbed material eluted in five bed volumes of buffer A.

(c) Absorbed material is eluted with five bed volumes of buffer A containing N-acetylglucosamine (2.5% w/v). Fractions are collected and monitored for N-acetylamino sugars and wall accessory polymer by appropriate methods (see above). Strongly absorbed material might require elution with a higher concentration of N-acetylglucosamine (10% w/v) or with di-N-acetylchitobiose (Lotan *et al.*, 1975).

D. LIPOTEICHOIC ACID AND OTHER LIPOCARBOHYDRATES FROM GRAM-POSITIVE BACTERIA

Lipoteichoic acids (LTAs) occur in nearly all Gram-positive bacteria, where they are located chiefly on the outer surface of the cytoplasmic membrane, with their poly(glycerolphosphate) chains intercalated with the cell wall. They can be important cell-surface antigens (Lambert *et al.*, 1977; Wicken and Knox, 1975). LTA is excreted during bacterial growth and may undergo interactions with cell wall components during this process. Thus, Ofek *et al.* (1982) have reported a specific interaction with M-protein in Group A streptococci, and Chiu *et al.* (1974) have isolated LTA from walls from ultrasonically disrupted *S. sanguis*. Where LTA is absent, it is replaced by other lipocarbohydrates (Hamada *et al.*, 1980), the most frequent of which is lipomannan (Powell *et al.*, 1975).

Although unpurified cell wall preparations and culture fluids usually contain some LTA, it is often in a deacylated form (Kessler & Shockman, 1979), and LTA is therefore best extracted from whole bacteria or an unfractionated broken cell suspension. The most effective extraction procedure uses aqueous phenol, which also extracts nucleic acids and lipids, especially glycolipids. Therefore, cells are usually defatted before extraction, and purification procedures are designed to remove contaminating nucleic acid, bearing in mind the micellar nature of LTA in aqueous solution. The presence of ester-linked alanine residues on the poly (glycerolphosphate) chain of the LTA is important for several of its properties, and the procedure described below, essentially that of Fischer *et al.* (1980), is designed to avoid alkaline hydrolysis of the sensitive D-alanine ester linkage. Coley *et al.* (1975) evaluated a number of extraction procedures with three different bacterial species and their paper provides useful examples of the variation that might be experienced, as does the comparative study of *Bacillus* LTAs by Iwasaki *et al.* (1986).

Because of the micellar nature of LTA and related polymers in aqueous solution, gel filtration on materials with very large exclusion limits, such as Sepharose or Biogel A, provides the best method of purification, as described below. Ion-exchange chromatography on DEAE-resins gives higher resolution, but only in the presence of non-ionic detergents that can be very difficult to remove completely from the purified LTA. It may, however, be useful for analytical purposes, and suitable procedures are described by Hancock *et al.* (1976) and Fischer *et al.* (1980). The most commonly used non-ionic detergent, Triton X-100, can be reduced to a low level by continuous ultrafiltration using an Amicon XM-50 membrane or other membranes, or hollow fibres with equivalent exclusion limits (Fiedler & Glaser, 1974).

(i) Extraction and purification of lipoteichoic acid

(a) Freshly harvested bacteria are suspended in 0.1 M sodium acetate (adjusted to pH 4.5 with acetic acid) at 800 mg wet wt/ml of buffer, and are defatted by mixing with two volumes of methanol and one volume of chloroform overnight at room temperature. The bacteria are recovered by filtration, washed with two volumes of methanol, and resuspended in 0.1 M sodium acetate buffer, pH 5.0 (500 mg of defatted bacteria per millilitre of buffer). Alternatively, freeze-dried bacteria are extracted with chloroform/methanol (2 : 1 v/v) (20 mg dry bacteria/ml) overnight, twice.

(b) The suspension of defatted cells is mixed with an equal volume of hot 80% w/v aqueous phenol* and stirred constantly for 45 minutes in a water bath at 65°C.

(c) On cooling, an emulsion forms which can be broken into two phases by centrifugation at 4°C at 5000g for 30 minutes. The upper, aqueous layer contains LTA and nucleic acids.

(d) Phenol is removed at pH 5.0 by dialysis in washed dialysis tubing against six changes of 40 volumes of 0.1 M sodium acetate, pH 5.0 at room temperature; the first three stages can be carried out over 4 h each, and the others over a further total of 24 h.

(e) Nucleic acids are degraded by incubation of the dialysed extract with nucleases (ribonuclease A, 5 units/ml; ribonuclease T_1, salt-free, 50 units/ml; deoxyribonuclease II, 30 units/ml; 1 mM magnesium chloride) at 20°C for 24 h with 1 ml toluene added to prevent microbial contamination.

*CAUTION—See section 4E for precautions when working with phenol.

(f) The phenol extraction procedure, as described in (b), (c) and (d)
     above, is used to remove the nuclease protein, and the aqueous phase
     is dialysed against 0.05 M sodium acetate, pH 4.0, and stored frozen.

(g) Purification is carried out by chromatography on Sepharose 6B. A
     column (2.6 × 45 cm) of Sepharose 6B is equilibrated at 4°C with
     0.05 M sodium acetate, pH 4, by upward pumped flow, and the crude
     LTA (50 mg/ml) is applied in the same buffer (up to 2.5 ml). Elution is
     carried out at 15 ml/h with the same buffer, by upward flow, and 3 ml
     fractions are collected. Every third fraction is assayed for absorption at
     260 nm and for organic phosphorus. Because of its micellar form, LTA
     is eluted close to the exclusion volume of the column, just behind a
     small peak of high-molecular-weight nucleic acid. The peak sample of
     LTA should have an A_{260} of less than 0.1 for a total phosphorus
     concentration of 100 μg/ml; degraded nucleic acids elute in a broad
     peak with an elution volume approximately 2.5 times that of the LTA.
     A smaller phosphorus-containing peak of partially or wholly deacy-
     lated LTA may be obtained roughly half-way between the LTA and
     nucleic acid peaks. If nucleic acid contamination, as indicated by A_{260},
     is significant, the material from the LTA peak of the eluate may be
     concentrated by ultrafiltration and rechromatographed (Coley *et al.*,
     1975).

Membrane-associated acidic lipomannans can be extracted from *Micrococcus* species and purified by the same procedure, except that phenol
extraction is carried out at room temperature, nuclease digestion is done at
pH 6.9 in 0.1 M sodium phosphate buffer, and Sepharose chromatography
is carried out in 0.2 M ammonium acetate (Powell *et al.*, 1975).

E. PREPARATION OF LIPOPOLYSACCHARIDE AND
ENTEROBACTERIAL COMMON ANTIGEN

The lipopolysaccharides (LPS) of Gram-negative bacteria and the related
but little understood molecule found in many enterobacteria—the
enterobacterial common antigen (ECA)—are components of the outer
membrane. Their amphipathic properties allow their selective extraction,
by solvents, from whole bacteria. Classically, LPS was extracted from
whole cells with aqueous phenol by the method developed by Westphal
and Luderitz (1954). This procedure was unsatisfactory for rough (R) LPS
and a method involving aqueous phenol, chloroform and petroleum spirit
was developed for extraction of R-LPS by Galanos *et al.* (1969). More
recently, a non-solvent method has been described for preparation of both
S and R forms of LPS for rapid analysis on polyacrylamide gels. The
following sections give details of the various methods for preparation and
purification of these molecules.

(i) Extraction of LPS by aqueous phenol*

(a) Finely divided lyophilized bacterial mass is suspended to a concentration of approximately 5% w/v in distilled water and heated to 67°C in a water bath.

(b) 90% w/w aqueous phenol is prepared by dissolving 90 g of phenol in 10 ml of water at 45°C, and a volume of this solution equal to that of the bacterial suspension is heated to 67°C in a water bath.

(c) The prewarmed bacterial suspension and the phenol solution are mixed and stirred at 67°C for 15 minutes.

(d) The mixture is carefully transferred to centrifuge tubes and cooled in ice until phase separation occurs. *Polycarbonate* tubes must *not* be used.

(e) The tubes are centrifuged at between 5000 and 10 000g for 15 minutes to complete separation of the phases. A swinging-bucket rotor is best for this purpose.

(f) The upper (aqueous) phase, containing the LPS, is carefully removed. In disposing of the phenol phase, the toxic nature of the phenol should be considered.

(g) The aqueous phase is transferred to washed dialysis tubing and dialysed against running tapwater for at least 18 h, until the smell of phenol cannot be detected.

(h) The solution is centrifuged for 15 minutes at 10 000g to remove any insoluble deposit.

(i) The dialysed extract is concentrated by rotary evaporation to approximately one-fifth of its original volume.

(j) The solution is centrifuged at 100 000g for 3 h, when the LPS should sediment to form a clear, gelatinous pellet. If the pellet has an opaque base it is because not all insoluble material was removed in stage (h). The pellet is resuspended in distilled water with the aid of a syringe fitted with a 23-gauge needle and centrifuged again as above. The final pellet is suspended in a small volume of distilled water and lyophilized.

(k) If it is necessary to deionize the LPS and prepare a particular salt form, see section E (iv) below.

(ii) Extraction of 'rough' LPS

(a) The extraction solvent consists of 90% phenol (see section E(i) above), chloroform and petroleum spirit (40–60°C boiling point) in the proportions 2 : 5 : 8 by volume. If the mixture is cloudy a small amount of

---

*CAUTION—Phenol solution is extremely caustic and toxic and should be handled with great care. Suitable decontamination material should be available in case of accidental spillage.

solid phenol should be dissolved in it until a clear solution is obtained.

(b) About 5 g of lyophilized bacteria are thoroughly homogenized in 20 ml of the extraction mixture and stirred for 2 minutes below 20°C.

(c) The mixture is centrifuged at 10 000g for 15 minutes.

(d) The supernate is filtered through Whatman No. 1 filter paper into a round-bottomed flask.

(e) The centrifuge pellet is extracted again as in (b–d) and the extract added to the first.

(f) Chloroform and petroleum are removed by rotary evaporation. If phenol crystallizes out it should be redissolved by addition of the minimum amount of water.

(g) After removal of the organic solvents water is added dropwise to the solution until LPS precipitates. No more water should be added when LPS starts to settle on standing for 1–2 minutes; approximately 0.4 ml of water is required when extraction is carried out in 20 ml of solvent. If too much water is added, phase-separation will occur.

(h) LPS is sedimented by centrifugation at 5000g for 10 minutes.

(i) The centrifuge tubes are drained and wiped out with tissue; the pellet is washed two or three times with 80% phenol and recovered by centrifugation each time.

(j) The final pellet is washed twice in ether and dried under vacuum.

(k) The dry material is taken up in 3–5 ml of water, with warming if necessary, using a syringe and 23-gauge needle, and LPS is sedimented by centrifugation at 100 000g for 4 h.

(l) The LPS is taken up in water and lyophilized. It can be deionized as described in section E(iv).

An alternative method, developed by Qureshi *et al.* (1982), can be substituted for step (g) as some R-LPS species are not fully precipitated by the addition of water. In this variant, six volumes of diethyl ether/acetone (1 : 5 by volume) are added to one volume of the phenol solution. After standing for 1 h the LPS is recovered by centrifugation as in (h) and washed two or three times with ether-acetone. The procedure from (k) above is then followed. This method is recommended for the preparation of R-LPS from *Pseudomonas aeruginosa* R mutants.

(iii) Proteinase K digestion method for the preparation of LPS

This method, developed by Hitchcock and Brown (1983), allows both S and R type LPS to be prepared in amounts suitable for analysis by PAGE and provides information on the chemotype of the LPS before large-scale preparation by one or other of the above methods. The principle of the

technique is that bacterial lysates can be digested with a powerful proteolytic enzyme leaving LPS as the only soluble macromolecular species
present.

(a) Bacteria are harvested by centrifugation from 10 ml of overnight
culture, or from washings from a plate, and washed once in
phosphate-buffered saline (PBS).
(b) The bacteria are suspended in PBS to an A_{525} between 0.5 and 0.6.
(c) The bacteria from 1.5 ml of suspension are sedimented in a microcentrifuge at 1000g for 3 minutes.
(d) The pellet is suspended in 0.05 ml of SDS-PAGE sample buffer
(2% SDS, 4% 2-mercaptoethanol, 10% glycerol, 0.002% bromophenol
blue in 1 M Tris-HCl buffer, pH 6.8) and heated at 100°C for 10
minutes.
(e) 25 μg of proteinase K (Protease Type XI, Sigma) in 0.01 ml of sample
buffer is added, and the mixture is incubated at 60°C for 60 minutes.
(f) 0.01 ml samples are taken for electrophoretic analysis on 14% polyacrylamide gels by the Laemmli (1970) system, omitting SDS from both
the stacking and separating gels.
(g) Gels are stained with silver (see Appendix 1). An example is shown in
Figure 4.2.

This technique can also be applied to envelope or outer membrane
preparations for analysing LPS. A cleaner picture results from the use of
such starting materials. The PAGE-separated proteinase K LPS can also be
analysed by immunoblotting (see Chapter 6 and Pyle & Schill, 1985).

(iv) Deionization of LPS by electrodialysis

The solubility of LPS in aqueous solutions depends on two major
factors. Firstly, the relative proportions of lipid and polysaccharide have
an influence; the greater the polysaccharide content, the higher the solubility. Secondly, LPS is an acidic molecule because of the presence of
phosphate and carboxyl groups; cations of various types neutralize these
negatively charged groups and affect solubility.

Galanos and Luderitz (1975) described a three-chambered electrodialysis cell for deionizing LPS. The apparatus developed by ISCO for
the electrophoretic concentration of proteins works well for deionizing
LPS. Figure 4.3 shows such a cell. It is used as follows.

(a) The LPS solution/suspension (5–10 mg/ml) in distilled water is placed
in the cell and distilled water is placed in the two electrode chambers.
The apparatus is ideally kept cool (below 10°C) in a cold-room.
(b) A voltage of up to 500 V is maintained across the cell.
(c) The pH in the cathodic chamber rises; when it reaches 9–10 the

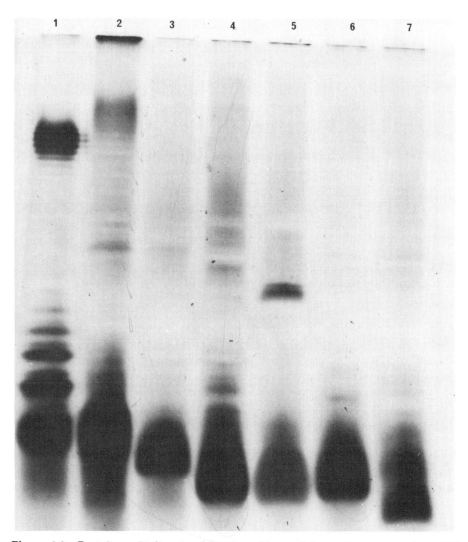

Figure 4.2   Proteinase K digests of rough and smooth bacteria, separated on 14% polyacrylamide gels (without SDS) and stained with silver. The patterns of the LPS are clearly seen against a faint background of undigested protein. Track 1, *E. coli* serotype 086 (a smooth type); track 2, *Salmonella typhimurium* Ra mutant; track 3, *Klebsiella aerogenes* rough mutant 10B (an Rb type); track 4, *E. coli* J5 mutant (an Rc type); track 5, *Pseudomonas aeruginosa* PAC 605 (an Rc type); track 6, *S. typhimurium* Rc; track 7, *S. minnesota* R595 (an Re type).

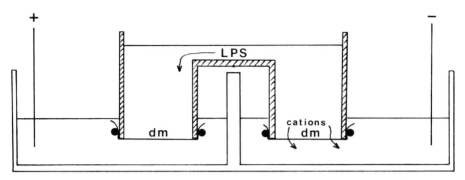

Figure 4.3 The apparatus used for the deionization of LPS. dm: dialysis membrane.

contents should be replaced with fresh distilled water. This should be repeated several times over 3–4 h.

(d) The free acid form of the LPS (deionized) precipitates from solution and is deposited as a gel on the membrane at the anodic end of the cell. It is recovered from the cell by draining off the excess water from the chamber, dismantling the apparatus, gently scraping off the gel and suspending it in distilled water.

The deionized LPS is almost insoluble in water and suspensions of 2 mg/ml produce a pH of 3.2–3.6 (Galanos & Luderitz, 1975). It can be made soluble by neutralization with various bases. Sodium or potassium hydroxide, pyridine, ethanolamine and triethylamine all produce soluble salt forms. The triethylamine salt is the most soluble, allowing the preparation of solutions of over 200 mg/ml. Lowest solubility, or insolubility, is shown by bivalent cation salts.

(v) Preparation of lipid A

Lipid A can be isolated readily from pure LPS by hydrolysis in mild acid at 100°C.

(a) LPS (preferably deionized) at a concentration of 2–10 mg/ml in 1% acetic acid is the usual starting material, although deionized LPS in distilled water produces a low enough pH to allow hydrolysis of the lipid–KDO linkage.

(b) The suspension or solution is heated at 100°C for 60–90 minutes.

(c) The lipid A precipitates from solution and can be removed by low-speed centrifugation. It is washed three times with hot distilled water.

(d) Lipid A can be made soluble by neutralization with potassium hydroxide or triethylamine and heating to 50–60°C.

The material left in solution after hydrolysis is the polysaccharide fraction of the LPS. Details of fractionation of this material into core and O-antigen fractions is described in Chapter 5C.

(vi) Preparation of enterobacterial common antigen (ECA)

This antigen has been recognized since 1962, but only recently has it been isolated and its structure partially determined. In possibly all strains of enterobacteria it is present as a hapten, but in only a few strains (notably rough forms) is it immunogenic in whole bacteria. Its repeating unit has been characterized as a disaccharide of N-acetylglucosamine and N-acetylmannosaminuronic acid with O-acetyl and possibly some fatty acid substituents. Several methods exist for its preparation, but the one described below, which was developed by Mannel and Meyer (1978), is probably the easiest to carry out in most laboratories and parallels the preparation of smooth and rough LPS. The species most commonly used is *Salmonella montevideo* wild type (e.g. strain SH94 from Central Public Health Laboratory, Helsinki, Finland).

(a) An aqueous phenol extract of whole bacteria is prepared as described for the isolation of smooth LPS in E(i)(a)–(g) above, and lyophilized.
(b) The lyophilized material is extracted with phenol–chloroform–petroleum spirit (2 : 5 : 8 by volume) as for the preparation of rough LPS. The LPS is precipitated from the extract with water and removed by centrifugation, as in E(ii)(a)–(h) above.
(c) The phenolic solution is placed in a dialysis tube, with allowance for considerable volume increase by osmosis, and dialysed extensively against running tapwater to remove phenol. The dialysed solution is lyophilized.
(d) The dry material is resuspended in water, then centrifuged at 105 000g for 4 h to remove all traces of LPS. The supernate is lyophilized again.
(e) The dry material is dissolved in 0.5 M ammonium acetate in methanol (pH 6.2) and applied to a column of DEAE cellulose equilibrated in the same solution.
(f) The column is eluted stepwise, first with the 0.5 M ammonium acetate in methanol, then with 1.0 M ammonium acetate in methanol. The ECA elutes in the 1.0 M solution and is lyophilized.
(g) The ECA is applied to the ion-exchange column again, as in (e), and eluted with 0.9 M ammonium acetate in methanol. The effluent is monitored by differential refractometry. The principal peak is lyophilized.

F. ISOLATION AND PURIFICATION OF SURFACE PROTEINS

As described in Chapter 1, proteins may occur in the cell wall covalently linked to the peptidoglycan, non-covalently bound in the peptidoglycan matrix, in organized layers such as the Gram-negative outer membrane and crystalline S-layers, and as surface appendages such as pili and flagella. The isolation of outer membrane proteins and of surface appendages are described in sections G and A of this chapter. In this section we deal with the proteins that occur as S-layers, and those that are tightly integrated into the peptidoglycan layer of the wall. In many cases these proteins are initially detected either by electron microscopy or immunochemically, so too little is known about their properties to permit completely reliable selection of an isolation technique. This section can therefore only be a guide to techniques that have proved successful in particular cases; some experimentation may be necessary to find one appropriate for a newly discovered protein, or for a bacterial strain not previously examined.

(i) Separation and purification of S-layers from Gram-positive and Gram-negative bacteria

P. Messner and U. B. Sleytr

Chemical analysis of a variety of eubacterial and archaebacterial S-layers has shown that most are composed of a single protein species. Occasionally, covalently linked carbohydrate chains have been found. SDS-polyacrylamide gel electrophoresis has demonstrated that the molecular masses of the monomers vary from 40 000 to 220 000 Da (Figure 4.4). Frequently pure S-layer preparations reveal more than one band on the SDS-PAGE gel. This can be explained by (a) the presence of two closely associated S-layers composed of different subunit species (Tsuboi *et al.*, 1982; Abe *et al.*, 1983; Kist & Murray, 1984); (b) stoichiometric amounts of more than one protein species in the surface array (Smit *et al.*, 1981); (c) proteolytic cleavage of S-layer subunits *in vitro*, or in the course of the isolation procedure (Baumeister *et al.*, 1982; Koval & Murray, 1984); or (d) microheterogeneity of the carbohydrate residues (Wieland *et al.*, 1982; Sleytr *et al.*, 1986b). Amino acid analyses of individual purified S-layers have shown that they are acidic proteins with a considerable amount of hydrophobic amino acids and practically no sulphur-containing ones (Sleytr & Messner, 1983; Kandler & Konig, 1985). Freeze-etching techniques provide the most accurate picture of the structure and orientation of S-layers on intact cells. By application of these techniques it has been

(a) (b) (c)

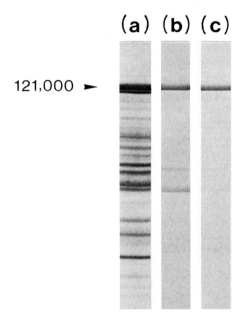

121,000 ►

Figure 4.4 SDS-polyacrylamide gel electrophoresis analysis of the S-layer preparation procedure of *Bacillus stearothermophilus* NRS1536/3c. (a) SDS-soluble whole cell extract; (b) cell wall preparation before the extraction with 5 M guanidine hydrochloride; (c) purified S-layer material. The arrow indicates the molecular weight.

Figure 4.5 Ultrathin sections (left) and schematic diagram (right) of the three main categories of bacterial cell envelopes containing crystalline surface layers. The schematic diagram displays, on the left, the thin-section profiles, in the middle, the molecular architecture showing the major components, and on the right, the freeze-fracture behaviour of cell envelopes. (a) Cell envelope structure of archaebacteria, which lack a rigid cell wall component—micrograph of *Pyrodictium occultum*, strain PL19; (b) Gram-positive cell envelope—micrograph of *Bacillus hydrosulfuricum*, strain L111-69; (c) Gram-negative cell envelope—micrograph of *Aquaspirillum serpens*, strain VHA. CM: Cytoplasmic membrane; CW: rigid cell wall layer in Gram-positive organisms composed of peptidoglycan and/or other polymers; OM: outer membrane; PG: peptidoglycan layer; S: crystalline S-layer. Arrows indicate locations for possible additional S-layers (S_A). Modified diagram from Sleytr & Messner (1983); micrograph of (a) by permission of H. König; (c) reproduced by permission of R. G. E. Murray.

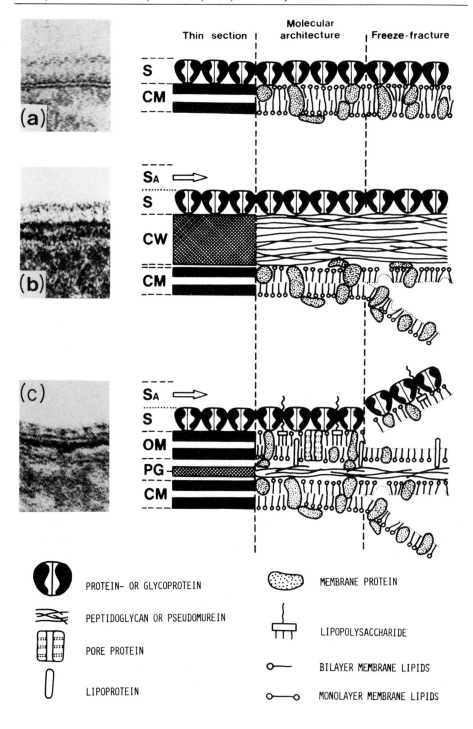

Thin section | Molecular architecture | Freeze-fracture

(a) S / CM

(b) S$_A$ / S / CW / CM

(c) S$_A$ / S / OM / PG / CM

PROTEIN- OR GLYCOPROTEIN	MEMBRANE PROTEIN
PEPTIDOGLYCAN OR PSEUDOMUREIN	LIPOPOLYSACCHARIDE
PORE PROTEIN	BILAYER MEMBRANE LIPIDS
LIPOPROTEIN	MONOLAYER MEMBRANE LIPIDS

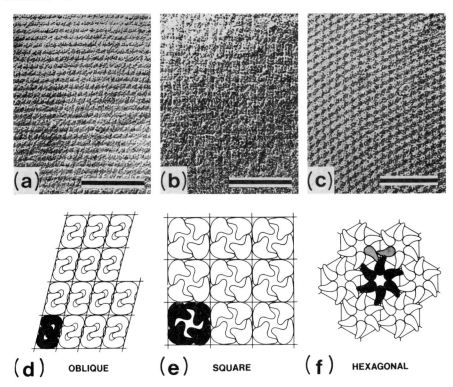

(d) OBLIQUE   **(e)** SQUARE   **(f)** HEXAGONAL

Figure 4.6 Electron micrographs (a)–(c) of freeze-etched preparations of intact cells, and (d)–(f) schematic presentation of the corresponding lattice types of the S-layers. (a) and (d) Oblique (p2) lattice of *Bacillus stearothermophilus* NRS2004/3a; (b) and (e) square (p4) lattice of *Bacillus stearothermophilus*, strain H4-65; (c) and (f) hexagonal (p6) lattice of *Clostridium thermohydrosulfuricum*, strain L111-69. Bar marker: 100 nm.

shown that S-layers cover the entire cell. Freeze-etched, freeze-dried, or negatively stained preparations have shown that S-layers have oblique (p2), square (p4) or hexagonal (p6) lattices (see Figures 4.5 and 4.6). The morphological units, composed of protein dimers, tetramers or hexamers have centre-to-centre spacing in the range 5–35 nm (Sleytr & Messner, 1983; Sleytr *et al.*, 1986a). Chemical analyses of a variety of eubacterial and archaebacterial S-layers have shown that most are composed of single protein species. Occasionally, covalently linked carbohydrate chains have been found. SDS-polyacrylamide gel electrophoresis has demonstrated that the molecular weights of the monomers vary from 40 000 to 220 000.

Isolated S-layer subunits have the ability to assemble *in vitro* into lattices identical to those observed on intact cells on removal of the

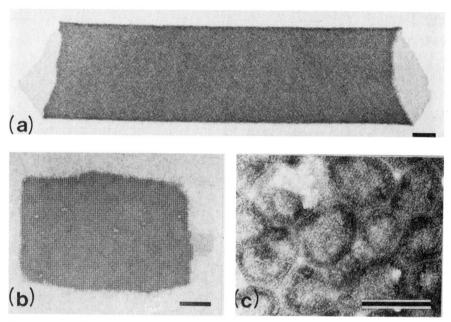

Figure 4.7 Negatively stained preparations of differently shaped *in vitro* self-assembly products. (a) Open-ended cylinder of *Bacillus stearothermophilus* NRS2004/3a; (b) flat sheet of *Desulfotomaculum nigrificans*, NCIB 8706; (c) closed vesicle of *Clostridium thermohydrosulfuricum*, strain L111-69. Bar markers: 200 nm.

disrupting agent that was used for the isolation of the subunits. Depending on the morphology and bonding properties of the subunits (Figure 4.7), open-ended cylinders, flat sheets or closed vesicles have been found as a result of the entropy-driven assembly process (Aebi *et al.*, 1973; Hastie & Brinton, 1979; Masuda & Kawata, 1980; Sleytr & Plohberger, 1980; Sleytr, 1981; Tsuboi *et al.*, 1982).

Recrystallization, reattachment and the functional aspects of S-layers have been discussed in great detail in a number of reviews (Sleytr, 1978; Sleytr, 1981; Sleytr & Messner, 1983). S-layers can be seen as the simplest biological membranes. They provide an excellent model system for studying relationships between structure and function and the dynamic aspects of membrane morphogenesis and biosynthesis.

S-layers are generally isolated from cell envelope fragments prepared by mechanical cell disruption (sonication, French Press, homogenization with glass beads) as described in Chapter 3 or, in special cases, by freeze-thawing. Treatment with nucleases may be necessary before final isolation of the cell walls by differential centrifugation. Treatment with

Table 4.4 General isolation procedures for S-layers.

Procedure	References
Gram-positive bacteria	
Removal of peptidoglycan layer with lysozyme	Nermut & Murray (1967); Sleytr (1976); Beveridge (1979); Masuda & Kawata (1979); Abe *et al.* (1983)
Extraction of S-layer proteins with low concentrations of chaotropic agents (e.g. 4 M urea, 1.5 M guanidinium hydrochloride)	Nermut & Murray (1967); McNary *et al.* (1968); Sleytr & Plohberger (1980)
Extraction of S-layer proteins with high concentrations of chaotropic agents or detergents (e.g. 8 M urea, 5 M guanidinium-HCl, 1% SDS, formamide)	Goundry *et al.* (1967); Howard & Tipper (1973); Sleytr (1976); Sleytr & Thorne (1976); Hastie & Brinton (1979); Masuda & Kawata (1979); Tsuboi *et al.* (1982)
Extraction with chaotropes may be followed by column chromatographic purification (e.g. Sepharose CL-6B in 5 M guanidinium-HCl)	Kupcu *et al.* (1984)
Gram-negative bacteria	
S-layer proteins are extracted from the outer membrane/peptidoglycan layer, or outer membrane, free of cytoplasmic membrane contamination	
Extraction with low concentrations of chaotropic agents	Buckmire & Murray (1970); Thornley *et al.* (1974)
Extraction with metal-chelating agents (e.g. 10 mM EDTA, EGTA)	Thornley *et al.* (1974); Thorne *et al.* (1975); Beveridge & Murray (1976b)
Extraction with detergents and perturbants (e.g. 0.1% non-ionic or zwitter-ionic detergents, 1% sodium lauryl sarcosinate, 2% sodium dodecylsulphate, lithium diiodosalicylate)	Baumeister *et al.* (1982); Evenburg & Lugtenberg (1982); Koval & Murray (1983); Phipps *et al.* (1983)
Extraction by cationic substitution, acid or alkali	Beveridge & Murray (1976a,b); Lapchine (1976)
Further purification may be carried out by column chromatography (e.g. Sephadex G-200, DEAE Sephadex CL-6B)	Buckmire & Murray (1973); Thornley *et al.* (1974); Phipps *et al.* (1983)

detergents such as Triton X-100 to selectively solubilize adherent cytoplasmic membrane fragments contaminating the walls (Schnaitman, 1971) has been used for both Gram-positive and Gram-negative bacteria prior to solubilization of the S-layer (Sleytr & Glauert, 1976; Beveridge, 1979). Table 4.4 summarizes procedures that have been used for extraction of S-layers.

1. Extraction of S-layer proteins from Gram-positive bacteria

The procedure described below has been applied to the isolation of S-layers from several Gram-positive bacteria, including *Bacillus stearothermophilus*, *B. sphaericus*, *Clostridium thermohydrosulfuricum*, *C. thermosaccharolyticum*, and *Desulfotomaculum nigrificans*. The reagents are as follows:

50 mM Tris-HCl buffer, pH 7.2 (buffer 1)
50 mM Tris-HCl buffer, pH 7.2, containing 0.5% w/v Triton X-100 (buffer 2)
5 M guanidinium hydrochloride in distilled water (solution 3)
5 mM calcium chloride in distilled water containing 0.01% w/v sodium azide (solution 4)

(a) 10 g (wet weight) of bacteria are washed twice with 50 ml of buffer 1 at 4°C, then resuspended in 50 ml of ice-cold buffer 1.
(b) The bacteria are disrupted by treatment in a Branson B15-P ultrasonic disintegrator at maximum power output, with continuous cooling in an ice-bath, for a total of 3 minutes in 30 second periods with intervals of cooling. The extent of disruption is assessed by microscopic examination.
(c) The broken cell suspension is centrifuged at 45 000g for 20 minutes at 4°C. The upper, opaque part of the pellet, consisting of crude cell envelope, is carefully scraped off the lower pellet of unbroken cells. The unbroken cells are resuspended and treated again as in (a) and (b).
(d) The combined crude envelopes (peptidoglycan layer with attached S-layer, contaminated with soluble proteins and cytoplasmic membrane) are washed by resuspension in cold buffer 1 and centrifugation as in (c). This is repeated three times, and each time the upper layer of the pellet is retained, while any residual whole cells in the lower pellet layer are discarded.
(e) The washed envelope preparation is thoroughly resuspended and mixed for 15 minutes at room temperature in buffer 2 to solubilize any contaminating cytoplasmic membrane.
(f) The envelopes are recovered by centrifugation at 45 000g for 20 minutes, then washed in cold buffer 1 as in (c), four times.
(g) The purity of the envelope preparation and the retention of S-layer is checked by examination in the electron microscope following negative staining with 1% uranyl acetate with and without fixation of the protein with 2.5% glutaraldehyde in 0.1 M cacodylate buffer, pH 7.2.
(h) S-layer subunits are extracted from the envelopes by homogenization of the pellet in 10 volumes of solution 3 and stirring gently overnight at room temperature (or 60°C for thermophiles). The envelopes are recov-

vered by centrifugation at 45 000*g* for 30 minutes and re-extracted as above. The combined extracts are dialysed for up to 24 h against several changes of solution 4 or 0.01% sodium azide, at 4°C, room temperature or 60°C to remove the chaotropic agent.

(i) The S-layer assembly products that have precipitated in the dialysis bag are collected by centrifugation at 6000*g* for 15 minutes, and stored frozen at −20°C (or at 4°C for electron microscopic studies). Additional purification may be obtained by redissolving the precipitate in solution 3 and repeating (h) and (i).

Column chromatography. The S-layer subunits, extracted as in (h) above, but not dialysed, may be purified by column chromatography.

(a) The extract (about 20 mg protein in 5 ml) is applied to a 2.6 × 100 cm column of Sepharose CL-6B, previously equilibrated in solution 3.
(b) Elution is carried out in solution, 3, elution of protein is monitored at 280 nm and 5 ml fractions are collected.
(c) Appropriate fractions are combined and dialysed as described in (h) above, then treated as in (h) and (i).

2. Extraction from Gram-negative bacteria

A variety of methods have been developed for separating S-layers from the adhering outer membrane (see Table 4.4). Many S-layers from Gram-negative bacteria show a specific ion-dependence for integrity, and consequently these environmental conditions have to be maintained during the purification and self-assembly steps.

(ii) Isolation and purification of proteins linked to the cell wall in Gram-positive bacteria

R. R. B. Russell

For many years work with Gram-positive cell walls employed material that had been treated with proteases to remove contaminating cytoplasmic membrane components. As a result proteins were not observed to be associated specifically with the cell wall unless, as in the case of S-layer proteins, they could be detected on the surface of the bacterium by electron microscopy. If, however, proteolysis is avoided during wall preparation (see Chapter 3), protein components can often be recognized. For example, Doyle *et al.* (1977) subjected *Bacillus subtilis* to sequential extraction with 5 M LiCl, 3% SDS at 100°C, 6 M guanidinium chloride and 8 M urea, agents that should disrupt all except covalent bonds, then assayed for protein remaining linked to peptidoglycan. Because routine colorimetric assays for proteins are inappropriate for use with particulate wall preparations, they used methods for detecting various non-peptidoglycan aminoacids to indicate the level of protein present: chloramine T-mediated radio-iodination which labels tyrosine and histidine, ^{35}S-labelled sodium sulphate which is metabolically incorporated into cystine, cysteine and methionine, and Koshland's reagent (2-hydroxy-5-nitrobenzyl bromide) which specifically alkylates exposed tryptophan residues in proteins. Complete hydrolysis of wall preparations followed by amino acid analysis could also indicate the presence of amino acids not characteristic of peptidoglycan and allow some estimate of the total amount of protein present. Doyle *et al.* (1977) calculated that about 0.6 per cent of the wall dry weight consisted of covalently linked protein, but whether such a value is typical of other micro-organisms is not known. Reusch (1982) applied a similar series of extractions to produce sacculi of *Streptococcus sanguis* and found the protein content to average 16 per cent. It therefore seems likely that a considerable variation in protein content will be found in different organisms.

1. Recognition of wall proteins

In order to progress towards elucidation of the structure and function of wall proteins it is essential to develop a means of monitoring their presence through purification procedures. There are several ways of achieving this:

By labelling. If wall proteins are extrinsically labelled by methods such as those described by Doyle *et al.* (1977), or by metabolic incorporation of radio-isotopes during growth, it is possible to track them during extraction from walls and during subsequent purification steps. The major problem with this approach is that of distinguishing between multiple forms of a single protein (perhaps generated by chemical or enzymic degradation) and multiple distinct wall components. Unless there is some way of recognizing the relationship of different fractions, one may arrive at a totally erroneous perception of the complexity of the wall's protein complement.

By function. The best-characterized proteins from Gram-positive bacterial walls are those that bind immunoglobulins by the Fc region: protein A of *Staphylococcus aureus* and the Fc receptors of streptococci of Lancefield groups C and G. These can be detected readily in their cell-bound form by using agglutination assays or by determining the ability of intact bacteria or wall preparations to bind radiolabelled immunoglobulin. In its soluble form, protein A can be assayed by radial diffusion in agar containing IgG, by ELISA, RIA, or Western blotting, all these techniques depending on the special property of binding to the Fc region rather than to the antibody region usually involved in such tests. The M-protein of streptococci of Lancefield group A is known to have antiphagocytic function, though its presence is generally detected with type-specific antibody. Further comparable characteristic functions include fibronectin-binding and fibrinogen-binding. Others may yet be discovered for other Gram-positive species.

By antibody. In group A streptococci the M-proteins are the immuno-dominant, type-specific antigen. Over 80 different M-types have been recognized and antisera capable of recognizing each type have been raised in rabbits by immunization with intact streptococci. With most species of bacteria, however, such immunization will induce antibodies against a range of surface-exposed antigens including polysaccharides, teichoic acids and lipoteichoic acid, as well as proteins. Furthermore, antibodies to peripherally located proteins, as well as covalently bound ones, will be formed. In order to obtain antiserum specific for wall proteins, therefore, it is necessary to immunize with 'clean' walls. For example, when *Streptococcus mutans* was extracted with SDS at 100°C in order to remove all but covalently bound proteins and the resultant wall preparations used to immunize a rabbit, the serum obtained had antibodies to just two wall-associated proteins (Russell, 1979).

Monoclonal antibodies to wall proteins can be a valuable tool for identifying individual wall components. One must continually bear in mind, however, that monoclonal antibodies raised against native antigens may not recognize proteins that have been denatured during extraction or electrophoretic separation. Furthermore, monoclonal antibodies may not recognize all fragments of a protein and so will fail to detect all derivative degradation products. This problem of heterogeneity of composition due to degradation, together with the possible presence of several different proteins or multiple forms of a single protein, is one reason why monoclonal antibodies on their own (or specific functional assays) are inadequate for resolving the complexity of wall proteins. The availability of a polyvalent antiserum against a wall preparation can be of great help, by allowing a range of highly sensitive antigen detection methods to be used.

2. Isolation of wall proteins

Once a protein has been defined as being associated with the wall, therefore, the researcher attempting purification has the following options:

(a) Enzymatic release

(b) Extraction of protein which is associated with the wall, but not co-valently linked, using chemical agents
(c) Purification of released proteins from the culture supernatant. These may be released by normal wall turnover, but various means are available for enhancing release—cloning the gene for a wall protein into a new host bacterium in which wall-linkage does not occur is a special example of this.

Enzymatic release of wall proteins. If one wishes to obtain a preparation of a protein which is covalently attached to peptidoglycan, one can either enzymatically digest the peptidoglycan so that only a minimal fragment remains linked to the protein or, alternatively, cleave the protein close to the point of linkage. In either case the result cannot be an intact, pure protein. Furthermore, the use of such digestion techniques is fraught with problems arising from the uncontrolled degradative activity of proteases often present in lytic agents. The history of research into staphylococcal protein A and streptococcal M-protein shows that for many years there was considerable disagreement about such fundamental facts as the size and charge of the molecules, because different workers using different extraction methods encountered heterogeneity in their preparations. The problems were only resolved when the proteins were purified from culture supernatants.

Wall-lytic enzymes. A wide range of enzymes catalysing the degradation of peptidoglycan and hence of cell walls has been described, and an account of them can be found in Chapter 4C. It must be emphasized, however, that even for the widely used and commercially available enzymes information is lacking on their contamination by proteolytic enzymes.

Lysozyme—Although this is probably the best-known agent for lysing bacteria, it is not suitable for work on the wall proteins of staphylococci (Sjoquist *et al.*, 1972) or streptococci (Reis *et al.*, 1984). It has, however, been employed by Mobley *et al.* (1983) for the release of protein from the walls of *Bacillus subtilis*. 50 μg of lysozyme for 50 mg of cell walls in 50 mM phosphate buffer (pH 6.5) was used. The susceptibility to lysozyme of some bacteria can be greatly enhanced by prior culture in a medium containing a high concentration of threonine or glycine (Scholler *et al.*, 1983).

Lysostaphin—This enzyme is particularly active against staphylococci and is used in some commercial rapid diagnostic kits. It was successfully used by Sjoquist *et al.* (1972) to release protein A but reproducibility was hard to obtain, presumably because lysostaphin has three separate enzyme activities which attack different linkages in peptidoglycan and the proportions of these vary from batch to batch (Schindler & Schuhardt, 1964).

Bacteriophage lysins—Since virulent bacteriophages lyse their hosts, the enzymes determined by them are potentially useful for wall digestion. Reis and her colleagues (Reis *et al.*, 1984) were able to purify the Fc receptor protein from a culture of group C *Streptococcus* after it had been infected with bacteriophage C1, and found that the released product was less heterogeneous in size and charge than when extracted by several other methods. The lytic enzyme of bacteriophage C1 can also attack walls of streptococci of groups A and G and has been used extensively for studies of the former by Fischetti *et al.* (1971) who have described a procedure for its purification.

Mutanolysin—This lytic agent, produced by *Streptomyces globisporus*, was originally characterized by its activity against *Streptococcus mutans* (Hamada *et al.*, 1978) but is also active against many other streptococci and other Gram-positive bacteria, though not staphylococci. It contains several enzymes which attack the cell wall, though it appears that the N-acetylmuramidase activity (M-1 enzyme) is most important. Many preparations contain an active protease and so further purification by ion-exchange chromatography is generally needed (Siegel *et al.*, 1981; Tille *et al.*, 1986). Alternatively, protease inhibitors can be used (see below). Digestion of walls with mutanolysin at 200 µg/ml in 0.01 Tris–maleate–NaOH buffer (pH 7.0) has proved successful in releasing wall proteins from *Streptococcus mutans* (Russell *et al.*, 1983).

Proteolytic enzymes. In cases where a wall protein has a limited number of sites susceptible to an enzyme such as trypsin or pepsin, it is possible by controlled digestion to release fragments of more or less consistent size. While such fragments may help the early stages of investigation of wall protein, it must be stressed that important features may have been destroyed by the proteolytic cleavage. For a recent example of the use of trypsin, see Reis *et al.* (1985).

Reagents that disrupt non-covalent interactions. Wall proteins may occur in a loosely associated state in the cell wall (even if only transitorily) from which they may be released with reagents such as detergents or chaotropes. As always, concurrent degradation is possible—van de Rijn and Fischetti (1981) obtained evidence consistent with the cleavage of M-protein by a wall or membrane-bound enzyme which was activated only during extraction of the protein with non-ionic detergent.

It seems likely that there is no single agent that will be of general applicability for extraction of wall-associated proteins, and each case may require the investigation of a number of approaches. Among the agents that have proved useful are ionic detergents (sodium dodecylsulphate, lithium dodecylsulphate, cetyl trimethylammonium bromide), non-ionic detergents (Triton X-100, Brij-58), zwitter-ionic detergent (Empigen BB), strong salt solutions (LiCl, NaCl), chelating agents (EDTA), extremes of pH and chaotropic agents (guanidinium chloride, sodium thiocyanate, urea).

The use of protease inhibitors. Degradation of wall proteins by endogenous proteases or by proteases present in lytic enzyme preparations can, in some cases, be overcome by the addition of one or more of the following inhibitors: phenylmethylsulphonylfluoride (PMSF) (2 mM), *N*-*p*-tosyl-ʟ-lysine chloromethylketone (2 mM) benzamidine hydrochloride (2 mM), aprotinin (two trypsin-inhibitor units per millilitre). PMSF should be made up as a concentrated stock solution (0.1 M) in isopropanol and diluted into the required aqueous solution immediately before use, as it has a half-life of only 35 minutes in water at pH 8. It is even more rapidly inactivated in Tris buffers, with which it should therefore be regarded as incompatible.

3. 'Free' wall-associated proteins

Many proteins which have a wall-bound location can also be found free in the culture supernatant. This might be due either to direct secretion, or to release of peptidoglycan-linked proteins by wall turnover. There is considerable variation in the extent to which wall proteins may be found in the free form. Within a bacterial species, different strains may vary in the extent of release; a single strain may show variations depending on how many subcultures it has undergone since primary isolation (Russell & Smith, 1986) or on the conditions of growth (Russell *et al.*, 1983; Knox & Wicken, 1985). In a number of cases release occurred throughout the growth cycle in batch culture, but further release of wall-bound protein may occur owing to autolysis in the stationary phase (Movitz, 1976; Russell, 1979).

Dissolution of cell walls will release wall-bound protein, but if similar procedures are applied to living bacteria in the presence of an osmotic stabilizer so that protoplasts are formed, protein will continue to be synthesized and will be released directly into the supernatant. Stable ʟ-forms, which lack cell walls, have also been successfully used as a source of free wall proteins (Forsgren, 1969; van de Rijn & Fischetti, 1981).

The introduction of techniques for cloning genes in foreign host bacteria has led to rapid progress in our understanding of wall proteins. The full sequences of the genes for protein A (Uhlen *et al.*, 1982; Guss *et al.*, 1984) and type 6 M-protein (Hollingshead *et al.*, 1986) have been published and the sequences of several other types of M-protein are expected soon. It is not the purpose of this book to consider details of genetic manipulation methods, but it may be noted that plasmid, cosmid and bacteriophage vectors have all been successfully used for the initial cloning of wall protein genes in *Escherichia coli*, their products being detected in recombinant gene banks by use of appropriate antisera. No linkage of the proteins to the *E. coli* peptidoglycan seems to occur and they generally accumulate in the periplasm. From this location they can be obtained by ultrasonic disruption or by detergent treatment of the host bacteria.

4. Purification of wall proteins

The purification of wall proteins does not pose any special difficulties and a wide range of biochemical fractionation techniques has been used. The pioneering work in this field has been done by researchers studying major proteins which are believed to be virulence factors of certain pathogens; but with the development of methods for analysing wall proteins, it is likely that proteins will be identified as components of the walls of a wide range of other Gram-positive bacteria.

G. ISOLATION AND PURIFICATION OF OUTER MEMBRANE PROTEINS FROM GRAM-NEGATIVE BACTERIA

P. A. Lambert

(i) Introduction

Outer membrane proteins (OMPs) are usually identified and defined in terms of their mobility on sodium dodecylsulphate–polyacrylamide gel electrophoresis (SDS-PAGE) (see Appendix 1). Staining with Coomassie blue or silver of outer membranes fractionated on SDS-PAGE reveals the OMP profile of the bacterium, which normally comprises several heavily stained bands (the major OMPs) and numerous minor bands, some of which may be contamination of cytoplasmic or inner membrane origin.

The OMP composition of a given species is subject to major genotypic and phenotypic variation. Consequently, careful consideration should be given to strain selection and growth conditions (see Chapter 2) to ensure maximum production of the OMPs of interest before embarking on extraction and purification procedures. Several important decisions must be made in devising a procedure for isolation and purification of any OMP.

1. Selection of an appropriate strain

The strain used should be easy to grow on a large scale and must produce the OMP of interest in satisfactory amounts. Several factors are important here. Firstly, yields of the protein sought can be greatly enhanced by selecting or constructing an overproducing strain. Secondly, it can be worthwhile to choose a strain that lacks, or produces in very small amounts, components that are persistent contaminants during extraction and purification. Such components might be other OMPs of similar properties or lipopolysaccharide. Finally, molecular cloning is valuable both for gene amplification and for insertion of genes from a pathogen or nutritionally fastidious organism into a host that is safer and easier to grow.

2. Choice of growth conditions

Although many of the major OMPs are constitutively expressed, their production can be modulated by growth conditions. For example, the OmpF porin of *Escherichia coli* is greatly reduced in level in media of high osmolarity. Other OMPs are induced by specific nutrient depletion. For example, a group of iron-regulated membrane proteins, including a number of siderophore receptors, is synthesized in most Gram-negative bacteria grown under iron restriction. The PhoE protein, an ion-selective porin, is produced in *E. coli* grown under phosphate limitation, and the ButB protein receptor is produced under vitamin B_{12} limitation. Synthesis of many other OMPs is induced in the presence of specific nutrients; for example, the Fec protein in response to the presence of citrate and a maltose-specific porin, the LamB protein, in response to the presence of maltose. Overall, the OMP profile of any organism must be regarded as extremely variable and dependent on the growth environment. Medium composition, pH, temperature, aeration and growth phase are all factors that might be varied to optimize OMP production.

3. Choice of harvesting and washing conditions

The obvious objective here is to remove unwanted components such as loosely associated slime or capsular polysaccharides that might cause problems during extraction and purification. Buffered saline is the usual choice for washing bacteria; with certain buffers, such as Tris, and especially in the presence of chelating agents such as EDTA, there is a risk of loss of OMPs owing to destabilization of the outer membrane.

4. Choice of methods for preparation of envelopes or outer membranes

Most Gram-negative bacteria can be broken easily by mechanical means (see Chapter 3). Envelopes recovered by centrifugation can be washed clean of cytoplasmic contaminants and either used directly for extraction of OMPs or subjected to further purification. Treatment of the envelope with detergents such as N-lauroyl sarcosinate ('Sarkosyl') can sometimes be used to remove cytoplasmic membrane, leaving the outer membrane and peptidoglycan intact (see Chapter 3). These outer membrane–peptidoglycan fractions are a particularly useful starting point for isolation of OMPs because they can be washed clean of contaminating material and are easily sedimented by centrifugation. An ultracentrifuge is not essential and large amounts of outer membrane can be prepared.

Some OMPs, such as the porins and some lipoproteins, are tightly but non-covalently associated with the peptidoglycan layer. This property can be exploited for their purification using mild extraction procedures with detergents (e.g. SDS) to remove all other envelope components. OMPs can then be released from the peptidoglycan by suitable extraction methods. In some cases it might be possible to use EDTA, detergents or saline wash to release outer membrane fragments from whole bacteria. These fragments, which are released from outer membrane blebs on the cell surface, form vesicles which can be recovered from the medium after removal of the bacteria. They can form a useful starting point for protein purification. Whatever system is chosen, consideration should be given to protecting OMPs from degradation by endogenous proteases. Protease inhibitors are discussed in more detail in section F(ii) of this chapter.

5. Choice of extraction procedures

A sequence of treatments with buffers, salts, detergent, chelating or chaotropic agents under particular conditions of pH, ionic strength and temperature must be devised to yield fractions extensively enriched in the OMP of interest. This is the key step in a successful OMP purification process. Inevitably, a compromise must be reached between the conflicting demands of efficiency of extraction, contamination with other outer membrane components and preservation of properties. Strong anionic detergents such as SDS are particularly useful for dissociation of outer membranes and for separating individual OMPs during gel permeation chromatography. However, SDS is strongly denaturing and difficult to remove from proteins. Zwitter-ionic detergents such as N-tetradecyl-N,N-dimethyl-3-amino-1-propanesulphonate (Zwittergent-3,14; Calbiochem-Behring, La Jolla, California, USA) and non-ionic detergents such as Triton X-100 and β-octylglucoside, have found useful applications in OMP isolation. They are generally less denaturing towards proteins.

6. Choice of purification techniques

The choice of a suitable system for separation and purification of proteins in an extract depends on the properties of the OMP, the major contaminants and the agents used to release them from the outer membrane. Gel permeation and ion-exchange column chromatography are most commonly employed. It is usually essential to maintain the presence of a detergent throughout the process to keep the OMPs in solution. Gel permeation is compatible with all detergents, although detection of eluting proteins is difficult in the presence of UV-absorbing detergents such as Triton X-100. Anion-exchange chromatography cannot be used with anionic detergents.

The methods described below have been developed from the experience of many groups working with purified OMPs. Methods differ according to the degree of purity sought and the requirements for retention of native configuration and function. Methods involving extraction with SDS at elevated temperatures are likely to yield denatured OMPs with fewer impurities. Conversely, milder methods should retain native conformation at the expense of some degree of contamination with lipopolysaccharide.

(ii) Preparation of porins

1. Method 1

This method for porin isolation is applicable to a wide range of strains. It relies on the non-covalent association of porins with peptidoglycan. This property is exploited using SDS treatment of sarkosyl-prepared outer membrane-peptidoglycan complexes to isolate peptidoglycan with associated proteins but essentially free of other outer membrane components. The porins, together with other peptidoglycan-associated proteins, are then released by increasing the salt concentration and temperature. Purification of the solubilized proteins is achieved by gel permeation chromatography or by electroelution from the polyacrylamide gel following SDS-PAGE (see method 2).

(a) Bacteria are grown to mid-exponential phase. 4 litres of culture containing between 5×10^8 and 10^9 bacteria per millilitre will provide adequate biomass.

(b) The bacteria are harvested by centrifugation and washed twice in phosphate-buffered saline (PBS; NaCl 8 g/l; KCl 0.2 g/l; KH_2PO_4 0.2 g/l; $Na_2HPO_4.2H_2O$ 2.9 g/l; pH 7.4) at 4°C.

(c) The bacteria are resuspended in 50 ml of PBS containing 0.1 mM phenylmethylsulphonylfluoride (PMSF) and broken by sonication or another mechanical technique. Unbroken cells are removed by centrifugation at 5000g for 5 minutes and discarded.

(d) The supernatant is mixed with 5 ml of 22% w/v sodium N-lauroyl sarcosinate (Sarkosyl). Rapid clearing of the mixture indicates solubilization of the cytoplasmic membrane. The mixture is incubated for 30 minutes at 20°C.

(e) The outer membrane–peptidoglycan complex is sedimented by centrifugation at 40 000g for 45 minutes. The pellet is resuspended in 50 ml of 2% Sarkosyl containing 0.1 mM PMSF and the outer membrane complex again recovered by centrifugation.

(f) The pellet is resuspended in 50 ml Tris-HCl buffer (pH 8.0, 10 mM) containing 2% w/v SDS at 20°C, and insoluble material is recovered by centrifugation at 100 000g for 45 minutes at 20°C. The pellet is washed twice more in the same way. The final insoluble material, peptidoglycan with associated proteins, is retained.

(g) The pellet is resuspended in 2 ml Tris-HCl buffer (pH 8.0, 10 mM) containing 2% SDS and 1.0 M NaCl at 37°C and incubated at this temperature for 90 minutes to release the porins and some peptidoglycan-associated lipoproteins. The peptidoglycan residue is removed by centrifugation at 100 000g for 45 minutes at 20°C, and the supernatant retained.

(h) The supernatant porin solution is applied to a Sephacryl S-200 (Pharmacia) column (1.5 × 90 cm) previously equilibrated in 10 mM Tris-HCl, 1% SDS, 0.4 M NaCl (pH 8.0), and proteins are eluted at room temperature with the same buffer. The absorbance of the effluent is monitored at 280 nm.

(i) Individually collected fractions of the eluate are examined by SDS-PAGE to determine the nature of the proteins eluted. The molecular masses determined by SDS-PAGE will not necessarily correspond to those predicted from the elution sequence since many porins exist as oligomers that are only dissociated on heating to 100°C during PAGE sample preparation.

Surprisingly, the pore-forming function of porins is retained during gel permeation in the presence of SDS. In the case of the *Pseudomonas aeruginosa* porin F, activity is retained on subsequent dialysis in 5 mM Tris-HCl, pH 8, containing 0.1% SDS and 0.1 M NaCl, but not in water or buffers of low ionic strength.

2. Method 2 (Parr *et al.*, 1986)

An alternative method for isolating the porins released from peptidoglycan involves electroelution from gels following SDS-PAGE:

(a) Follow stages (a) to (g) from method 1 to obtain 2 ml of porin extract.

(b) 0.5 ml of the protein extract is mixed with 0.5 ml SDS-PAGE sample buffer (see Appendix 1) omitting 2-mercaptoethanol and incubated at 20°C for 10 minutes.

(c) This mixture is applied as a single broad band to an SDS-polyacrylamide gel and electrophoresis is carried out as for normal OMP analysis.

(d) A narrow strip is cut from each side of the gel and stained rapidly with Coomassie blue to locate the protein bands. The main piece of gel is retained at 5°C during this process. The appropriate regions of the unstained gel are then cut out. (Under the non-denaturing electrophoresis conditions used here, porins run as globular oligomers, not as fully denatured monomeric polypeptides.)

The excised gel segments are crushed with a glass rod and suspended in 5 ml of Tris–glycine buffer (15 mM Tris, 192 mM glycine, pH 8.3) containing 20% methanol and 0.1% SDS. The crushed gel suspension is placed in a dialysis tubing sac, which is then suspended in an electroelution chamber (an electroblotting cell such as the Bio-Rad Trans Blot is suitable), and subjected to electroelution at 50 V for 2 h at 4°C. The buffer in the tank is changed and elution continued for a further 15 h at 10 V. The supernatant is removed from the dialysis sac and dialysed against 5 mM Tris-HCl, pH 8.0, containing 0.1% SDS and 0.1 M NaCl.

The advantage of the electroelution technique is the virtual removal of contaminating lipopolysaccharide. Porins prepared by gel permeation still contain significant quantities of associated lipopolysaccharide.

The methods described above work well for the *E. coli* OmpF, OmpC, LamB, PhoE and protein K porins, the OmpD porin of *Salmonella typhimurium* and the *P. aeruginosa* porins F and P. They are generally applicable to other organisms, but in some cases harsher conditions are required to remove the porins from the peptidoglycan or to separate other outer membrane proteins from those associated with the peptidoglycan. For example, the NmpC porin of *E. coli* remains bound to peptidoglycan even after treatment with 2% SDS, high salt and EDTA at 37°C for 2 h. The same procedure at 60°C is required to effect quantitative release. The OmpC and OmpF porins of *S. typhimurium* also require harsher conditions for solubilization. Inclusion of 5 mM EDTA or 4 M guanidine thiocyanate is required to release them from peptidoglycan. Many methods have been developed to extract and purify porins from specific bacterial species. Some of these, illustrating the use of different solubilization agents and chromatographic separation techniques, are described below.

3. Haemophilus influenzae porin (Vachon *et al.*, 1985)

(a) Bacteria are grown, harvested and washed as described in method 1.
(b) The bacteria are suspended in 50 ml of 2% w/v cetyltrimethyl-

ammonium bromide containing 0.1 mM PMSF and gently stirred at 20°C for 1 h.

(c) The suspension is centrifuged at 20 000g for 15 minutes, and the pellet resuspended in 30 ml of water containing 0.1 mM PMSF. 10 ml of 4M CaCl$_2$ is added and the mixture stirred for 1 h at 20°C. 10 ml of ethanol are added.

(d) The mixture is centrifuged at 20 000g for 15 minutes and the pellet of nucleic acids discarded. The supernatant is mixed with 150 ml of ethanol.

(e) The protein so precipitated is recovered by centrifugation at 20 000g for 15 minutes and washed twice with ethanol, then once with acetone, and dried.

(f) The protein is resuspended to a concentration of 2% w/v in Tris-HCl buffer (50 mM, pH 8.0) containing 10 mM Na$_2$EDTA, 0.5 M NaCl and 5% w/v Zwittergent Z-3,14 (Calbiochem). After stirring at 20°C for 1 h the suspension is centrifuged at 20 000g for 15 minutes. The supernatant is retained and the pellet is extracted with detergent again, in the same way, then discarded. The supernatant extracts are combined.

(g) The supernatants are dialysed exhaustively against Tris-HCl buffer (50 mM, pH 8.0) containing 10 mM Na$_2$EDTA, 0.5 M NaCl and 0.05%, w/v Zwittergent Z-3,14, to reduce the detergent concentration.

(h) The dialysed protein extract is applied to a column (2.5 × 20 cm) of DEAE-Sepharose CL-6B (Pharmacia) previously equilibrated in the buffer described in (g), and proteins are eluted with a total of three bed volumes of the same buffer, without a salt gradient. The eluate is continuously monitored for protein at 280 nm.

(i) The protein fractions are analysed by SDS-PAGE. The porin, which behaves as a 40 000 Da polypeptide on SDS-PAGE, elutes in a buffer volume of about 150 ml (on a column of 120 ml packed bed volume). Some other non-porin proteins elute at about the void volume.

4. *Neisseria gonorrhoeae protein I* (Blake & Gotschlich, 1982)

(a) Bacteria are grown, harvested and washed as described in method 1 above, and treated with Zwittergent as for *Haemophilus influenzae* stages (a) to (f).

(b) The combined Zwittergent extracts are dialysed against Tris-HCl buffer (pH 8.0, 50 mM) containing 10 mM Na$_2$EDTA and 0.05% Zwittergent Z-3,14.

(c) The dialysed solution is applied to a column (2.5 × 20 cm) of DEAE-Sepharose CL-6B equilibrated in the above buffer. The column is washed with three bed volumes of the same buffer, then proteins are eluted in a 600 ml linear gradient of NaCl, from 0 to 0.5 M, in the same buffer.

(d) Protein I elutes at a salt concentration of between 0.4 and 0.5 M. The protein-containing fractions are analysed by SDS-PAGE, and fractions containing protein I are pooled. The protein is precipitated by the addition of ethanol to 80% v/v. The precipitate is collected by centrifugation at 20 000g for 15 minutes and dissolved in 1 ml of Tris-HCl buffer (50 mM, pH 8.0) containing 10 mM Na$_2$EDTA and 5% Zwittergent Z-3,14.

(e) The solution is applied to a column (1.5 × 90 cm) of Sephacryl S-200 previously equilibrated in Tris-HCl (100 mM, pH 8.0) containing 10 mM Na$_2$EDTA, 0.2 M NaCl and 0.05% Zwittergent Z-3, 14. Elution is carried out with the same buffer. Protein 1 elutes with an apparent molecular mass of 160 000 Da, and runs on SDS-PAGE as a 34 000 Da polypeptide.

5. *Legionella pneumophila porin* (Gabey *et al.*, 1985)

Although the general porin preparation method (1) can be used for this bacterium, only 50 per cent of the porin is released from the peptidoglycan. The following procedure is more efficient.

(a) Bacteria are grown, harvested and washed as in method 1.

(b) The washed cells are resuspended in 5 ml of 1 M sodium acetate (pH 4.0) containing 1 mM 2,3-dimercaptopropanol, and mixed with 50 ml of 5% w/v Zwittergent Z-3, 14 in 0.5 M CaCl$_2$ at 4°C. The suspension in its container is immersed in a sonic water bath for 30 seconds at 4°C.

(c) 15 ml of ice-cold ethanol is added dropwise, with stirring, and the mixture is stirred for a further 30 minutes at 20°C. The precipitate is collected by centrifugation at 20 000g for 15 minutes.

(d) The pellet is treated again as in (b) and (c). This time the supernatant, containing the porin, is retained. The protein is precipitated by the addition of 2.5 volumes of ice-cold ethanol.

Purification is carried out by ion-exchange and gel-permeation chromatography in buffers containing Zwittergent.

6. *Rhodopseudomonas sphaeroides porin* (Weckesser *et al.*, 1984)

Porin from this organism can be prepared by the general method 1, but a simple saline wash of whole cells has also been shown to release lipopolysaccharide–phospholipid–protein complexes rich in porin. Porin can be conveniently purified from this material by the following procedure.

(a) Bacteria are grown to mid-exponential phase in 4 litres of culture, harvested by centrifugation and resuspended (approx. 10 g wet wt) in 50 ml of 0.9% w/v NaCl containing 1 mM PMSF. The suspension is shaken at 37°C for 1 h.

(b) After centrifugation at 30 000g for 1 h the supernatant is retained, dialysed exhaustively against water at 4°C and freeze-dried.

(c) The lyophilized material is dissolved at 25 mg/ml in Tris-HCl buffer (5 mM, pH 7.2) containing 0.3% w/v lithium dodecylsulphate, 0.4 M LiCl, 5 mM Na$_2$EDTA and 0.3 mM NaN$_3$. The solution is applied to a column (1.5 × 90 cm) of Sephacryl S-200 previously equilibrated with the same buffer, and eluted with this buffer at 4°C. The first peak of protein eluting after the void volume contains the porin, with a molecular mass of 47 000 Da on SDS-PAGE.

7. Glucose-inducible porin D1 of Pseudomonas aeruginosa (Hancock & Carey, 1980)

The D1 porin of *Pseudomonas aeruginosa* is induced by growth on glucose and is thought to be involved in glucose uptake. It can be isolated from cells grown on glucose by extraction of outer membranes with Triton-EDTA and ion-exchange chromatography.

(a) Bacteria are grown in a glucose minimal medium and outer membranes are prepared with Sarkosyl as described in method 1, (a) to (e).

(b) The outer membrane pellet is suspended in 50 ml Tris-HCl buffer (20 mM, pH 8.0) containing 2% w/v Triton X-100 and recovered by centrifugation at 100 000g for 1 h at 4°C.

(c) The pellet is resuspended in the buffer above, but containing 10 mM Na$_2$EDTA in addition to the Triton, to extract the porin. The supernatant is obtained after centrifugation at 100 000g for 1 h at 4°C.

(d) The porin is purified from the supernatant solution by ion-exchange chromatography in the presence of 0.1% Triton X-100.

(iii) Purification of lipoproteins

Method 1 for porin isolation also provides the basis for procedures designed to purify other outer membrane proteins that are non-covalently associated with the peptidoglycan layer. An increasingly wide range of lipoproteins of this type are now being recognized in Gram-negative bacteria. *E. coli*, for example, produces at least nine distinct lipoprotein species. The following method can be used for the preparation and study of lipoproteins in a range of bacteria.

(a) The outer membrane–peptidoglycan complex is isolated, for example as described in method 1 above (see also Chapter 3).

(b) The pellet from 4 litres of culture is resuspended in 200 ml of Tris-HCl buffer (10 mM, pH 7.8) containing 2% SDS, 10% v/v glycerol and 0.15 M NaCl and incubated at 30°C for 1 h.

(c) The pellet is recovered by centrifugation at 100 000g for 1 h (this and all subsequent centrifugation steps in which SDS is present are carried out at 20°C), and treated as in (b), but omitting NaCl from the buffer solution.

(d) The pellet is again recovered by centrifugation and peptidoglycan-associated proteins are extracted into 200 ml of the buffer as in (b) but omitting NaCl, at 50°C for 1 h. Insoluble material is removed by centrifugation at 100 000g for 1 h and the supernatant is retained.

(e) Solid MgCl$_2$ is dissolved in the supernatant to give a final concentration of 10 mM, then 400 ml of acetone at 4°C are added. The precipitated proteins are recovered by centrifugation at 20 000g for 30 minutes and washed once with 90% acetone at 4°C.

(f) The precipitate is dissolved in 2 ml of sodium phosphate buffer (25 mM in phosphate, pH 7.2) containing 1% SDS, 5 mM Na$_2$EDTA, 50 mM NaCl and 0.02% w/v NaN$_3$ ('SDS buffer'). The solution is dialysed against a large excess of the same buffer, overnight.

(g) The sample is applied to a column (1.5 × 90 cm) of Sephacryl S-200 previously equilibrated in SDS buffer, and proteins are eluted in the same buffer. Lipoproteins are located by analysis of protein-containing fractions by SDS-PAGE. For *E. coli* the free form of the Braun lipoprotein (7–10 000 Da), peptidoglycan-associated lipoproteins ('PALs', 18–25 000 Da) and porins (33–40 000 Da) are all eluted from the column.

If sufficient resolution of the components is not obtained after repeated chromatography of the individual fractions in the same system, further resolution may be obtained by ion-exchange chromatography in the presence of Triton X-100 and urea.

(h) Proteins are precipitated from appropriate combined fractions from the Sephacryl column by addition of solid MgCl$_2$ to 10 mM, followed by two volumes of ice-cold acetone. The precipitated proteins are collected by centrifugation as in (e) above and dissolved in 10 ml of Tris-HCl buffer (10 mM, pH 7.4) containing 2% w/v Triton X-100 and 6 M urea ('Triton–urea buffer'). The solution is dialysed overnight against the same buffer, at 20°C.

(i) The solution is applied to a column (1 × 40 cm) of DEAE-cellulose previously equilibrated with Triton–urea buffer. The column is washed with 90 ml of the same buffer, then eluted with a 200 ml linear gradient of 0 to 0.2 M NaCl in the same buffer. The 21 000 Da PAL of *E. coli* and the H2 lipoprotein of *P. aeruginosa* elute around 0.05 M

NaCl. Analogous PALs have been reported in *S. typhimurium*, *Klebsiella aerogenes*, *Serratia marcescens*, *Proteus vulgaris*, *Proteus mirabilis*, *Pseudomonas fluorescens* and *Pseudomonas putida*.

(iv) OmpA protein of *Escherichia coli* (van Alphen *et al.*, 1977)

Unlike the porins and lipoproteins, the OmpA protein is not peptidoglycan-associated. It is released from outer membrane–peptidoglycan complexes such as the Sarkosyl-insoluble preparation (see method 1 above) by treatment with SDS at 30°C and purified, as follows.

(a) Bacteria are grown, harvested and washed, and the Sarkosyl-insoluble fraction of the cell envelope is prepared as described in method 1 (a)–(e).

(b) The pellet is resuspended in 20 ml of Tris-HCl buffer (2 mM, pH 7.8) containing 2% w/v Triton X-100 and 10 mM $MgCl_2$ and incubated at 20°C for 10 minutes. The pellet is recovered by centrifugation at 40 000g for 1 h and washed once in 120 ml of Tris-HCl buffer (2 mM, pH 7.8).

(c) The washed pellet is resuspended in 20 ml of Tris-HCl buffer (10 mM, pH 7.5) containing 2% SDS, 5 mM Na_2EDTA and 0.05% v/v 2-mercaptoethanol, and stirred at 30°C for 30 minutes. The mixture is centrifuged at 40 000g for 1 h and the supernatant retained. The pellet is extracted again in 20 ml of the same buffer and the supernatants combined. The pellet is discarded.

(d) Extracted protein is precipitated from the combined supernatants by the addition of nine volumes of ice-cold acetone, and recovered by centrifugation at 20 000g for 20 minutes at 4°C. The protein pellet is washed twice in cold 90% v/v acetone and dried.

(e) The dried material is washed twice with 20 ml of chloroform–methanol (2 : 1 by volume) to remove free lipids, and dried again.

(f) Protein is solubilized from the pellet into 30 ml of Tris-HCl (2 mM, pH 7.8) containing 2% w/v Triton X-100, 2 mM Na_2EDTA and 8 M urea for 30 minutes at 60°C, and insoluble, non-protein material is removed by centrifugation at 20 000g for 30 minutes. The protein is precipitated from the extract with acetone as in (d) above.

(g) The dry protein is dissolved in 1 ml of Tris-HCl buffer (10 mM, pH 7.5) containing 1.5% w/v SDS and 5 mM Na_2EDTA and dialysed overnight against the same buffer.

(h) The sample is applied to a column (1.5 × 90 cm) of Sephacryl S-200 equilibrated in the buffer described in (g) and proteins are eluted in the same buffer. The OmpA protein runs as the major included component, with several other minor separate protein peaks. Note that the OmpA protein is heat-modifiable: it behaves on SDS-PAGE as a 28 000 Da polypeptide when denatured at 20°C and as a 35 000 Da protein when denatured at 100°C.

(v) Other outer membrane proteins

Proteins other than porins, lipoproteins and OmpA protein can occur in major quantities in outer membranes and fulfil vital functions. For example, the iron-regulated membrane proteins (IRMPs, 74–83 000 Da), most thoroughly studied in *E. coli*, act as receptors for iron siderophore complexes and form a high-affinity iron uptake system. Analogous IRMPs appear to occur in most Gram-negative bacteria grown under iron restriction. The BtuB protein (60 000 Da) in *E. coli* is a vitamin B_{12} receptor, whose synthesis is induced by the vitamin and which is involved in its uptake. Other important OMPs are encoded by plasmids—for example, the TraT protein of *E. coli* (25 000 Da) which is involved in surface exclusion and mediates serum resistance.

Although there is not yet much information on the isolation and purification of these proteins, modifications of the standard methods 1 and 2 above should be applicable. As a basic approach, steps (a) to (e) in method 1 could be used to prepare an outer membrane–peptidoglycan complex. Treatment with 2% SDS (method 1(f)) can then be followed by the procedure described for the OmpA protein (stages (d) onwards) if the desired protein is soluble, or by steps (g) onwards (method 1) if the protein remains associated with insoluble peptidoglycan. If separation of peptidoglycan-associated proteins from other proteins by method 1 is inadequate, and suitable mutants or growth conditions cannot be found to minimize contamination, method 2 should be followed.

H. ISOLATION OF EXOPOLYSACCHARIDES

The term exopolysaccharide refers to the polysaccharides that are found outside the cell wall/envelope of the bacterium. They include (i) the discrete capsule that can be revealed by a variety of negative staining techniques, such as Indian ink for light microscopy (Duguid, 1951) or ruthenium red for electron microscopy (see Electron Microscopy in Appendix 1); and (ii) the loosely attached or unattached slime that is found free in the culture supernate and is often difficult to visualize in microscopy. Exopolysaccharides are found in a wide range of bacteria, both Gram-positive and Gram-negative. In pathogens they are often associated with virulence and in enterobacteria they are usually referred to as K antigens.

Perhaps more than with any other bacterial polymer, their production is markedly influenced by environmental conditions. In the laboratory careful consideration must be given to culture conditions before the preparation of exopolysaccharides is undertaken. The ratio of carbon to nitrogen in the medium and the phase of growth are the two factors that have the greatest influence. A generalization is that more exopolysaccharide is produced under conditions of excess carbon and energy source and with nitrogen limitation at stationary phase. The incubation temperature also may influence the amounts produced in some species.

The production of exopolysaccharides of high purity is fraught with difficulties. These are associated with the physical properties of polysaccharides in solution: they tend to be very viscous and to have properties similar to those of the molecules with which they are likely to be contaminated—nucleic acids, lipopolysaccharides, intracellular glycogen and high-molecular-weight medium components. The production of large amounts of impure material is, however, relatively simple, especially if the polysaccharide is found free in the culture fluid. This is described in method 1, below.

For most species that produce large amounts of high-molecular-weight polysaccharide, culture should, if possible, be carried out on solid medium to avoid undue contamination by medium components. Large petri dishes or large glass or enamel cooking trays with lids can be used. However, if the medium is dialysable broth cultures can be used. Broth cultures are also recommended for the preparation of low-molecular-weight exopolysaccharides produced by some enterobacteria and neisseriae (see method 3, below).

It is also worthy of note that capsule production may be lost or diminished by repeated laboratory subculture. If this is suspected it is worth passaging a sublethal dose of the organism two or three times through a mouse peritoneum, with the expectation that the peritoneal phagocytic cells will select capsule-formers.

During the preparation of exopolysaccharides, heat treatments and alkaline hydrolysis procedures that have, in the past, been recommended for the release of capsules are to be avoided if at all possible because of the problems of modification and contamination of the product. Several more satisfactory methods have been developed for the isolation of exopolysaccharides. The first method described below is recommended for species such as *Klebsiella aerogenes* and *Streptococcus pneumoniae* that produce large capsules or copious slime. Method 2, or a combination of 1 and 2, provides an alternative. The third method is specifically for the low-molecular-weight (high-electrophoretic-mobility) exopolysaccharides of certain enterobacteria and neisseriae. For a bacterium of unknown capsule type it is probably worth attempting method 1 and investigating the impure product by chemical and immunochemical methods to estimate roughly its molecular weight and degree of contamination with other materials.

(i) General method for capsular polysaccharides

The following is a general starting procedure for a bacterium producing easily detectable capsules.

(a) Bacteria are harvested in stationary phase from solid medium by gently scraping with a glass rod, and are suspended in phosphate-buffered saline containing 0.5% formaldehyde.

(b) Exopolysaccharide is stripped off the bacteria by shearing forces. Gentle stirring is all that is required for slime producers, but blending in a liquidizer or blender is necessary for firmly bound capsules. The time of blending is dependent on the size of the capsule and how firmly it is attached. Removal of capsule from the cells can be monitored by the Indian ink method. Heating during blending should be avoided.

(c) Bacteria are removed by centrifugation. Relatively high g forces ($>10\ 000g$) may be required to sediment the cells through the viscous fluid. Removal of cells should be checked by phase-contrast microscopy. Several cycles of centrifugation might be required.

(d) Four volumes of ice-cold acetone are added to the supernate to precipitate the polysaccharide. Complete precipitation may take up to 18 h at 4°C. The precipitate is recovered by low-speed centrifugation. In other cases, such as with alginate from *Pseudomonas aeruginosa*, the polysaccharide may be wound on to a glass rod immediately after addition of the acetone.

(e) The precipitate is washed in acetone several times, dissolved in water, and lyophilized.

At this stage the preparation will be contaminated with a range of cellular products and medium components. The advantages of the method, however, are that it is simple, only physical separation methods are used and contamination with intracellular and cell-bound components is kept to a minimum. A variety of procedures is available at this stage but none of them has proved universally successful. Perhaps the best method of purification is to extract the lyophilized material with 45% aqueous phenol as in the preparation of smooth LPS (see section E) and to use the $100\ 000g$ supernate as the starting material as in step (ii)(c) below. Alternatively, if contamination is minimal—for example, where the medium constituents can be removed by dialysis—it is possible to proceed to step (ii)(c) without the phenol extraction step.

The following two methods have been developed by Jann and co-workers (Jann, 1985) for the isolation of exopolysaccharides from *Escherichia coli*.

(ii) High-molecular-weight exopolysaccharide

(a) Steps (a) to (i) of the procedure for isolation of smooth LPS by the aqueous phenol extraction method (see section E) are followed.

(b) The extract is centrifuged at 100 000*g* for 3 h to sediment LPS. The supernate is retained and lyophilized. This contains exopolysaccharide, RNA and possibly some intracellular glycogen.

(c) The lyophilized material is dissolved (10 mg/ml) in 0.25 M NaCl and stirred with half its volume of 0.25 M NaCl containing 4% w/v cetyltrimethylammonium bromide (CTAB) for 15 minutes at room temperature. The precipitated RNA–CTAB complex is removed by centrifugation at 10 000*g* for 1 h.

(d) Water (approximately three volumes are required) is slowly added to the supernate until a second precipitate forms. The mixture is allowed to stand in an ice-bath for 2 h and the precipitated exopolysaccharide is collected by centrifugation.

(e) The precipitate is dissolved in a minimum volume of 1 M NaCl (5–8 ml for 500 mg of starting material) and the acidic polysaccharide is precipitated again by the addition of 8–10 volumes of ethanol. The mixture is allowed to stand in an ice-bath for 2 h and the precipitate is collected by centrifugation.

(f) The procedure in (e) is repeated, and the final precipitate is dissolved in a minimum of distilled water. If the solution is opalescent it is clarified by centrifugation at 100 000*g* for 2 h.

(g) The solution is concentrated if necessary, by rotary evaporation, and lyophilized.

(iii) Low-molecular-weight exopolysaccharide

(a) Bacteria are grown to late exponential phase in 10 litres of a medium whose constituents can be removed by dialysis. The culture supernate is retained after removal of the bacteria by centrifugation.

(b) An equal volume of 0.2% w/v CTAB is added to the supernate and the mixture allowed to stand overnight at 4°C.

(c) The suspension is centrifuged at 10 000*g* for 30 minutes and the pellet is homogenized in 450 ml of distilled water to give a fine suspension, and cooled in ice.

(d) Solid $CaCl_2.6H_2O$ (110 g) is added and dissolved to give a final concentration of 1 M.

(e) The solution is cooled in an ice-bath and 132 ml of 96% ethanol is added (final concentration 20% by volume). The precipitate is removed by centrifugation at 16 000*g* for 20 minutes.

(f) Ethanol is added to the supernate to give a final concentration of 80% by volume, and the mixture is allowed to stand at 4°C for 16 h. The precipitated exopolysaccharide is collected by centrifugation at 16 000*g* for 20 minutes.

(g) The pellet is dissolved in 500 ml of 10% saturated sodium acetate solution and mixed with 585 ml of 80% phenol buffered to pH 6.5 with sodium acetate (500 ml of 90% w/v phenol mixed with 85 ml of 16% w/v aqueous sodium acetate). The mixture is stirred vigorously at 4°C for 3–5 h.

(h) The liquid phases are separated by centrifugation at 16 000*g* for 20 minutes, and the upper, aqueous phase is retained.

(i) Ethanol is added to the aqueous phase to a final concentration of 80% by volume, and the mixture is kept at 4°C overnight. The precipitated polysaccharide is collected and re-extracted with buffered phenol as in (g) and (h) above.

(j) The pellet is dissolved in a small volume of 10% saturated sodium acetate and any insoluble material is removed by centrifugation at 100 000*g* for 4 h.

(k) The supernate is dialysed against several changes of distilled water, concentrated by rotary evaporation and lyophilized.

I. ISOLATION AND PURIFICATION OF MYCOBACTERIAL WALL LIPIDS

D. E. Minnikin

(i) Introduction

As described in Chapter 1, mycobacterial cell envelopes contain both covalently bound and free lipids. It is often impossible to extract the free lipids separately from those of the cytoplasmic membrane, so preparations will usually contain phospholipids such as diphosphatidyl glycerol, phosphatidyl ethanolamine, phosphatidyl inositol and the family of phosphatidyl inositol di- and pentamannosides (Minnikin, 1982; Dobson *et al.*, 1985). The isolation and purification of the bound mycolic acids and wall free lipids will be considered separately.

1. Bound mycolic acids

Mycolic acids (see Figure 1.6 in Chapter 1) are covalently linked to an arabinose unit of the wall arabinogalactan and require vigorous chemical treatment for their release. In early studies, alcoholic alkaline treatment was followed by esterification of the free acids with diazomethane. This procedure can cause racemization at C2, resulting in diastereoisomers which separate during thin-layer chromatography (Minnikin & Polgar, 1966). The problem of racemization has been overcome by the use of potassium hydroxide in 2-methoxyethanol for hydrolysis, followed by diazomethane esterification (Daffe *et al.*, 1983), or a procedure involving digestion with methanolic tetramethylammonium hydroxide and addition of iodomethane in dimethylformamide (Minnikin *et al.*, 1984a). Both these procedures are mainly suitable for the isolation of mycolic acid methyl esters, and more flexible alternatives, involving phase-transfer catalysed esterification with various halides is preferred (Dobson *et al.*, 1985).

Acid methanolysis at 75°C was developed as a rapid procedure for distinguishing between mycolate patterns from mycobacteria, nocardiae, rhodococci and corynebacteria (Minnikin *et al.*, 1980), but a temperature of 50°C is recommended to avoid partial degradation of cyclopropane rings in mycobacteria mycolates (Dobson *et al.*, 1985). This method is of value in detecting the presence of epoxymycolates, which are degraded to easily recognizable methoxy-hydroxy derivatives.

All the above methods cleave wax ester mycolates (see Figure 1.6) to carboxymycolates and 2-alkanols, and if the intact wax ester mycolate is required it is necessary to use hydrolytic methods which do not split the link between the two parts of the molecule. Hydrolysis with dilute aqueous alkali is suitable for this purpose, but since the wax ester and ketomycolates co-chromatograph on TLC it is necessary to modify the latter by reduction to an alcohol before separation (D. E. Minnikin, S. M. Minnikin and J. H. Parlett, unpublished results).

2. Free lipids

Mycobacterial free lipids vary widely in structure and polarity, from apolar waxes to glycolipids which are comparable in polarity with the very polar cytoplasmic membrane phosphatidyl inositol mannosides (Dobson *et al.*, 1985) (see Figure 1.7 in Chapter 1). The least polar waxes are the dimycocerosates of the phthiocerol family and a related series of dimycocerosates of glycosyl phenol phthiocerol. A family of glycolipids composed of trehalose acylated with mycolipenic acid and related methyl-branched acids is characteristic of *Mycobacterium tuberculosis* and the same species produces sulphated trehaloses acylated with another family of methyl-branched acids, the phthioceranates.

Glycopeptidolipids are characteristic surface components of many mycobacteria. Those from *M. avium*, *M. intracellulare* and *M. scrofulaceum* have been studied in detail and they occur in non-polar and polar varieties, the latter being lipid surface antigens (Brennan *et al.*, 1978; 1981). Other mycobacteria have a different kind of glycolipid surface antigen, the best studied being the acylated trehalose-based oligosaccharides of *M. kansasii* (Hunter *et al.*, 1983). Finally, trehalose 6,6'-dimycolates, historically known as 'cord factors', are present in small amounts in most mycobacteria investigated.

These free lipid types do not all occur together in a single species but their complexity and range of polarities necessitate special extraction and purification strategies. The traditional extraction method used ethanol, which also killed the bacteria, and diethyl ether mixtures followed by chloroform to produce partially fractionated lipids (Asselineau, 1966). Complete extraction of all free lipids can be achieved by use of chloroform–methanol mixtures (Brennan *et al.*, 1978; Hunter *et al.*, 1983). The recommended method, however, originally designed to extract non-polar isoprenoid quinones separately from plasma membrane polar lipids (Minnikin *et al.*, 1984b), produces non-polar and polar mycobacterial lipids in essentially separate fractions (Dobson *et al.*, 1985). It involves preliminary mixing with a biphasic mixture of methanolic saline and petroleum spirit to give the non-polar lipids in the petroleum spirit layer. The lower aqueous methanol layer and the partially extracted cells are then processed according to the well-proved method of Bligh and Dyer (1959).

(ii) Mycolic acid extraction and purification

It is not convenient to purify free mycolic acids, so they are usually isolated as their methyl or other esters. Several variations of the recommended procedures are available according to the components required. Since the mycolates are principally covalently bound to the cell wall, either whole organisms or defatted cells may be used as a source, the latter having the advantage that the lipid-soluble components of the free lipids will no longer be present. The procedures described below are those of Dobson *et al.* (1985) and Minnikin *et al.* (1985a).

1. Alkaline extraction

(a) 50 mg (dry wt) of biomass is treated, in a glass tube with a Teflon-lined cap, with a solution of 1 ml of methanol, 30% aqueous potassium hydroxide and toluene (1 : 1 : 0.1 by volume) at 75°C, overnight.

(b) After cooling, the mixture is acidified to pH 1 with 1 M HCl, and free mycolic acids are extracted three times into 1 ml of diethyl ether.

Alternatively, and more efficiently, tetrabutylammonium hydroxide

(TBAH) may be used, to yield tetrabutylammonium salts of the mycolic acids.

(a) 50 mg biomass is treated as above with 2 ml of 5% or 15% aqueous TBAH overnight at 100°C.
(b) Where 15% TBAH was used, after cooling, 2 ml water is added.

The lower concentration of TBAH does not degrade wax ester mycolates.

2. *Phase-transfer catalysed esterification of mycolic acids*

Mycolic acid TBAH salts prepared as described above, or derived from the free mycolic acids from the methanolic KOH procedure by mixing up to 50 mg of mycolic acid with 1 ml of 0.8% NaOH w/v, 3.39% w/v tetra-butylammonium hydrogen sulphate in water, are used in this procedure.

(a) 1 ml of mycolic acid TBAH salt solution is mixed with 1 ml of dichloromethane and 25 μl of iodomethane or pentafluorobenzyl bromide* for 30 minutes at room temperature, then centrifuged at 1000g for 5 minutes.
(b) The upper phase is removed and replaced with 1 ml of 1 M HCl. The tube contents are mixed again and the upper phase discarded.
(c) The lower phase is mixed with 1 ml of water and again the upper aqueous phase is discarded.
(d) The lower phase is evaporated to dryness.

The acid wash is included in the procedure to remove contaminating material that might interfere with subsequent chromatography. The procedure permits the preparation of pentafluorobenzyl instead of methyl esters.

3. *Acid methanolysis*

Release of mycolates from dry biomass as their methyl esters is achieved by acid methanolysis. Epoxymycolates are degraded, but the method does allow the positive *detection* of epoxymycolates.

(a) 50 mg (dry wt) of biomass is heated with 3 ml of dry methanol–toluene–sulphuric acid (30 : 15 : 1 by volume) in a glass tube with a PTFE-lined cap at 50°C overnight.
(b) The mixture is cooled and shaken with 2 ml of petroleum spirit (60–80°C boiling point). The upper layer is retained.

*CAUTION—Pentafluorobenzyl bromide is a potent lachrymator. Work must be carried out in a fume hood and excess reagent must be decomposed in dilute alkali.

(c) The upper layer is passed through a column of ammonium hydrogen carbonate (1 cm long) in a Pasteur pipette plugged with cotton wool and prewashed with diethyl ether. The sample is washed through the column with a further 1 ml of petroleum ether. The combined eluates are evaporated to dryness.

The procedure can be scaled up economically and conveniently for preparation of large amounts by heating under gentle reflux for 3 h, extracting with petroleum ether, washing with 5% aqueous sodium bicarbonate and water, followed by evaporation to dryness.

4. *Chromatography of mycolates*

Preparations of mycolic acid methyl esters are best examined by TLC to establish which types are present in a particular mixture. The use of two-dimensional TLC (Minnikin *et al.*, 1980) provides patterns that show the types of mycolates released by both alkaline (Figure 4.8) and acid (Figure 4.9) treatment (Dobson *et al.*, 1985). As noted above, epoxymycolates (M in Figure 4.8(c)) are transformed by acid to characteristic polar derivatives (I,J,N,O in Figure 4.9(c)) and this is valuable because intact epoxymycolates co-chromatograph with ketomycolates in these TLC systems. Undegraded wax ester mycolates also co-chromatograph with ketomycolates (Figures 4.8 and 4.9), so that, to prepare a pure sample of the former, the ketomycolate must be transformed into a more polar derivative by reduction with sodium borohydride in tetrahydrofuran.

Milligram quantities of mycolic acid esters can be isolated by preparative TLC using 10 × 10 cm pieces of Merck 5735 plastic-backed silica-gel sheets; separated components are located by spraying with 0.01% ethanolic Rhodamine 6G and viewing under long-wave (366 nm) UV light. The silica gel is then scraped from the sheets, or pieces are cut out, and the mycolates are extracted by shaking with diethyl ether, in which Rhodamine 6G is insoluble. For quantities up to 1 g, single-dimension TLC on 20 × 20 cm glass plates coated with 1–2 mm layers of Merck 7748 $PF_{254+366}$ silica gel is used. The developing solvent mixtures shown in Figures 4.8 and 4.9 are suitable in many cases, but the similar chromatographic behaviour of alpha'- and methoxymycolates (Figures 4.8 and 4.9, (d) and (a) respectively) and their close proximity to alpha-mycolates requires special procedures. Separation of alpha-, methoxy- and ketomycolate mixtures from *M. tuberculosis* (Figures 4.8, 4.9(a)) and of alpha- and alpha'-mycolates from *M. chelonae* (Figures 4.8, 4.9(d)) can be achieved efficiently using unidimensional multiple-elution chromatography, developing six times with petroleum spirit–diethyl ether (19 : 1 by volume) (Davidson *et al.*, 1982). If alpha'- and methoxymycolates occur

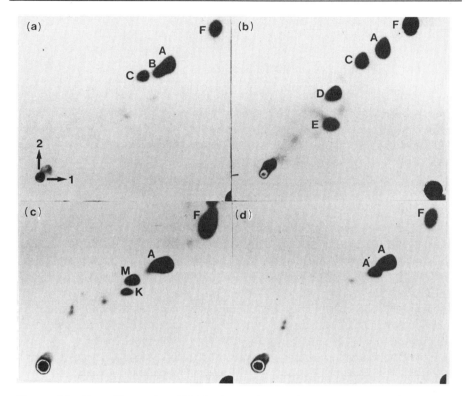

Figure 4.8   Two-dimensional TLC of mycolates released by alkali. First direction, petroleum ether (bp 60–80°C)–acetone (95:5, thrice); second direction,
toluene–acetone (97:3; once). Organisms: (a) *M. tuberculosis*, (b) *M. avium*, (c)
M. fortuitum, (d) *M. chelonae*. A: α-mycolate; A': α'-mycolate; B: methoxymycolate; C: ketomycolate; D: ω-carboxymycolate; E: 2-alkanols; F: fatty acid methyl
esters; K: unidentified mycolate; M: epoxymycolate. Detection: charring with
5% ethanolic molybdophosphoric acid. Reproduced by permission of Academic
Press from Dobson *et al.* (1985).

together, as in *M. agri* and *M. thermoresistibile* (Minnikin *et al.*, 1984a;
1985a) it is possible to use dichloromethane as a preliminary developing
solvent, since methoxymycolates migrate with alpha-mycolates, allowing
alpha'-mycolates to be isolated; the methoxymycolate can then be separated from the alpha-mycolate as described above.

An alternative approach to separating these complex mixtures is to
convert the total mixture into *tert*-butyldimethylsilyl (TBDMS) ethers, as
follows.

(a) A solution of mycolates in toluene (0.15 ml) is mixed with TBDMS

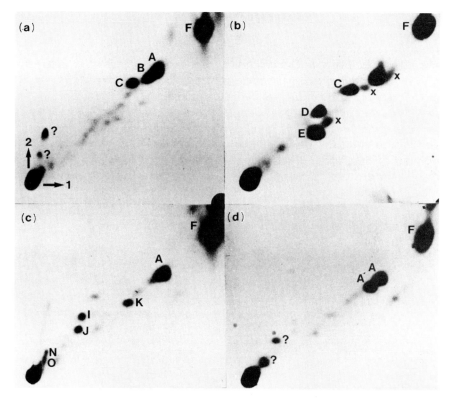

Figure 4.9 Two-dimensional TLC of mycolates released by acid methanolysis. The organisms and chromatographic details were as in Figure 4.8. Additional abbreviations—I,J,N and O: products from acid degradation of epoxymycolates; X and ?: unknowns. Reproduced by permission of Academic Press from Dobson *et al.* (1985).

chloride (0.15 g) and imidazole (0.17 g) dissolved in 1 ml of dimethyl-formamide and heated at 75°C overnight.

(b) The reaction mixture is extracted three times with petroleum spirit (0.3 ml).

(c) The combined extracts are passed through a 1 cm column of neutral alumina to remove traces of dimethylformamide.

(d) TLC of the mycolate TBDMS ethers is carried out using petroleum spirit–toluene (7:3 by volume) to give good separation of the derivatives of alpha-($R_f 0.75$), methoxy-($R_f 0.31$), keto-($R_f 0.21$) and omega-carboxy-($R_f 0.07$) mycolates.

The chromatographic mobilities of the TBDMS ethers of alpha- and alpha'-mycolates are similar since they differ only in the size and degree of unsaturation of the main carbon chain. The purified ethers are converted back to mycolates by acid methanolysis, using the procedure described above for extraction from biomass. The excellent separation of the TBDMS ethers allows the preparation of very pure mycolates; these derivatives are also very suitable for mass spectrometry and NMR spectroscopy.

HPLC on silica gel columns allows measurement of the relative amounts of the different mycolate types and reverse-phase HPLC separates the individual homologues. It is convenient to label mycolates with a chromophore suitable for UV-detection and the use of pentafluorobenzyl esters is recommended, since they can be readily prepared by the phase-transfer esterification procedure described above. These derivatives can also be analysed by two-dimensional TLC using petroleum spirit-acetone

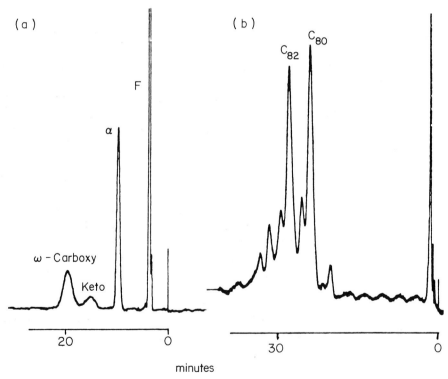

Figure 4.10 HPLC of pentafluorobenzyl esters of mycolic acids from *M. avium* D4. (a) Normal phase chromatography of total fatty acid derivatives using a Waters Radial-Pak B silica gel cartridge with 1-chlorobutane–tetrahydrofuran (99.75 : 0.25) at 0.5 ml/minute. (b) Reverse phase chromatography of α-mycolate derivative using a Merck 15654 RP-18 cartridge with tetrahydrofuran–acetonitrile (50 : 60) at 0.9 ml/minute. F: pentafluorobenzyl esters of non-hydroxylated fatty acids.

(98:2) in the first dimension and toluene–acetone (99:1) in the second. Examples of HPLC separations are shown in Figure 4.10. The identity of the separated mycolate derivatives can be confirmed by IR and NMR spectroscopy, and mass spectrometry, combined gas chromatography and mass spectrometry of trimethylsilyl ethers of methyl mycolates being particularly powerful (Yano, 1985).

(iii) Free lipid extraction and purification

The procedure of Dobson *et al.* (1985) extracts the complete range of mycobacterial lipids and provides an indication of the relative proportions of the different lipid types in a series of TLC patterns.

1. *Lipid extraction*

(a) 50 mg dry walls or biomass is mixed with 2 ml of methanol–0.3% aq.NaCl (10:1 by volume) and 2 ml of petroleum spirit (60–80°C boiling point) in a tube with a PTFE-lined cap, for 15 minutes at room temperature.
(b) The upper layer is removed and retained. 1 ml of petroleum spirit is added to the lower layer and mixed for a further 15 minutes.
(c) The upper layer is removed, combined with the upper layer retained from the first extraction, and evaporated to dryness at less than 37°C. This yields the **non-polar lipids**.
(d) The lower (aqueous) layer from (c) is heated in a boiling-water bath for 5 minutes, then cooled to room temperature.
(e) 2.3 ml chloroform–methanol–0.3% aq.NaCl (9:10:3 by volume) is added and the contents of the tube are mixed for 60 minutes.
(f) The mixture is centrifuged to separate the solvent extract from the biomass. The supernatant is retained and the residue is extracted again, with 0.75 ml of chloroform–methanol–0.3% aq.NaCl (5:10:4) for 30 minutes.
(g) The combined solvent extracts are mixed with 1.3 ml of chloroform and 1.3 ml of 0.3% aq.NaCl for 5 minutes, then centrifuged to separate the layers. The upper layer is discarded. The lower (chloroform) layer is evaporated to dryness below 37°C, giving the **polar lipids**.

2. *Lipid chromatography*

Five two-dimensional TLC systems are necessary to cover the full polarity range of mycobacterial lipids. Examples of the application of these systems are shown in Figure 4.11; others are described in Dobson *et al.* (1985), Minnikin *et al.* (1985b) and Ridell *et al.* (1986). It is useful to use

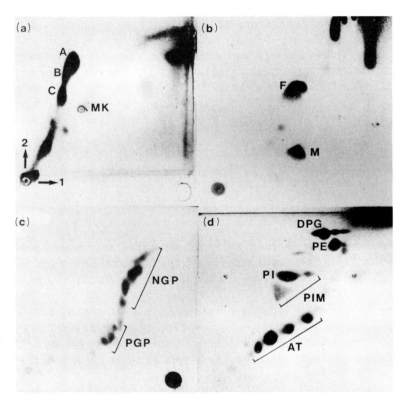

Figure 4.11 TLC of representative mycobacterial free lipids. (a) Phthiocerol dimycocerosates from *M. tuberculosis*. First direction, petroleum ether (bp 60–80°C)–ethyl acetate 95 : 5, thrice); second direction, petroleum ether–acetone (98 : 2, once). (b) Glycosylphenolphthiocerol dimycocerosates from *M. kansasii*. First direction, chloroform–methanol (96 : 4); second direction, toluene–acetone (80 : 20). (c) Glycopeptidolipids from *M. avium*, serovar 2. First direction, chloroform–methanol–water (100 : 14 : 0.8); second direction, chloroform–acetone–methanol–water (50 : 60 : 2.5 : 3). (d) Trehalose-based oligosaccharides from *M. kansasii*. First direction, chloroform–methanol–water (60 : 30 : 6); second direction, chloroform–acetic acid–methanol–water (40 : 25 : 3.6). Detection: charring with 5% ethanolic molybdophosphoric acid.

Key—A–C: dimycocerosates of the phthiocerol family; MK: menaquinone; TG: triacylglycerol; F: free fatty acid; M: glycosylphenolphthiocerol dimycocerosates; NGP and PGP: non-polar and polar glycopeptidolipids; AT: acylated trehalose-based oligosaccharides; DPG: diphosphatidylglycerol; PE: phosphatidylethanolamine; PI: phosphatidylinositol; PIM: phosphatidylinositol mannosides.

measured volumes of solvent for the lipids from known weights of biomass, and to chromatograph measured volumes of the lipid solutions in order to obtain an estimate of the relative proportions of lipids in different samples and of the amount of biomass required for the preparation of a given lipid. It should be noted that some of the surface lipid antigens, those from *M. kansasii* for example (Figure 4.11(d)), overlap with the usual plasma membrane polar lipids. Certain classes of lipid antigens can be directly detected by immuno-TLC (Ridell *et al.*, 1986).

Considering the variety of mycobacterial lipid types, it is not easy to recommend a universal procedure for lipid isolation. For milligram amounts, preparative TLC on plastic sheets, such as Merck 5735, is convenient; combinations of the developing solvents used on analytical plates (Figure 4.11) can be used. Separated non-polar lipids can be extracted with diethyl ether after detection with Rhodamine 6G as described above for mycolate purification. For more polar lipids that cannot be extracted from the silica gel with ether, chromatograms can be sprayed with 0.01% 1,6-diphenylhextriene in petroleum spirit–acetone (9 : 1 by volume) and bands observed under long-wave (366 nm) UV light (Hyslop & York, 1980). Lipid bands are marked, then the plates are redeveloped in petroleum spirit–acetone (9 : 1) to remove the detection reagent. The polar lipids are subsequently extracted from the silica gel into chloroform–methanol (9 : 1). Several hundred milligrams can be isolated on 1–2 mm layers of silica gel containing a fluorescent indicator, such as Merck 7748 $PF_{254+366}$, lipids being detected under long-wave UV light. These plates are not suitable for use with polar solvents, however, since the 366 nm detection agent is soluble; silica gel such as Merck 7747 PF_{254} may be used, with diphenylhexatriene detection, in these cases. Larger-scale preparations require the use of column chromatography on silica gel or DEAE-cellulose, the latter being particularly useful for the glycopeptidolipids (Brennan *et al.*, 1981b; Hunger *et al.*, 1983). References to the original papers must be made for precise details of the individual separations. The monographs of Christie (1982) and Kates (1986) are excellent practical handbooks for general lipid techniques.

Bacterial Cell Surface Techniques
I. Hancock and I. Poxton
© 1988 John Wiley & Sons Ltd.

5

Chemical Analysis of Envelope Polymers

The surface polymers of bacteria consist of peptidoglycan, proteins, carbohydrates and lipids, or complexes of these polymers (see Chapter 1). The analytical techniques employed for determining chemical composition and structure tend to be common to several of these macromolecular types. Traditionally the polymers were broken down to their component oligomers or monomers by hydrolytic or enzymic cleavage, and the composition and structure determined by a range of chromatographic and colorimetric methods.

In this chapter, conditions and guidelines are given for the hydrolysis of bacterial surface components and for more selective chemical degradation. Methods for the preparation of the necessary derivatives for subsequent analysis, and the techniques of gas chromatography (GC) and high-performance liquid chromatography (HPLC) for determining sugar composition are described. The commonly used assays for determining quantitative composition are listed in Appendix 1.

A. ANALYSIS OF POLYSACCHARIDES

All bacterial cell walls contain carbohydrate polymers and, while these macromolecules may bear a variety of alternative names, the application of the techniques of polysaccharide analysis is central to their characterization and measurement. In this section we describe the most important of these techniques; the literature on polysaccharide chemistry is huge and we have been extremely selective. Every laboratory working in the field will have its own versions of the general methods described; we have not attempted to review them. For detailed critiques, the reader should consult comprehensive works on carbohydrate chemistry.

In addition to the general features of polysaccharides—the presence of sugars in glycosidic linkage—bacterial cell wall carbohydrate polymers contain a variety of additional features such as acylated sugars, phosphodiester linkages, alditols and alditol phosphates, and glycopeptide

137

components. Characterization of these structures requires special techniques which are described in succeeding sections of this chapter.

With the exception of NMR, the main techniques of structural determination are destructive, usually employing acid- or base-catalysed hydrolysis. Most wall polymers contain a variety of linkages of differing susceptibilities to hydrolysis, and it is essential that these differences are taken into account during analysis. Badly chosen conditions may lead to degradation and loss of individual sugar components; careful selection of hydrolysis conditions and accurate quantitative analysis of the procedure can give valuable information about the types of components and linkages present.

Even more selective degradation can be obtained by oxidation of polysaccharides and related polymers such as teichoic acids (see section 5D) with periodate, since this process only leads to cleavage of the bond between two carbon atoms, each of which carries a free hydroxyl group or an unacylated amino group. In the most commonly used development of this procedure, Smith degradation, the aldehyde groups so formed are reduced with sodium borohydride and the oxidized/reduced polymer is subject to acid hydrolysis, permitting identification of the sugar residues that are susceptible to periodate oxidation and hence yielding valuable information about the positions of glycosyl substitution. With heteropolysaccharides containing only one periodate-susceptible sugar residue per repeating unit, the method often permits the isolation of the oxidized/reduced repeating oligosaccharide. An example of the use of the technique is given in section 5A(vi).

(i) Hydrolysis of carbohydrate polymers

For most carbohydrates hydrolysis of a solution at 1–2 mg/ml in 2 M HCl, *in vacuo*, for 2–3 h at 100°C is used. It is carried out in a thick-walled glass tube, 15 × 125 mm, previously constricted in a flame about 25 mm from the open end. After addition of the sample, the tube is connected to a vacuum line and the constricted neck is sealed in a flame while under vacuum. For volumes exceeding about 0.5 ml it is necessary to freeze the sample before applying the vacuum to prevent 'bumping'. Alternatively, purpose-made glass reaction vials with PTFE-lined septa in screw caps can be used. The head-space can be flushed with nitrogen through the septum. The tube must be placed directly into boiling water and not allowed to warm up slowly. This prevents prior de-N-acetylation of amino sugars which would make subsequent cleavage of the glycosidic bond very difficult.

When proteins, peptidoglycan and amino sugar-containing polymers are hydrolysed, more concentrated acid (4–6 M) and longer hydrolysis times (up to 16 h) are required. Considerable degradation, especially of sugars, is likely, and if quantitation is important, control compounds should be run under similar conditions.

It is convenient to remove HCl, without loss of material, by desiccation over NaOH pellets and phosphorus pentoxide *in vacuo*. This should be carried out as quickly as possible to minimize further degradation. After initial drying a few drops of water should be added, then drying repeated. Acid may also be removed by anion exchange resin (in the OH^- or CO_3^- form), but anionic material may be lost. If the presence of salt in the product does not matter, the hydrolysate may be neutralized with sodium hydroxide or ammonia solution. In both these neutralization procedures heat may be generated, causing further degradation. They should be carried out at as low a temperature as possible. If sulphuric acid is used for hydrolysis it may be removed by ion-exchange resin or by neutralization with barium carbonate, the barium sulphate being removed as an insoluble precipitate. Insoluble barium salts of phosphorylated products may also be precipitated in this procedure.

Partial acid hydrolysis with weak acid or at lower temperatures is sometimes useful. An example of this is the cleavage of lipopolysaccharide into lipid A and polysaccharide by 1% acetic acid (see Chapter 4E, and later in this chapter).

The hydrolysis of teichoic acids and other accessory wall polymers containing sugar phosphates requires special consideration. This is dealt with later in this chapter (section D), together with alkaline hydrolysis and selective cleavage of phosphate ester linkages with hydrofluoric acid. For some analytical procedures, such as HPLC, it is useful to *N*-acetylate amino sugars in the hydrolysate. This is conveniently done as follows.

Samples (0.05 ml) containing 10–50 nmol total hexosamine are mixed with 0.01 ml of saturated sodium bicarbonate solution and 0.01 ml acetic anhydride at room temperature for 10 minutes. Excess acetic anhydride is destroyed by heating in a boiling-water bath for 3 minutes, then cooling quickly.

(ii) Smith degradation of polysaccharides (Abdel-Akher *et al.*, 1952)

In this procedure, oligo- and polysaccharides and related polymers are subjected to periodate oxidation, followed by borohydride reduction and acid hydrolysis. It is possible to vary both the duration of periodate oxidation and the conditions of acid hydrolysis and thereby, in some cases, isolate smaller oligosaccharide fragments for further investigation (see section (vi)7 below).

(a) Polysaccharide (4 mg) is dissolved in 1 ml of 0.1 M NaIO$_4$ and incubated in the dark for 3–4 days at room temperature. It is essential that the solution is neutral or acid, though buffering is not usually necessary. Selective oxidation can sometimes be achieved by the use of a lower concentration of periodate at a lower temperature and for a shorter time. For example, terminal non-reducing galactofuranose residues, or internal ones linked at C2 or C3, can be selectively degraded to arabinose in the Smith procedure if oxidation is carried out with 0.05 M periodate, for 2 h at 3°C. In this case, only the theoretically required amount of periodate need be added, thus avoiding over-oxidation.

(b) Solid NaBH$_4$ (20 mg) is added, and the mixture incubated overnight at room temperature.

(c) Acetic acid is added dropwise to decompose residual borohydride, and when hydrogen evolution has ceased the solution is passed through a column (2 ml) of Dowex 50 ion-exchange resin (H$^+$ form). A further 20 ml of water is passed through the column and the combined eluates are evaporated to dryness under vacuum. Borate is removed by the repeated addition and evaporation of 1 ml of methanol at 60°C.

(d) The residue is hydrolysed in acid as described above and the products examined chromatographically. Complete hydrolysis to monosaccharides can be carried out in 2 M HCl at 100°C; more selective hydrolysis of acetals in the oxidized portions of the polymer, to yield oligosaccharides from heteropolysaccharides, may be achieved in many cases by treatment at room temperature with 0.5 M H$_2$SO$_4$, for 24 h. The progress of hydrolysis may be followed by the appearance of new periodate-oxidizable groups, periodate consumption being monitored by the colorimetric method of Avigad (1969) or by UV absorption at 223 nm. 10^{-4} M solutions of periodate and iodate have A$_{223}$ of 1 and 0.167 respectively.

(iii) Separation and quantitation of sugars as their alditol acetates, by gas chromatography

Sugars have been analysed by GC in a wide range of derived volatile forms (see Drucker, 1981) but most derivation procedures give rise to multiple peaks for single sugars on GC, complicating analysis and quantitation. A simple and convenient procedure, that produces only one peak per sugar, is the separation of the reduced forms of the aldoses as their peracetyl derivatives, as devised originally by Sawardeker *et al.* (1965). The original method, and many adaptations of it, suffered the disadvantage that borate generated during the reduction step with borohydride interfered with subsequent acetylation, and therefore had to be removed by repeated evaporation as trimethyl borate in anhydrous methanol. The method described here uses a catalyst for acetylation, 1-methylimidazole, that is active in the presence of borate (Blakeney *et al.*, 1983). Some aminosugars may not be completely acetylated in this procedure.

(a) An acid hydrolysate of polysaccharide (up to 15 mg) in 1.5 ml is made alkaline by addition of 0.32 ml of 15M ammonia solution. 0.05 ml of myo-inositol solution (20 mg/ml) is added if necessary as internal standard.

(b) Samples of the hydrolysate (about 1 mg total sugars in 0.1 ml) are mixed with 1 ml of sodium borohydride solution (2 g of sodium borohydride dissolved in 100 ml of anhydrous dimethylsulphoxide at 100°C and cooled) and incubated at 40°C for 90 minutes. (Dimethylsulphoxide may be dried by storage over molecular sieve Type 4A— Union Carbide.)

(c) 0.1 ml of glacial acetic acid is added to destroy residual borohydride.

(d) 0.2 ml of 1-methylimidazole and 2 ml of acetic anhydride, in that order, are added and the mixture is left at room temperature for 10 minutes, for complete acetylation. 5 ml of water is added to decompose excess acetic anhydride.

(e) When the mixture is cool, peracetylated sugars are extracted into 1 ml of dichloromethane by vigorous mixing on a vortex mixer. The lower, dichloromethane, phase can be stored at −20°C.

Gas–liquid chromatography was originally carried out on a column of 3% ECNSS-M on a suitable support at 170–175°C. More thermally stable stationary phases, such as SP2340 (Supelco) and OV225 are now available, and the temperature can be raised to 240°C after elution of the hexitol acetates, to elute and separate *N*-acetylhexosaminitol acetates. Silar 10C, either in a capillary column or on Chromosorb W in a packed column, is an alternative thermally stable packing that gives good results with temperature programming from 190 to 230°C. The derivatives elute in the following order from polar stationary phases: glycerol, erythritol, rhamnitol, ribitol, arabinitol, xylitol, mannitol, galactitol glucitol, *N*-acetylgalactosaminitol, *N*-acetylglucosaminitol. Blakeney *et al.* (1983) give data on elution from Silar 10C. Integrated peak areas may require correction for individual sugar derivatives when used for quantitative measurements of sugar composition.

(iv) Separation of sugars by HPLC

Sugars may be separated without derivation by HPLC on either strong cation-exchange columns or by reverse-phase chromatography. In ion-exchange HPLC resolution is improved by the use of a specific counter-ionic form of the resin, such as the lead form. All manufacturers of HPLC columns and packings supply material specifically designed for carbohydrate separations. The main problem encountered in analysing underived sugars is detection. Sugars can be detected by their UV-absorption, but only at very low wavelengths, where background noise is high and sensitivity low. Refractive index monitoring with an in-line refractometer is commonly used, but despite recent improvements in instrumentation sensitivity is, again, not very high.

A wide range of HPLC techniques and conditions for carbohydrate analysis has been reviewed by Honda (1984), who provides a thorough bibliography. A typical HPLC system is that described by Sutherland & Kennedy (1986) for the analysis of sugars in hydrolysates of lipopolysaccharides. The hydrolysates are neutralized with anion-exchange resin in the bicarbonate form, lyophilized and redissolved in water, and filtered. Samples are applied to an HPX-87P ion-exchange column, 7.8 × 300 mm (Biorad Ltd) and eluted with water at 0.6 ml/minute at 85°C.

Sensitivity of detection can be enhanced greatly by the use of fluorescent or UV-absorbing derivatives. The latter are particularly useful as a UV detector is the commonest form of instrumentation for HPLC. Derivatives can be prepared by dansylation (Alpenfels, 1981) or as secondary amines with *p*-aminobenzoic ethyl ester by reduction of the Schiffs base formed between the sugar and the aromatic amine with cyanoborohydride (Wang *et al.*, 1984). In both cases the aldehydic group of the reducing sugar is derived, so separation by ion-exchange is not possible and reverse-phase HPLC is used.

(v) Methylation analysis

Before the introduction of ^{13}C nuclear magnetic resonance spectrometry (NMR), the only effective method of determining the linkage and substitution patterns of polysaccharides was by methylation analysis. It remains a powerful tool for structural investigations of oligosaccharides, polysaccharides and glycoconjugates and is usually used in conjunction with NMR or mass spectrometry. However, the technique does not give information about the anomeric configurations of the sugars, or about their sequence.

The principle of the technique is initially to convert all the free hydroxyl

Figure 5.1 Stages in methylation analysis of a hypothetical polysaccharide with a trisaccharide repeating unit. Step 1: permethylation of polysaccharide in DMSO with dimsyl sodium; step 2: acid hydrolysis; step 3: reduction of partially methylated monosaccharides with sodium borohydride; step 4: acetylation; step 5: analysis by GC–MS. Modified from Jansson *et al.* (1976).

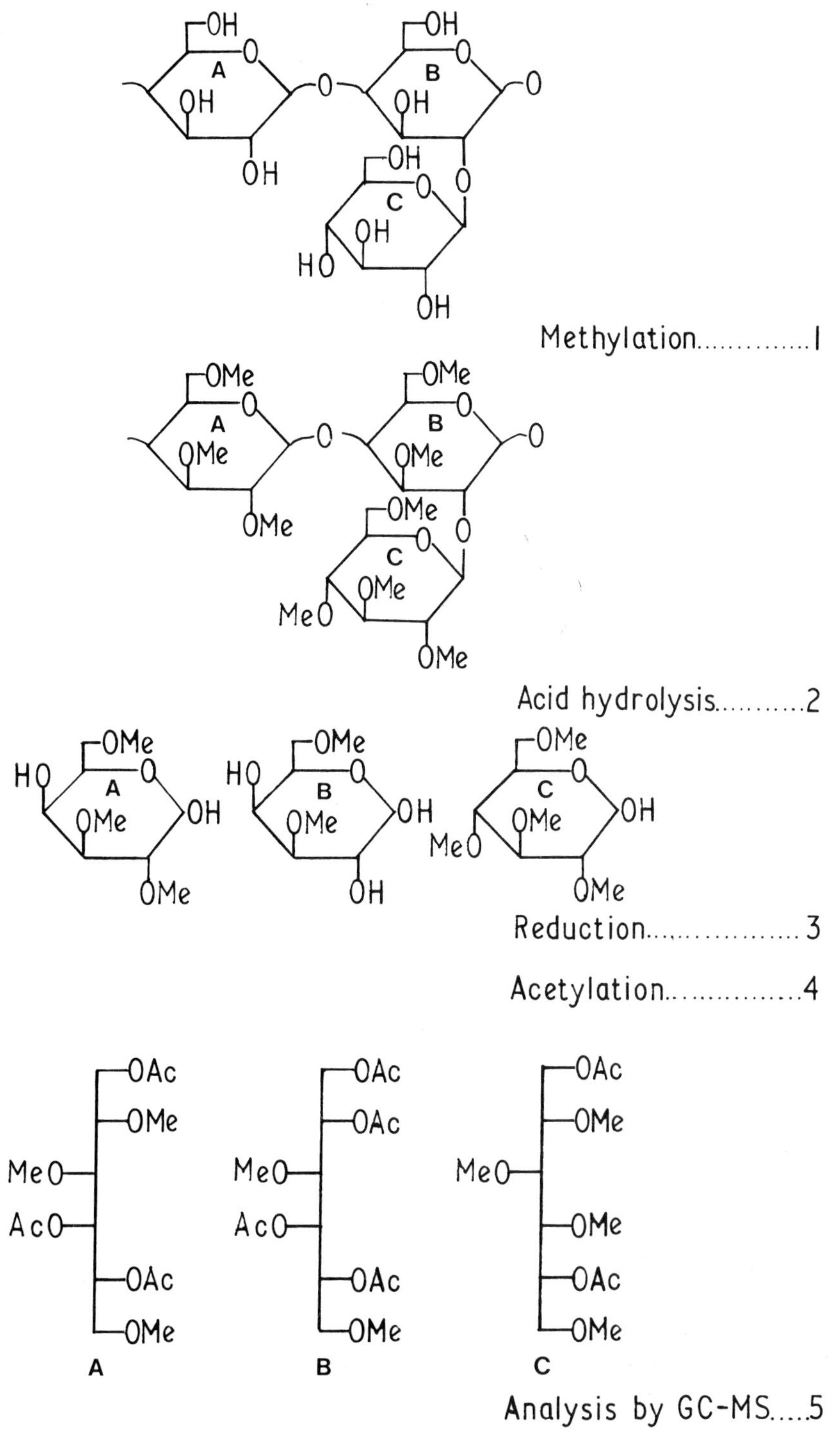
A
B
C
Methylation..............1
Acid hydrolysis...........2
Reduction.................3
Acetylation...............4
Analysis by GC-MS.....5

groups in the intact polymer to acid-stable methyl ethers. This is followed by cleavage of the glycosidic linkages by acid hydrolysis and subsequent reduction to yield partially methylated alditols. The free hydroxyls, originating from the glycosidic linkages of the sugars, are then acetylated. Analysis of the derivatives by combined gas chromatography and mass spectrometry (GC–MS) gives information about the linkage and substitution patterns of the native polysaccharide. Figures 5.1 and 5.2 show a hypothetical example of the technique.

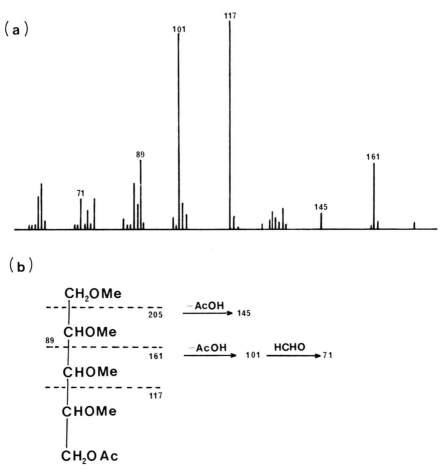

Figure 5.2 Analysis of 5-*O*-acetyl-1,2,3,4-tetra-*O*-methyl ribitol. (a) Mass spectrum with major diagnostic ions labelled in *m/e*. (b) Fragmentation pattern. For more details of the production and analysis of this product, see Poxton *et al*. (1978).

Much of the pioneering work on methylation analysis of bacterial polymers was performed by Bengt Lindberg and his group in Stockholm, and in 1976 they published a practical guide to the methylation analysis of carbohydrates (Jansson *et al.*, 1976). This remains the best source of mass spectra of partially methylated alditol acetates; copies can be obtained from Professor Lindberg.

1. *Procedure for methylation of polysaccharides*

This is based on the original method of Hakamori (1964).

(a) 2 M dimsyl sodium* is prepared as follows. 5 g of oil-free sodium hydride* is prepared in a glass vial by washing a 50% oil suspension three times with light petroleum and flushing dry with nitrogen. 50 ml of dimethyl sulphoxide (DMSO), dried over a molecular sieve, is added. The vial is closed with a septum fitted with a hypodermic needle. The vial is incubated in a sonic bath at 50–60°C for 4 h. Liberated hydrogen escapes through the needle. The greyish solution is stable for at least a month at 4°C.

(b) A dry sample of polysaccharide (0.5–3.0 mg) is dissolved in 0.5–3.0 ml of dry DMSO in a vial fitted with a septum and two hypodermic needles, and the vial is flushed with nitrogen through one needle. It is often necessary to sonicate the sample in an ultrasonic bath to aid dissolution. It is most important that the material is completely dissolved if methylation is to be complete. If the polysaccharide will not dissolve, a number of techniques, described by Jansson *et al.* (1976), must be tried.

(c) 0.5–1.5 ml of 2 M dimsyl sodium in DMSO is added to the vial dropwise, with a syringe, excess pressure being vented through the second needle. The mixture is ultrasonicated for 30 minutes at room temperature, then left overnight.

(d) The vial is cooled in an ice-bath and 0.5–1.5 ml of methyl iodide is added dropwise with a syringe. The mixture is ultrasonicated for 1 h, then excess methyl iodide is removed by flushing with nitrogen or by rotary evaporation below 40°C.

(e) For polymeric products the reaction mixture is dialysed in washed dialysis tubing against running water and the contents are dried by rotary evaporation.

(f) For oligomeric products that might be lost in the diffusate during dialysis, the mixture is diluted with 5 ml of water and the aqueous

---

*CAUTION—Dimsyl sodium and sodium hydride are both extremely caustic and react violently with water.

phase, which contains the permethylated product, is extracted four times with chloroform or methylene chloride. The combined organic phases are washed four times with water and then dried by rotary evaporation.

2. *Acid hydrolysis and alditol acetate formation*

The glycosidic linkages in the permethylated product are hyrolysed in acid under the mildest possible conditions (see section A(i) above). The partially methylated monosaccharides are then converted to alditol acetates as described in A(iii) above.

GC analysis of the derivatives can be performed on a variety of moderately polar columns such as OV 225 or SP 1000. Wall-coated capillary columns give good results. GC alone is not sufficient for analysis as a full range of standards for comparison of retention times with the test samples may not be available. It is best to use combined GC–MS; the mass spectra can be compared readily with those shown in Jansson *et al.* (1976).

The review by Lonngren and Svensson (1974) is worth consulting before an attempt to interpret any mass spectra not shown in Jansson *et al.* (1976). An example of how a mass spectrum can be analysed, from an unknown methylated alditol acetate produced from an oligosaccharide fragment of pneumococcal C-polysaccharide (a teichoic acid), is shown in Figure 5.2.

(vi) Structural analysis by high-resolution ^{1}H and ^{13}C NMR

D. R. Bundle

1. *Introduction*

Proton and ^{13}C NMR spectroscopy now complements methylation analysis as the principal technique for structural studies on polysaccharides. The introduction of two-dimensional NMR experiments (Bax *et al.*, 1981) has made it possible, using only NMR methods at 500 MHz, to determine unambiguously the structure of a polysaccharide repeating unit containing as many as five hexose residues. In practice, most research groups in biological departments do not have access to the large slices of instrument time necessary to accomplish an analysis of this kind. This section will therefore consider NMR techniques that complement conventional analysis and are faster and simpler to execute. At the same time, newer and more elaborate methods are discussed.

High-resolution NMR spectra are, today, exclusively obtained by the pulsed Fourier transform technique. The sample, dissolved in a deuterated solvent and contained in a precision glass tube, is spun in a homogeneous magnetic field. A series of high-power radio-frequency pulses is applied to the sample and the free induction decay of magnetic spin states is recorded after each pulse. Coherent addition of the responses is performed in a computer which is used to perform the transform after sufficient intensity of the accumulated free induction decay has been attained. In this way the familiar intensity versus frequency spectrum is obtained. The magnetic field and the radio frequency are locked together via the deuterium resonance of the solvent, thus ensuring the long-term instrument stability necessary for coherent accumulation of several thousand transients. This is particularly important for less sensitive nuclei such as ^{13}C, which possess an intrinsic sensitivity which is 1.6 per cent relative to that of protons and suffers further from low natural abundance (1.1 per cent). Therefore, while fewer than 100 transients usually suffice to give a useable ^{1}H spectrum, several thousand are needed for a ^{13}C spectrum of the same sample. Basic principles of Fourier transform NMR spectroscopy are described in Shaw (1984) and applications of ^{13}C NMR and ^{1}H NMR to polysaccharides has been reviewed (Perlin & Casu, 1982).

The remarkable successes that have resulted from the application of NMR to bacterial polysaccharides are based on the regular and periodic nature of these polymers. Because of these multiple repeating units the effective concentration of the structural entity under scrutiny, the repeating unit, is such that the resonances of the major structural elements are readily observed. Microheterogeneity of repeating units inevitably complicates analysis and in extreme cases may prevent unambiguous solution of the structure. Aperiodic structures such as the glycan chains of glycoproteins have proved far more difficult to study by ^{13}C methods owing to the low effective concentration of each residue. In this field ^{1}H NMR spectroscopy has been the principal method used.

2. *Relative merits of ^{1}H and ^{13}C NMR data*

The first successful determination of an unknown bacterial cell wall polysaccharide structure by NMR methods employed ^{13}C spectroscopy (Bundle *et al.*, 1974). This technique was effective despite the low field strengths used because ^{13}C spectra, unlike ^{1}H spectra, are often well resolved on account of the wider range of chemical shifts—200 p.p.m. compared with 10 p.p.m. for protons. In the intervening period advances in instrumentation have dramatically improved spectral dispersion and sensitivity. Not only has this allowed more complex structures to be solved, but also higher sensitivities allow data to be collected in a fraction of the time. Furthermore, the intrinsic value of ^{1}H NMR spectra may be most effectively exploited at the higher field strengths now available (Bundle *et al.*, 1986; Perry *et al.*, 1987). At magnetic field strengths corresponding to ^{1}H resonance frequencies between 80 and 200 MHz, most ^{1}H spectra are too complex to give information on more than the prominent constituents. However, at proton frequencies between 300 and 500 MHz, proton spectra are extremely valuable and able to unravel anomeric configurations, sequence and conformational details (Bundle *et al.*, 1986; Perry *et al.*, 1987). Therefore, at the lower field strengths ^{13}C spectra are more likely to provide valuable structural information, while at higher magnetic field strengths increased spectral dispersion and sensitivity enhance the power of ^{13}C NMR, particularly when used in conjunction with complementary proton data. Since proton spectra are almost always obtained during instrument preparation prior to recording ^{13}C data on high field instruments, this data may be obtained at no extra cost in instrument time.

3. Sample preparation

The solvent of choice for polysaccharides is deuterium oxide and only in rare cases is it desirable or necessary to use others such as DMSO-d_6. In order to record ^{13}C spectra of polysaccharides with good signal-to-noise ratio (S/N) within a reasonable time, adequate amounts of sample must be available. The quantity required is a function of the spectrometer operating frequency, the size or number of monosaccharides comprising the polymer repeating unit, and other parameters such as viscosity, molecular weight and motional properties of the solution. Typically, at ^{13}C frequencies in the range 20–50 MHz, samples of 50–100 mg will give good S/N after 12 h of signal averaging, for those polysaccharides containing three or four hexose residues per repeating unit. At higher ^{13}C frequencies S/N increases so that at 100–125 MHz the same sample would yield comparable spectra within 30–60 minutes.

Irrespective of the spectrometer frequency, the sample should be dissolved in a volume of D$_2$O slightly greater than that defined by the receiver coil of the probe head. For 5 mm sample tubes this volume is usually 0.5 ml, while for 10 mm tubes 1.0–1.5 ml will be required.

(a) Crude salt-free polysaccharide is obtained from gel-permeation chromatography or by dialysis against water, and lyophilized.

(b) The material is dissolved in 1 ml of 99.8 at.% D$_2$O and lyophilized again. This should be done in a small desiccator or vacuum centrifuge reserved for the purpose, to avoid contact with water vapour. This step reduces residual water content and exchangeable protons, which would otherwise appear as an intense HOD peak at about 4.7 p.p.m. in the ^{1}H spectrum, possibly obscuring anomeric protons. If only ^{13}C spectra are to be recorded, this stage may be omitted.

(c) The sample is dissolved in the minimum volume of D_2O, as defined above, and, if necessary, filtered into the NMR tube through a small cotton-wool plug in the stem of a Pasteur pipette. When minimum volumes are used in sample tubes larger than 5 mm, Teflon vortex plugs must be used.

Signal-to-noise will be directly related to the number of nuclei (or spins) situated in the receiver coil volume, and large sample tubes increase the amount of sample in this space. Large diameter tubes (>5 mm) can be exploited most effectively for samples of low solubility or when a large amount of sample is available, thereby reducing the instrument time needed to obtain a high S/N. However, a sample dissolved in 0.5 ml of solvent in a 5 mm diameter tube will always yield better S/N than the same sample dissolved in 1.0 ml and run in a 10 mm tube. Although the S/N may be increased by prolonged signal averaging such improvements are proportional to the square root of the number of transients collected, and consequently a fourfold increase in the number of transients is required to improve S/N by a factor of two. Therefore, if an adequate S/N cannot be achieved within 12 h it is not practicable to continue signal averaging. Instead, more sample should be obtained, or other experiments planned exploiting the much more sensitive ^1H nuclei.

Although ^{13}C spectra are more forgiving of poorly prepared samples containing undissolved solids than proton spectra, optimum resolution is obtained when samples are free of insoluble impurities and contain no paramagnetic impurities such as dissolved gases and heavy metal ions. Dissolved oxygen may be removed by boiling the sample for about 1 minute in the NMR tube or by purging with nitrogen or argon. Complexation with EDTA or treatment with ion-exchange resins removes paramagnetic metal ions.

4. *Effects of concentration, temperature and pH*

Concentration may affect the quality of spectra obtained from a given sample. Increased sample concentration results in higher solution viscosity, which in turn decreases relaxation times of the magnetic nuclei and hence causes line broadening. Elevated temperatures reduce viscosity and line widths and therefore improve resolution, but temperature increases also reduce the differential population of the two magnetic spin states, the maintenance of which is essential for observation of an NMR resonance. Consequently temperature increases may degrade S/N ratios, but in most cases, particularly for polysaccharides, this effect is offset by the sharper lines that result from reduced line broadening.

Temperature changes significantly affect the chemical shifts of carbohydrate signals in ^{13}C spectra. Consequently, when data are to be compared accurately, ^{13}C spectra should be recorded at the same temperature and on samples which have reached equilibrium in the probe (Bock *et al.*, 1980).

Carbon-13 spectra are recorded with proton decoupling, and in this way each carbon atom, unless substituted by another paramagnetic nucleus such as ^{31}P (Bundle *et al.*, 1974), appears as a singlet. In order to remove proton coupling the sample is subjected to broadband irradiation at the corresponding proton resonance frequency. As magnetic field strength increases, the energy required to remove proton and ^{13}C coupling increases, and for polysaccharide samples dissolved in D_2O considerable heating of the sample occurs, especially for spectra recorded above 50 MHz. This microwave heating is exacerbated by increases in the dielectric constant of the solvent and the presence of ionic groups or salts. The temperature gradients set up degrade resolution and the S/N. The technique of pulse decoupling minimizes these effects, but nevertheless it is desirable to avoid the presence of salts in samples. In addition, it is generally advisable to record spectra at temperatures somewhat above the ambient probe temperature, and this should be standardized at the beginning of an investigation.

The effect of sample pH on chemical shift is profound for samples containing basic or acidic groups such as uronic acids or amines. When these components are present in a polysaccharide the sample pH should be carefully controlled. Acid- or base-induced shifts can be highly informative when such functional groups are present (Bock & Pedersen, 1983).

5. Referencing of chemical shift

Carbon-13 chemical shifts are defined relative to the ^{13}C resonance of tetramethylsilane (TMS) but, since it is insoluble in water, it is most usual to reference ^{13}C shifts for samples in D_2O relative to internal dioxane (δ_C 67.400 p.p.m.) or acetone (δ_C 30.5 p.p.m.). Proton chemical shifts are also referenced to TMS but again a water-soluble reference is employed. In recent years acetone (δ_H 2.225 p.p.m.) has become the most widely adopted reference. A large body of literature cites proton shifts relative to one of two water-soluble TMS derivatives, 4,4-dimethyl 4-silapentane sodium sulphonate (DSS) or sodium 3-trimethylsilyl)3,3,2,2-tetra deutero-propionate (TSP). It is very important to check the relative reference scale especially when literature comparisons are to be used.

6. Resonance assignment

Until the advent of two-dimensional shift-correlation, assignment of ^{13}C chemical shifts was largely based on empirical rules derived from comparison with model compounds. Oligosaccharide and polysaccharide resonances were assigned by comparison with the chemical shifts of previously assigned constituent saccharides (Gorin, 1980; Jennings & Smith, 1978). Although this method has been successful, it is empirical and error-prone, though few assignment errors have been critical to structure determination, probably because of the caution placed upon such tentative assignments. The extensive use of ^{13}C NMR has provided a large database for chemical shifts of monosaccharides, oligosaccharides and polysaccharides, and these have been tabulated in the literature for the first two classes of compound (Bock & Pederson, 1983; Bock *et al.*, 1984). Unless recourse is made to unambiguous assignment techniques, such data will provide the simplest route for assignment. Eventually the correlation of ^1H and ^{13}C shifts will make empirical methods redundant, especially as these techniques become more sensitive and less demanding in material. General and simple rules for the position of characteristic ^{13}C resonances of carbohydrates are summarized as follows.

1. The anomeric carbon atoms of pyranoses and furanoses are located in the lowest field section of the spectrum, 90–110 p.p.m. Acetal carbon atoms of pyruvate or other acetal-linked moieties are also found within this range.
2. Carbon atoms bearing unsubstituted primary hydroxyl groups are found at 60–64 p.p.m.
3. Carbon atoms of secondary hydroxyl groups in pyranoses and furanoses give signals at 65–85 p.p.m. Signals of *O*-alkylated carbon atoms are shifted to lower field by 5–10 p.p.m. This observation forms the basis of identification of glycosylation sites in poly- and oligosaccharides.
4. Deoxy carbon atoms may be positioned anywhere between 16 and 55 p.p.m. depending on the specific functionality. For example, amino, deoxy-carbon atoms resonate in the range 45–55 p.p.m., whilst 2-, 3- or 4-deoxy sugars exhibit a methylene carbon at 30–40 p.p.m. and 6-deoxy carbon atoms have chemical shifts between 16 and 20 p.p.m.
5. Non-carbohydrate substituents also possess highly characteristic chemical shifts. *O*- and *N*-acetyl groups exhibit CH_3 signals at 21 and 23 p.p.m. respectively (Bundle *et al.*, 1973). As already mentioned, acetal carbon atoms of groups such as pyruvate resonate at 100–110 p.p.m. and the chemical shift is indicative of whether the ring is of the dioxane (6-membered) or dioxolane (5-membered) type (Garegg *et al.*, 1979).

These simple rules are useful in the initial survey of a new polysaccharide of unknown structure. It is possible to make assignments of the number and type of monosaccharide constituents in the repeating unit and also determine whether there are non-carbohydrate substituents present.

7. Determination of anomeric configuration

One of the most useful aspects of NMR methods is the ability to determine anomeric configuration, the practical aspects of which have been briefly reviewed (Bundle & Lemieux, 1976).

The single most useful parameter for this purpose is the one-bond heteronuclear C–H coupling constant, measured at the anomeric centre. Pyranose sugars with equatorially oriented protons (generally alpha-linkages) have $^1J_{C,H}$ values of 170 Hz, whilst the other anomer has a value of 160 Hz (Bock & Pedersen, 1974). This parameter is most easily obtained from the anomeric carbon resonances of a proton-coupled carbon spectrum. Since these signals are well separated from the resonances of other carbon atoms they provide a convenient spectral window for measuring this value. This is illustrated for the O-polysaccharide of *Salmonella godesberg* which has a pentasaccharide repeating unit (see Figure 5.3(c)) as revealed by the five discrete anomeric carbon resonances (Figure 5.3(b)) and which in its proton coupled ^{13}C spectrum shows clearly the well resolved doublets of each anomeric signal (Figure 5.3(a)). Experimentally this spectrum is recorded under conditions of gated decoupling to take advantage of the n.O.e. contribution to S/N (Wehrli & Wirthlin, 1976; Martin *et al.*, 1980). Recently, however, it has been shown that ^{13}C satellite peaks of anomeric protons may also be used to obtain this coupling constant with the higher sensitivity of proton spectra. This experiment requires good S/N and high spectral dispersion, and is consequently most effective at 500 MHz (Bock & Pedersen, 1985). ^{13}Carbon chemical shifts are also indicative of anomeric configuration, since for any pair the β-anomer generally gives the lower field signal. However, this difference is often small and cannot be used as a reliable guide, particularly in polysaccharide samples (Bundle & Lemieux, 1976). If the ^1H spectra are sufficiently resolved, the chemical shift of the anomeric resonance of component pyranosidically linked sugars provides a useful guide to anomeric configuration. Generally the anomeric protons that have chemical shift values higher than 5.0 p.p.m. are assigned as equatorially oriented, which for most common sugars corresponds to alpha-linked residues (Bundle & Lemieux, 1976).

Three-bond coupling constants between H-1 and H-2 protons in the pyranose ring are related to the torsional angle for the segment H1–C1–C2–H2 by the Karplus equation. Consequently, for hexopyranosides with an axial H-2, a coupling constant to H-1 of about 3 Hz is characteristic of an alpha-linkage and 7–8 Hz is characteristic of a β-linkage.

8. Structure determination

The elucidation of the *Salmonella landau* O-polysaccharide structure (Figure 5.4) is discussed as an example of the combination of simple NMR

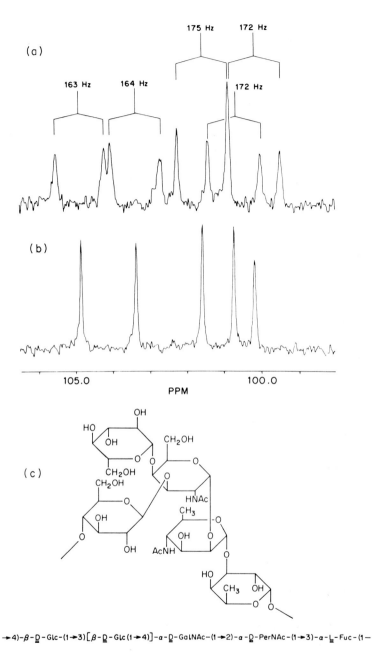

$\rightarrow$4)-β-$\underline{D}$-Glc-(1$\rightarrow$3)[β-$\underline{D}$-Glc(1$\rightarrow$4)]-α-$\underline{D}$-GalNAc-(1$\rightarrow$2)-α-$\underline{D}$-PerNAc-(1$\rightarrow$3)-α-$\underline{L}$-Fuc-(1$-$

Figure 5.3 (a) Proton-coupled ^{13}C spectrum illustrating the five anomeric doublets resulting from one-bond ^{13}C, ^{1}H coupling constants, *ca.* 160 Hz indicating two β-pyranosyl residues and *ca.* 170 Hz indicating three α-pyranosyl residues. (b) Proton-decoupled ^{13}C spectrum of the same spectral region showing the five anomeric resonances of the *Salmonella godesberg O*-polysaccharide (Perry *et al.*, 1986), the structure of which is shown in (c). (c) Chemical repeating unit of the *S. godesberg O*-polysaccharide.

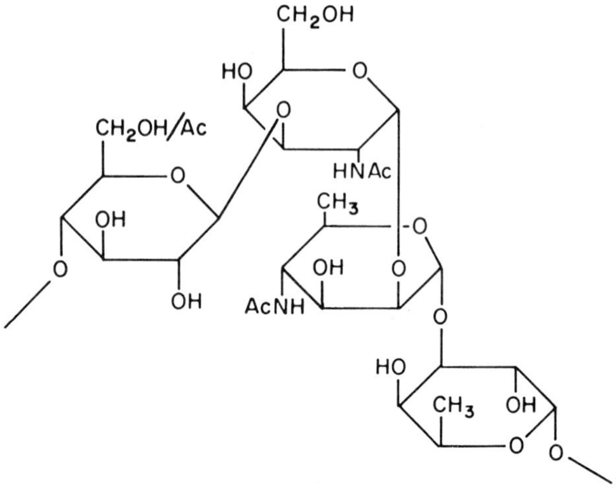

→4)-β-$\underline{D}$-Glc-(1→3)-α-$\underline{D}$-GalNAc-(1→2)-α-$\underline{D}$-PerNAc-(1→3)-α-$\underline{L}$-Fuc-(1—

Figure 5.4 Tetrasaccharide repeating unit of the *Salmonella landau* O-polysaccharide. Reproduced by permission of the *Canadian Journal of Chemistry* from Bundle *et al.* (1986).

techniques with classical structural methods. Although the structure of this polysaccharide was solved by two-dimensional NMR methods at 500 MHz (Bundle *et al.*, 1986) it was also studied by methylation analysis and partial degradation (Perry *et al.*, unpublished results). The combination of the latter methods with one-dimensional NMR techniques constitutes an effective approach to structural analysis. Complete structural analysis of a polysaccharide requires unambiguous determination of the following:

1. The identity and absolute configuration of the component monosaccharides.
2. The ring size and anomeric configuration of each monosaccharide.
3. The position of the glycosylation sites of the monosaccharides.
4. The sequence of monosaccharide residues in the linear or branched structure.
5. The nature, quantity and position of non-carbohydrate substituents.

Compositional analysis by GC mass spectroscopy, the method of choice for component sugar analysis (Gunner *et al.*, 1961), including determination of absolute configuration (Gerwig *et al.*, 1978) showed that the *S. landau* O-polysaccharide contained four monosaccharides, D-glucose (Glc), L-fucose (Fuc), 2-acetamido-2-deoxy-D-galactose (GalNAc) and

4-acetamido-4,6-dideoxy-D-mannose   (PerNAc).   Methylation   analysis
showed the following substitution pattern: Glc substituted at O–4 and O–5,
Fuc substituted at O–3 and O–5, GalNAc substituted at O–3 and O–5,
and PerNAc substituted at O–2 and O–5. This indicated that all four
monosaccharides were present in the pyranose ring form, though the
possibility of the furanose ring form of Glc could not be ruled out on the
basis of methylation analysis alone. The polysaccharide was released from
the lipid A of the lipopolysaccharide in a high-molecular-weight form by
dilute acid hydrolysis, ruling out the possibility of acid-labile linkages in
the O-chain. Subsequent NMR data were consistent with this conclusion.
The determination of anomeric configuration, monosaccharide sequence
and the nature of the non-carbohydrate substituents are questions amen-
able to solution by NMR alone, or in conjunction with chemical and
enzymic methods.

Proton and ^{13}C spectra of the native polysaccharide showed the presence
of non-stoichiometric amounts of ester-bound acetate (Figures 5.5(a) and
5.6(a)). The chemical shift of the acetate methyl group is diagnostic of
amide or ester acetate groups, and it can be seen that the two expected
amide acetate resonances at 23 and 22.9 p.p.m. are of equal intensity,
while the characteristic ester resonance at 21.2 p.p.m. has only half this
intensity. This conclusion can also be reached by considering the corresponding proton resonances in the region of 2 p.p.m. Mild alkali treatment
removes the ester group and the spectrum of the modified polysaccharide
is considerably simplified (Figures 5.5(b) and 5.6(b)). Examination of the
anomeric region of both the ^{13}C and ^{1}H spectra indicated the presence of
four anomeric resonances of equal intensity. Therefore, the polysaccharide
repeating unit contains equimolar amounts of the four component monosaccharides in a linear sequence, since methylation data did not indicate
the presence of a branching residue. It can also be seen that the presence of
one O-acetyl residue for every two repeating units causes considerable
spectral complexity in the anomeric and ring carbon regions of the spectra.
Removal of this microheterogeneity is essential for successful application
of NMR techniques. Comparison of the ^{13}C spectra of native and
O-deacetylated polysaccharides (Figures 5.5(a) and (b)) shows that the two
expected CH$_2$OH resonances at about 61.0 p.p.m. are present in the latter
but in the former an additional resonance occurs at 63.1 p.p.m. This could
be shown to be due to a carbon atom bearing two protons by performing a
DEPT experiment (Doddrell *et al.*, 1982) which distinguishes carbon atoms
on the basis of the number of attached protons. Upon removal of acetate
this resonance disappears, while that at 60.7 p.p.m. increases in intensity
relative to the other hydroxymethyl resonance at 62.1 p.p.m. Chemical
shift changes of this magnitude are consistent with the former C-6 group
(60.7 p.p.m.) being the acetylation site. The one-bond coupling constants

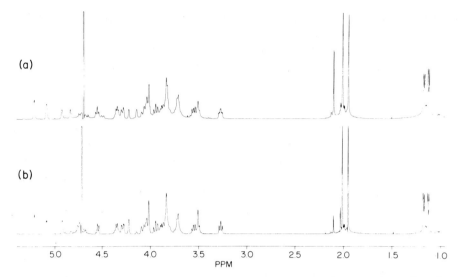

Figure 5.5   500 MHz ¹H NMR spectra of (a) native *S. landau O*-polysaccharide, and (b) de-*O*-acetylated polysaccharide. Reproduced by permission of the *Canadian Journal of Chemistry* from Bundle *et al*. (1986).

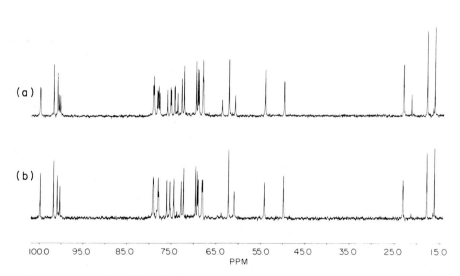

Figure 5.6 125 MHz ¹³C NMR spectra of (a) native *S. landau O*-polysaccharide and (b) the de-*O*-acetylated polysaccharide. Reproduced by permission of the *Canadian Journal of Chemistry* from Bundle *et al*. (1986).

at the anomeric carbon atoms obtained from a proton-coupled carbon spectrum recorded in the gated mode (see section on determination of anomeric configuration, above) showed values of 104.7 p.p.m (^1J = 163 Hz),101.7 p.p.m. (^1J = 172 Hz), 100.8 p.p.m. (^1J = 174 Hz) and 100.2 p.p.m. (^1J = 170 Hz), indicating the presence of a single beta linkage and three alpha linkages. Consistent with this, the ^{1}H spectrum showed a single anomeric proton with chemical shift δ 4.55 p.p.m. and a large (7.9 Hz) three-bond coupling constant, indicating that one sugar possessed *trans*-diaxial protons, thereby excluding the sugar with the *manno* configuration. Thus the 4-acetamido-4,6-dideoxymannose must be alpha-linked. The monosaccharide with the β-configuration cannot be identified from the existing data and further progress would either use conventional degradation techniques monitored by NMR or analysis of NMR data by techniques requiring extensive spectrometer and computer time (Bundle *et al.*, 1986; Jansson *et al.*, 1986). The use of NMR to correlate each anomeric proton with the aglyconic proton or carbon atoms of the adjacent pyranose ring, and hence determine the sequence and site of glycosylation, also requires large amounts of instrument time and expertise and is beyond the scope of this book (Dabrowski *et al.*, 1981; Bax *et al.*, 1984). However, the *S. landau* structure was also easily established by Smith degradation (see section A(ii) above) and one-dimensional NMR experiments. Since the glucose residue was the only component sugar susceptible to periodate oxidation, Smith degradation gave a trisaccharide glycoside of erythritol and NMR showed that the β-glycosidic linkage was now absent as judged by the disappearance of the proton and ^{13}C signals at δ 4.55 and δ 104.7 p.p.m. The β-linked residue was thus glucose. A second Smith degradation destroyed the terminal non-reducing residue of the glycoside to give a disaccharide glycoside of glycerol. Since in the process GalNAc was destroyed and PerNAc became the terminal non-reducing residue, the sequence of the polysaccharide was therefore established as Glc–GalNAc–PerNAc–Fuc. At each successive Smith degradation a single anomeric ^{1}H signal disappeared and by correlation with the position of the anomeric resonances of the precursor substance the anomeric resonances of the polysaccharide could be assigned. In addition the loss of characteristic signals of groups such as the C2 of the GalNAc residue were correlated with the progress of the Smith degradation. In this way the structure of Figure 5.5 was confirmed with the exception of the location of the *O*-acetyl group. Consideration of typical ^{13}C shifts for the C6 signals of β-linked glucose and alpha-linked GalNAc indicated that the resonance at 60.7 p.p.m is typical of β-glucose (Bock & Pedersen, 1983; Bock *et al.*, 1984). Since it was this signal that decreased in intensity in the presence of the *O*-acetyl group, the acetyl group must be located at C6 of the glucose residue.

Application of both ^{1}H and ^{13}C NMR methods has become a crucial technique in polysaccharide chemistry. Routine use of one-dimensional spectra facilitates the isolation and purification phases of an investigation, as well as providing an unrivalled, non-destructive, tool for structural elucidation.

B. PEPTIDOGLYCAN

(i) Investigation of murein structure and metabolism by high-pressure liquid chromatography

B. Glauner and U. Schwartz

1. Introduction

As described in Chapter 1, most bacteria are armoured by a sacculus, an exoskeleton which maintains the mechanical stability of the envelope and cell shape. The sacculus is tailored from a polymer, murein, in which glycan strands are cross-linked by short peptide side-chains. An understanding of murein structure and metabolism (the target of penicillin action) depends on exact knowledge of murein chemistry. For a long time the only method available for analysing murein structure was paper or thin-layer chromatography of murein subunits obtained by controlled enzymic degradation (Primosigh *et al.*, 1961). Recently, a high-pressure liquid chromatography (HPLC) technique has been developed which can separate muropeptides with much better resolution and sensitivity than paper chromatography (Glauner & Schwartz, 1983; Glauner *et al.*, 1987a). The new method has disclosed an unexpected complexity of murein composition. When applied to the murein of *E. coli* it revealed about 80 different components, instead of the 10 shown by paper chromatography. Seven different types of peptide side-chains and two types of peptide cross-linkage (A${}_2$pm–Ala–A${}_2$pm and A${}_2$pm–A${}_2$pm) have been detected. The observed diversity of muropeptides originates in the more or less free combination of the different side-chains in peptide cross-linked dimeric, trimeric and tetrameric compounds and in the occurrence of the 1,6-anhydro derivative of muramic acid (Glauner *et al.*, 1987b). With HPLC, all the compounds can be obtained on a preparative scale and used as probes for the investigation of murein metabolism. In the following sections, the separation and quantification of muropeptides by HPLC is exemplified using *E. coli* murein. The methods, however, can be adapted to the analysis of murein from other micro-organisms with only slight modification.

2. Analysis of murein composition

Growth of bacteria. As murein metabolism is a dynamic process, murein composition depends strongly on culture conditions such as medium composition, temperature, aeration and culture density. The results of different experiments can only be compared if all these conditions are kept constant. These effects are discussed in more detail in Chapter 2, where the advantages of chemostat culture for peptidoglycan studies (Driehuis & Wouters, 1987) are described.

Preparation of unlabelled muropeptides

(a) 400 ml of exponential culture of *E. coli* (A_{578} 0.6) is rapidly cooled to 0°C by shaking in an ice–salt bath for about 2 minutes to prevent uncontrolled attack of endogenous murein hydrolases. The bacteria are harvested by centrifugation at 12 000g for 15 minutes.

(b) The cell pellet is suspended in 3 ml of ice-cold water and added dropwise to 3 ml of boiling 8% SDS within 10 minutes, with vigorous stirring throughout. Degradation of high-molecular-weight DNA and complete solubilization of membranes is achieved by boiling for a further 30 minutes.

(c) The suspension is allowed to stand at room temperature, overnight, then sacculi are collected by centrifugation at 130 000g for 60 minutes at room temperature.

(d) The pellet is freed of SDS by four washes with 8 ml of water. After each wash, great care should be taken to resuspend the sacculi homogeneously. This can be done by agitating on a mechanical mixer with a round-ended glass rod in the tube containing the sacculi and a small volume of water.

(e) Glycogen trapped inside the sacculi (Leutgeb & Weidel, 1963) is degraded with alpha-amylase—the sacculi are suspended in 1 ml of Tris-HCl buffer (10 mM, pH 7.0) containing alpha-amylase (from *Bacillus subtilis*, 100 μg). Care should be taken that the enzyme used is free of contaminating murein hydrolases.

(f) Covalently linked lipoprotein is released by incubation with pronase (from *Streptomyces griseus*, 200 μg/ml) in the same buffer for 1 h at 60°C. Contaminating murein hydrolases in the enzyme can be inactivated by prior incubation of the enzyme for 2 h at 60°C, at a concentration of 10 mg/ml.

(g) The sample (1 ml) is mixed with 1 ml of 8% SDS and incubated at 100°C for 15 minutes, the mixture is diluted with 6 ml of water, and sacculi are recovered by centrifugation and washed with water three times as described above.

(h) The purified sacculi are suspended in 0.5 ml of sodium phosphate buffer (20 mM in phosphate, pH 4.8) containing 0.02% sodium azide, and digested with Chalaropsis muramidase (20 μg/ml) at 37°C overnight. The Chalaropsis enzyme is prepared as described by Hash & Rothlauf (1967) and stored frozen at −18°C at a concentration of 1 mg/ml in the phosphate buffer. The enzyme reaction is stopped by boiling the sample for 3 minutes.

(i) Insoluble material is removed by centrifugation in a microfuge for 5 minutes. The muropeptide concentration in the supernatant will be about 1 mg/ml and may be determined by amino acid analysis. It may be stored frozen at −18°C.

Preparation of radiolabelled murein. The modification of murein structure during growth and division can be studied by pulse- and pulse-chase experiments with radiolabelled diaminopimelic acid (A_2pm) as a murein-specific label. The decarboxylation of the diaminopimelic acid to lysine and its incorporation into protein does not occur in strains that carry a mutation in Lys A such as *E. coli* W7 (Dap A, Lys A; Hartmann *et al.*, 1972) or *E. coli* MC4100 (Lys A; Wientjes *et al.*, 1985). The following protocol is useful for labelling strain W7 and subsequent HPLC.

(a) *E. coli* W7 is grown in the presence of unlabelled A_2pm (2 μg/ml) to A_{578} 0.4.

(b) For short (10–60 s) pulses of labelling 25 ml of culture is filtered at the growth temperature by suction through nitrocellulose membrane filters (0.45 μm).

(c) The bacteria on the filters are washed with prewarmed A_2pm-free medium, then covered with 0.2 ml of medium containing 50–200 μCi[^{3}H]-A_2pm (30 m Ci/mM). After the desired time of labelling the medium is removed by suction and the filter is boiled in 6 ml of 4% SDS. For longer pulses the radiolabel is added directly to the growing culture (up to 4 μCi/ml); the labelled cells are harvested by filtration and boiled as above.

(e) In pulse-chase experiments the filtered cells (see (b) above) are resuspended in 0.4 ml of prewarmed A_2pm-free medium and 200 μCi [^{3}H]-A_2pm are added. After 2 minutes incubation with vigorous shaking the cells are recovered by filtration, washed with prewarmed medium containing unlabelled A_2pm (400 μg/ml) and then resuspended in 100 ml of the same medium.

(f) Samples (10–25 ml) are taken at appropriate times, filtered and boiled in 4% SDS.

(g) After addition of 50 μg unlabelled sacculi as carrier, the muropeptides are prepared as described earlier in this section, scaled down appropriately after the first centrifugation step.

Reduction of muropeptides with sodium borohydride. Muropeptides with a free reducing sugar exist in different anomeric configurations, which are eluted separately during HPLC and thereby complicate the muropeptide pattern. Different anomeric configurations are avoided by reduction of the muramic acid to the corresponding sugar alcohol.

(a) Muropeptide solution prepared as described earlier (0.1 ml, 1–2 mg/ml) is mixed with 0.1 ml of sodium borate buffer (0.5 M in borate, pH 9) and 1–2 mg of solid sodium borohydride are added immediately.

(b) After incubation for 30 minutes at room temperature, excess borohydride is destroyed by addition of 5ml of 20% phosphoric acid.

(c) When evolution of hydrogen is complete, the sample is adjusted to pH 3 or 4 with phosphoric acid.

Muropeptides are not completely stable under these conditions but can be stored for a few weeks at $-18°C$.

Analytical separation of reduced muropeptides. The analytical separation of the reduced muropeptides prepared as above is done by HPLC on a 3 μm Hypersil ODS column (250 × 4.6 mm) with a performance of at least 20 000 theoretical plates.

(a) The column is equilibrated with phosphate buffer (50 mM in sodium, pH 4.31; prepared by titrating NaOH with phosphoric acid) containing 0.8 mg of sodium azide per litre, at a flow rate of 0.5 ml/minute at a temperature of 55°C for 40 minutes.

(b) The sample, 10–100 μl (10–50 μg or 50–250 × 10³ c.p.m.), is injected and the column eluted with a linear gradient starting with the equilibrating buffer and reaching 75 mM sodium phosphate (pH 4.95, 75 mM in sodium; prepared by titration as above) containing 15% v/v methanol (see Figure 5.7). The pH of the eluting buffers should be accurate to at least ± 0.02 units and the temperature must be kept constant within ±0.5°C.

The sodium azide in the initial eluant buffer compensates for the increasing absorption of the effluent caused by the methanol in the second buffer. The amount of azide needed depends on the geometry of the flow cell.

(c) For the detection of radiolabelled muropeptides, fractions (100 μl) are collected, mixed with a liquid scintillation cocktail and measured in a liquid scintillation counter.

The above conditions are highly optimized with respect to temperature, pH, ionic strength and steepness of the gradient, to ensure the complete separation of all *E. coli* muropeptides (Glauner *et al.*, 1987a). For the adaptation of these conditions to other equipment or columns, the following six conditions should be kept (for abbreviations, see Figure 5.7):

1. Tri Anh should elute directly before Tetra–Tri A_2pm Gly_4.
2. Tetra–Tetra Gly_4 should elute between Tetra–Tri A_2pm and Tetra–Tri.
3. Tetra Anh should elute directly after Tetra–Tetra.
4. Tri–Tri A_2pm LysArg should elute directly after Tetra–Penta.
5. Tetra–Tri LysArg should elute between Tetra–Tetra–Tri and Tetra–Tetra.
6. Tetra–Tetra–Tri LysArg should elute directly before Tetra–Tetra Anhl.

As a practical guide for setting up the system, the changes in separation caused by column temperature, the pH of the eluant buffer and the steepness of the gradient are shown in Figure 5.7. The arrows show the change of peak position relative to the neighbouring peaks. Small deviations in the pattern may often be compensated for by changing the temperature by ±3°C.

Quantification of muropeptides. The separated muropeptides may be detected by UV-absorption at 205 nm or by the radioactivity of radiolabelled samples. The radioactivity in collected fractions can be used directly to calculate the molar proportions of the different components. The large amount of data obtained from one analysis (about 600 fractions) makes it advisable to use a computer for the integration of elution patterns.

In the case of UV detection, different extinction coefficients of muropeptides, depending on their chemical structures, have to be considered. For a correction of the area percentage values obtained by integration of the UV absorption signal, the following formula can be applied (Glauner, 1987a):

$$C = D/[(0.6D + 0.1A) - (0.1E + 0.2L)], \qquad (1)$$

where C is the conversion factor, D is the number of disaccharide units, E is the number of 1,6-anhydro end groups, and L is the number of Lys–Arg residues. The values of C for the different muropeptides are in the range 0.8–1.3. The area percentage values of the peaks obtained by UV detection have to be multiplied by these factors. The conversion factors have been normalized for murein from *E. coli* wild-type such that the sum of all converted values is 100 per cent. In cases of murein with a composition different from that of *E. coli*, the converted values have to be renormalized.

Evaluation of the analytical data. In principle, the molar amounts of the muropeptides in a sample can be compared directly between different

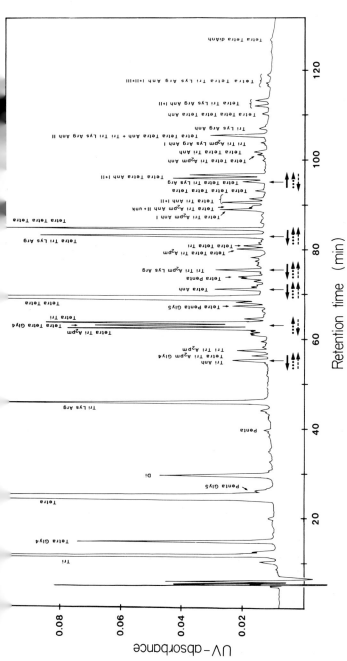

Figure 5.7 Separation of *E. coli* muropeptides by HPLC. Muropeptides (30 μg) prepared from *E. coli* W7 and reduced by sodium borohydride were separated on a 3 μm Hypersil ODS column (250 × 4.6 mm) with a linear gradient from methanol-free 50 mM sodium phosphate of pH 4.31 with 8 mg/l sodium azide to 15% methanol in 50 mM sodium phosphate of pH 4.95 (flow rate 0.5 ml/minute, run time 135 minutes, column temperature 55°C) using an equipment from Waters (two pumps M6000A, gradient controller M660, automatic injector M710B, variable UV-detector M450, data module M730). The column was operated in combination with a guard column (20 × 3.9 mm) packed with Bondapack C18 Corasil. Two pulse dampers were used, causing a dead volume of about 8 ml (16 minutes) between the pumps and the column. For temperature control, a water jacket was used. The effluent was monitored at 205 nm.

Key—Di: disaccharide dipeptide; Tri: disaccharide tripeptide; Tetra: disaccharide tetrapeptide; Penta: disaccharide pentapeptide; Gly4/Gly5: the alanine in position 4 or 5 is replaced by glycine; Anh: 1,6-anhydro muramic acid; LysArg: Lysine–Arginine residue; A_2pm: A_2pm–A_2pm cross-bridge; unk: unknown. The arrows below the chromatogram show the shift of the considered peak relative to the neighbouring peak(s) with increasing column temperature (⟶), increasing pH of the eluent buffer (•••▶) or increasing steepness of the gradient (- - -▶).

experiments. Considering the large number of compounds, this is not recommended. It is particularly difficult to determine changes in the percentage and cross-linkage of the different disaccharide peptides because they occur in various combinations as elements of oligomeric compounds. The muropeptide patterns of different samples are easier to compare after calculation of the percentage of the two types of peptide cross-bridges (A_2pm–A_2pm and A_2pm–Ala–A_2pm), the proportion of the different peptide side-chains (Di, Tri, Tri LysArg, Tetra, Tetra Gly$_4$, Penta, Penta Gly$_5$) and their participation in the two types of cross-linkage, the percentage and cross-linkage of 1,6-anhydro chain ends, the average length of the glycan strands and the sum of monomeric, dimeric, trimeric and tetrameric compounds, as exemplified in Table 5.1.

The cross-linkage is defined as the percentage of cross-bridges related to the total number of disaccharide peptides, as in the following formula:

$$\text{Cross-linkage } (\%) = 100 \times (\tfrac{1}{2}\Sigma \text{ dimers} + \tfrac{2}{3}\Sigma \text{ trimers}$$
$$+ \tfrac{3}{4}\Sigma \text{ tetramers})/\Sigma \text{ all muropeptides.} \tag{2}$$

With the exception of the value obtained for the cross-linkage of 1,6-anhydro chain ends, these results equal the percentage of disaccharide peptides serving as donor sites for cross-bridge formation (Figure 5.8). In the case of trimeric compounds containing both an A_2pm–A_2pm and an A_2pm–Ala–A_2pm cross-bridge, only one of the three side-chains is considered for the calculation of the amount of each type of cross-linkage.

The relative amount of a particular peptide side-chain (X) is defined as the percentage of disaccharide peptides carrying X and possessing a free terminal carboxyl group, that is, side-chains which are free and those which serve as the acceptors in peptide cross-bridge formation (Figure 5.9):

$$X(\%) = 100 \times (\Sigma \text{ X-monomers} + \tfrac{1}{2}\Sigma \text{ X-dimers}$$
$$+ \tfrac{1}{3}\Sigma \text{ X-trimers} + \tfrac{1}{4}\Sigma \text{ X-tetramers})/\Sigma \text{ all.} \tag{3}$$

By this definition tri- and tetrapeptide donors (that is, with no free carboxyl) are taken as a group of side-chains that have been considered already in calculating the extent of cross-linkage. Therefore the values for the different side-chains and the cross-linkage add up to 100 per cent.

The average glycan chain length can be calculated from the percentage of 1,6-anhydro disaccharide peptides, which form the blocked reducing ends of the glycan strands. Other chain termini do not exist in *E. coli* (Holtje *et al.*, 1975). The following formula is used to calculate the average chain length:

$$\text{Average chain length} = \Sigma \text{ all muropeptides}/\Sigma \text{ 1,6-anhydro}$$
$$\text{disaccharide peptides} \tag{4}$$

Table 5.1 The murein composition of E. coli KN126 and E. coli W7 grown at 30°C in PB-medium, analysed with HPLC as shown in Figure 5.7. The data collected were corrected for the different extinction coefficients of muropeptides at 205 nm and used to calculate the percentage of the different peptide side-chains, the proportion of 1,6-anhydro end groups, the average length of the glycan strands, the cross-linkage (A_2pm–Ala–A_2pm as well as A_2pm–A_2pm) and the relative amounts of monomers, dimers, trimers and tetramers (bold-faced). The free and cross-linked portions of the different disaccharide peptides are presented individually. The standard deviations were obtained from double determinations of eleven independent muropeptide samples and are given as percentages of the indicated values.

	E. coli KN126 (%)	E. coli W7 (%)	Average standard deviation (%)
Dipeptide	2.72	2.61	9
Tripeptide	11.01	9.78	10
free	80.5	81.5	2
acceptor (Ala)	18.0	16.6	11
acceptor (A_2pm)	1.5	1.9	15
Tri LysArg	5.47	4.97	5
free	68.7	66.8	4
acceptor (Ala)	26.9	27.3	7
acceptor (A_2pm)	4.3	5.9	15
Tetrapeptide	54.25	54.59	4
free	69.1	67.7	3
acceptor (Ala)	28.7	29.3	4
acceptor (A_2pm)	2.2	3.0	10
Pentapeptide	0.15	0.16	18
free	40	50	15
acceptor (Ala)	60	50	13
Tetra Gly4	2.50	3.33	8
free	70.0	72.7	2
acceptor (Ala)	28.6	25.4	3
acceptor (A_2pm)	1.4	2.0	23
Penta Gly5	0.39	0.44	17
free	62	75	11
acceptor (Ala)	38	25	20
1,6-anhydro ends	3.00	3.30	7
uncross-linked	40.7	36.4	4
cross-linked (Ala)	46.2	47.1	5
cross-linked (A_2pm)	13.1	16.6	7
Average chain-length	33.3	30.3	5
Cross-linkage			
total	23.06	23.73	3
Ala	21.40	21.54	3
A_2pm	1.66	2.19	8
Monomers	**55.30**	**54.07**	2
Dimers	**40.52**	**41.42**	2
Trimers	**4.06**	**4.36**	9
Tetramers	**0.13**	**0.15**	18

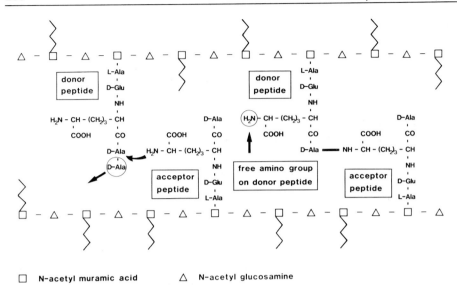

☐ N-acetyl muramic acid   △ N-acetyl glucosamine

Figure 5.8   Mechanism for the formation of the A$_2$pm–Ala–A$_2$pm cross-bridge in *E. coli*.

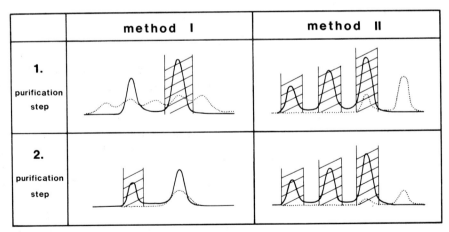

Figure 5.9   Two methods for the purification of unreduced muropeptides using the anomerization reaction. In a first step the hatched fractions are pooled. After readjustment of the anomeric equilibrium, a second purification step follows under the same conditions. —— anomers of the considered muropeptide, ····· contaminations.

where

Σ 1,6-anhydro disaccharide = Σ anhydro monomers + $\frac{1}{2}\Sigma$ monoanhydro
      peptides   dimers + $\frac{1}{3}\Sigma$ monoanhydro trimers +
                             $\frac{1}{4}\Sigma$ monoanhydro tetramers + Σ dianhydro
                             dimers + $\frac{2}{3}\Sigma$ dianhydro trimers.   (5)

(The dianhydro trimers are not detectable in *E. coli* wild-type.)

As an example, the results obtained for the murein composition of two *E. coli* wild-type strains are presented in Table 5.1. The mean standard deviations for determination of the different values are also given.

3. Preparative separation of muropeptides

HPLC also permits the isolation of muropeptides on a preparative scale. However, the separation of reduced muropeptides is much easier than resolving unreduced, native compounds. Fortunately, reduced muropeptides serve, in all cases so far tested, as substrates for enzymes for which the unreduced muropeptides are the natural substrates. For example, they are substrates for both endopeptidases and for carboxypeptidase II of *E. coli*. For the semi-preparative separation of reduced muropeptides the analytical conditions described in section B(i)2 above are satisfactory. About 1 mg of muropeptides (up to 5 mg for preparation of the major components) can be applied to a 250 × 4.6 mm column of 3 μm Hypersil ODS. For the purification of larger quantities, a wider column (e.g. 250 × 20 mm) packed with 5 μm Hypersil ODS should be used. For the preparation of highly purified compounds it is necessary to repeat the separation steps, using a methanol gradient from 0 to 15 per cent in 50 mM sodium phosphate buffer, the first at pH 5.5, and the second at pH 3.5.

Unreduced muropeptides can also be prepared by HPLC, but with more difficulty. The different anomeric configurations of a muropeptide are present in equilibrium before separation commences, but as conditions change during separation mutarotation occurs, leading to the formation of a smear between individual peaks and poor recovery. Mutarotation can be suppressed by reduction of the column temperature, but at the cost of column performance. A separation at room temperature is the best compromise.

Advantage can be taken of the anomerization reaction (Figure 5.8). Two methods can be used. In the first method only one anomer of the desired compound is eluted in the first separation step. After readjustment of the anomeric equilibrium overnight, an additional anomer can often be obtained in pure state in a second chromatographic step (Figure 5.8(a)). In the second method, all anomers of a desired muropeptide are collected in the first separation step. Subsequent readjustment of the anomeric equilibrium then permits separation of some contaminants not separated during the first step (Figure 5.8(b)). These techniques should be combined with changes in the pH of the eluant buffers. For the initial separation of unreduced muropeptides a linear gradient from 0 to 15 per cent methanol in 50 mM sodium phosphate, pH 5.5, is recommended. This may be followed by chromatography at pH 6.5 and 4.0.

4. *Determination of radiolabelled donor and acceptor peptides*

A peptide cross-bridge consists of a donor and an acceptor side-chain. The terminal carboxyl group of the donor side-chain, which provided the chemical energy during cross-link formation, is connected to the epsilon-amino group of the diaminopimelic acid residue of the acceptor side-chain (Figure 5.9). Donor and acceptor peptides are distinct from each other since only the donor has a free amino group. After modification of this amino function, for example by dansylation, both types of side-chains can be assayed independently. HPLC separates the dansylated and non-dansylated diaminopimelic acid residues after total acid hydrolysis of the derivatized muropeptides.

During pulse and pulse-chase labelling of murein, the two A_2pm residues in the peptide cross-bridges are labelled differently. After short periods of labelling, the radioactivity is detected predominantly in the donor peptides, because the newly inserted units contain most of the pentapeptide side-chains that can provide energy for transpeptidation. Investigation of the time-course of incorporation into donor and acceptor peptides and of the fate of the radiolabel during a 'cold' chase, give some insight into the mechanism by which the sacculus grows. For example, it is possible to distinguish between the incorporation of new material into a narrow growth zone, where mainly new (labelled) subunits are cross-linked with each other, and a more randomly distributed incorporation where new subunits are connected mainly to existing, unlabelled, murein (Burman & Park, 1984). Furthermore, a possible cleavage of cross-bridges during the ageing of murein (Goodell & Schwartz, 1983) can be revealed. In combination with HPLC separation of the muropeptides, the different oligomeric compounds can be analysed independently, to study their contribution to the structure of the sacculus and their role in murein metabolism.

Procedure. For the analysis of the donor/acceptor ratio in oligomeric subunits, at least 5000 c.p.m. of an individual muropeptide is required. The sample should have a low content of inorganic salts, which interfere with the dansylation reaction. If necessary, the compound may be desalted by HPLC on any ODS support.

(a) A muropeptide solution to be desalted is injected on to a 5 μm Hypersil ODS column (125 × 4.6 mm) equilibrated with 50 mM sodium phosphate, pH 4 (flow rate 0.5 ml/minute at 50°C) and eluted with the same buffer. After 2 minutes the column is washed with 15% methanol in water. Under this condition the muropeptides elute almost free of salt in a narrow zone at the boundary between the buffer and the methanol. This material is used for dansylation.

(b) The sample is adjusted to pH 9.5–10 with sodium hydroxide and mixed with 0.02 ml of 0.2 M sodium borate, pH 10. The sample is lyophilized and redissolved in 0.04 ml of water.

(c) The sample is mixed with 0.02 ml of dansyl chloride solution (22.5 mM in acetonitrile) and kept in the dark for 90 minutes at 37°C.

(d) The dansylation mixture is lyophilized, dissolved in 0.5 ml of 5 M HCl and heated for 14 h at 100°C in a sealed tube or vial. The hydrolysate is dried and freed of acid by evaporation *in vacuo* over P_2O_5 and KOH in a desiccator. The sample is redissolved in 0.1 ml of water and adjusted to pH 4 for HPLC.

(e) The sample is applied to a column (125 × 4.6 mm) of 5 μm Hypersil and eluted with a linear gradient of 35–80 per cent (by volume) methanol in 35 mM sodium phosphate, pH 6.5, at a flow rate of 1.5 ml/minute at room temperature. The running time is 5 minutes. 1 ml fractions are collected. A_2pm (acceptor) elutes after about 1 minute; dansylated A_2pm (donor) elutes after about 4.5 minutes.

(f) The radioactivity in the fractions is determined.

The data must be corrected for a release of dansyl groups during acid hydrolysis. This is conveniently done by isolating and analysing in the same way muropeptides from uniformly labelled murein, in which the radioactive donor:acceptor ratio will be unity. The correction factor is then calculated as follows:

$$k = [\text{dans.}A_2\text{pm (c.p.m.)} + \text{non-dans.}A_2\text{pm (c.p.m.)}]/2 \times \text{dans.}A_2\text{pm} \tag{6}$$
$$\text{(c.p.m.).}$$

The experimental data obtained for the percentage of donor (D) or acceptor (A) peptides are corrected by this factor as follows:

$$D_{\text{corr}} = Dk; \quad A_{\text{corr}} = A - D(k - 1) \tag{7}$$

5. *Determination of the activity of murein hydrolases*

The enzymic activity of several murein hydrolases can be determined with isolated muropeptides as substrates. The separation of substrate and reaction products can be achieved quickly and conveniently by HPLC. For this purpose a short column (125 × 4.6 mm) packed with Hypersil ODS, 5 μm, is suitable.

Endopeptidase. [^{3}H]-A$_2$pm-labelled bis(disaccharide tetrapeptide) (Tetra–Tetra), reduced with sodium borohydride and purified by HPLC, is used as substrate. Endopeptidases cleave this compound into two molecules of disaccharide tetrapeptide (Tetra). The substrate (20 pmole, 10 000 c.p.m.) is incubated with the enzyme in Tris–maleate buffer (10 mM in Tris) at pH 8.0 for the penicillin-sensitive DD-endopeptidase (Tamura *et al*., 1976) or at pH 6.0 for the penicillin-insensitive LD/DD-endopeptidase (Keck & Schwartz, 1979) containing 10 mM MgCl$_2$, 0.2% Triton X-100 and 0.02% sodium azide in a total volume of 0.04 ml. After incubation at 37°C for 10 to 60 minutes, the reaction is stopped by boiling for 3 minutes. Precipitated protein is removed by centrifugation in a microfuge. The supernatant is mixed with 0.01 ml of 2 M sodium phosphate buffer, pH 3, and applied to the HPLC column. It is eluted with 15% methanol in 50 mM sodium phosphate, pH 2.8, at a flow rate of 0.6 ml/minute at room temperature. The reaction product (Tetra) elutes with a retention time of about 3.5 minutes. In the presence of carboxypeptidase II, Tri may also be formed. This elutes at 2.5 minutes. The undegraded substrate elutes at 8 minutes. Radioactive products are measured in fractions of 0.3 ml.

Carboxypeptidase II. Carboxypeptidase II of *E. coli* (Izaki & Strominger, 1968) can use the reduced disaccharide tetrapeptide (Tetra) as substrate. About 40 pM (10 000 c.p.m.) of the [^{3}H]-A$_2$pm-labelled compound are incubated with the emzyme in 25 mM Tris–maleate buffer, pH 8.0, containing 0.1 mM dithiothreitol and 0.02% sodium azide. After 10 to 60 minutes at 37°C the reaction is stopped by boiling. The pH of the sample is adjusted to between 3 and 4 with phosphoric acid. Insoluble material is removed by centrifugation. The supernatant is applied to the HPLC column and elution carried out with 3% methanol in 50 mM sodium phosphate buffer, pH 4.3, at a flow rate of 0.6 ml/minute at room temperature. The reaction produce, disaccharide tripeptide (Tri), elutes at 4.5 minutes, the disaccharide tetrapeptide (Tetra) at 7.5 minutes. Radioactivity is measured in 0.3 ml fractions.

Carboxypeptidase IA. Carboxypeptidase IA can also be assayed by HPLC, using the nucleotide precursor of peptidoglycan synthesis, UDP-muramyl pentapeptide, as substrate. The nucleotide is prepared from *Bacillus cereus* as described by Lugtenberg *et al*. (1971) and Tamura *et al*. (1976), and purified by HPLC. 0.2–1 nM of the substrate is incubated with the enzyme for between 30 and 120 minutes at 37°C in 0.05 ml of Tris-HCl buffer (5 mM, pH 7.4) containing 0.1 mM dithiothreitol, 0.1% Triton X-100 and 0.02% sodium azide. After boiling and centrifugation the sample is applied to the HPLC column and eluted with 50 mM sodium phosphate buffer, pH 6.0. The reaction product, UDP-muramyltetrapeptide, elutes at 2.5 minutes, the substrate at 4 minutes. The samples are quantified by their UV absorbance.

(ii) Determination of the average chain length of peptidoglycan

If a suitable HPLC system is available, peptidoglycan chain lengths can be estimated as described above (Glauner & Schwartz), but this method is only applicable to those few peptidoglycans that are known to have 1,6-anhydro blocked reducing ends. However, a similar principle can be applied more generally by quantitation of the reduced form of the reducing-terminal sugars of the glycan chains following reduction with sodium [^{3}H]-borohydride, by periodate oxidation of the non-reducing terminal amino sugar, or by selective β-elimination of the lactylpeptide unit from the reducing-terminal muramic acid residue (Ward, 1973).

1. *Borohydride reduction**

(a) Isolated peptidoglycan or clean cell wall containing 5–20 μmol hexosamine (see Appendix 1) is suspended in 1 ml of 12.5 mM NaOH at 4°C and 1 ml of freshly prepared sodium [^{3}H]-borohydride (10 Ci/mole; 0.2 M in 12.5 mM NaOH) is added. The mixture is incubated for 6 h.
(b) The mixture is acidified with acetic acid and left at room temperature until evolution of gas is complete, then frozen and dried under vacuum in a desiccator over fresh, solid NaOH with two ethanol- or acetone-solid CO_2 traps between the dessicator and the pump.
(c) The product is alternately dissolved in 10 mM acetic acid and dried as above, or on a rotary evaporator, until all exchangeable radioactivity is removed, as indicated by a constant level of total radioactivity in the sample.
(d) The final dry sample is hydrolysed in 4 M HCl at 100°C for 4 h, and the acid is removed by evaporation under vacuum.
(e) The hydrolysis products—radioactive muramitol and glucosaminitol,

---

*CAUTION—Reduction with ^{3}H-borohydride in aqueous solution is accompanied by the liberation of radioactive tritium gas, and more is liberated during destruction of excess borohydride with acid. Local radioisotope regulations may govern how much tritium gas may be released in this way; in any case a fume hood vented to the atmosphere independently of the main ventilation system is essential.

It should also be noted that a substantial proportion of the radioactivity initially incorporated into the product will be in the form of exchangeable protons in hydroxyl and amino groups. This will equilibrate, rapidly under acidic conditions, with solvent water. This is the reason for the repeated evaporation of dilute acid from the reduced product. The condensate from this process will also be radioactive.

unlabelled muramic acid and glucosamine—are separated by chromatography on a column (0.8 × 20 cm) of AG50 × 2, H⁺ form, 200–400 mesh ion-exchange resin previously equilibrated with 0.1 M pyridine-acetic acid, pH 2.8. Products are eluted first with 70 ml of the same buffer followed by 70 ml of 0.133 M pyridine-acetic acid, pH 3.85; 2 ml fractions are collected throughout. Muramitol, followed by muramic acid elute in the first buffer, glucosaminitol followed by glucosamine in the second. Elution is monitored by measurement of radioactivity (muramitol and glucosaminitol) and hexosamine (muramic acid and glucosamine) as described in Appendix 1. Variable amounts of radioactive, hexosamine-free artifacts may elute in a peak before muramitol; a peak eluting after glucosamine has been identified as alaninol, presumed to be derived from ester-linked alanine in wall teichoic acid (Ward, 1973).

(f) Mean chain lengths are calculated from the amount of sugar alcohol (estimated from the amount of radioactivity and the known specific radioactivity of the sodium borohydride) and the total hexosamine, in the hydrolysed sample. Reducing terminal muramic acid residues (i.e. those that yield radioactive muramitol) represent the biosynthetic reducing termini of the glycan chains. Reducing terminal glucosamines presumably arise by post-synthetic endoglucosaminidase activity.

2. Periodate oxidation

Glycans consisting of 1–4 linked N-acylhexosamine residues are not susceptible to periodate oxidation except for the terminal non-reducing N-acetylglucosamine group. The latter is oxidized between C3 and C4 to give a product which liberates glycerol, originating in the C4–C6 moiety of the sugar, on borohydride reduction followed by dilute acid hydrolysis. This has been used to estimate non-reducing terminal N-acetylglucosamine residues in peptidoglycan in clean cell walls by measuring radioactive glycerol produced following reduction with ³H-borohydride. It should be noted that N-acetylhexosaminyl substituents of teichoic acid, or terminal N-acetylhexosamines in other accessory wall polymers, would also give radioactive glycerol.

(a) Walls are oxidized by treatment with 25 mM sodium metaperiodate (freshly prepared) in 0.1 M sodium acetate buffer, pH 4.5, in the dark at 4°C for 24 h. The walls are recovered by brief centrifugation, then reduced with sodium [³H]-borohydride as described above.

(b) After destruction of residual borohydride with the minimum amount of 0.1 M HCl, the walls are recovered by centrifugation and suspended in 0.02 M HCl. 100 μg of carrier glycerol is added and the mixture is incubated at room temperature for 16 h.

(c) The walls are removed by centrifugation and washed with a small volume of water. The initial supernatant and the washings, containing the radioactive glycerol, are combined and transferred quantitatively to the origin of a chromatogram on Whatman 3MM paper. After development in ethanol–0.5 M aqueous ammonium acetate, pH 3.8 (5 : 2 by volume) overnight, the radioactivity in the band corresponding to authentic glycerol is measured.

Evaporation under reduced pressure, particularly at elevated temperatures, should be avoided during this procedure, as glycerol is slightly volatile under such conditions.

3. *β-elimination of lactyl peptides from reducing terminal muramyl peptides*

The O-lactyl peptide group of a muramyl peptide is subject to β-elimination under alkaline conditions when there is a free reducing group on the muramic acid residue (Perkins, 1967). This process has been useful in studying the growing end of the peptidoglycan during biosynthesis (Ward & Perkins, 1973).

Peptidoglycan is treated with 4 M aqueous NH_3 at 37°C for 6 h. The released lactyl peptide(s) can be isolated by preparative paper chromatography on Whatman 3MM paper, previously washed in 1 M ammonium acetate, then water. The eluting solvent isobutyric acid–0.5 M aqueous NH_3 (5 : 3 by volume) is suitable.

4. *Labelling of non-reducing terminal N-acetylglucosamine residues using galactosyl transferase*

Galactosyl transferase from bovine milk catalyses the transfer of galactose from UDP-galactose to C4 of N-acetylglucosamine or N-acetylglucosaminyl residues. It can therefore be used to radioactively label non-reducing peptidoglycan chain ends (Schindler *et al.*, 1976). The enzyme is a glycoprotein and is subject to autogalactosylation, which could give misleading results in some procedures. Therefore it is advisable to autogalactosylate the enzyme using unlabelled UDP-galactose before use (Torres & Hart, 1984), if work is to be done on a very small scale.

(a) Bovine milk galactosyl transferase (Sigma) is incubated in 0.05 M Tris-HCl, pH 7.3, with UDP-galactose (0.4 mM), $MnCl_2$ (5 mM), 2-mercaptoethanol (1 mM) and aprotinin solution (Sigma, 15–30 TIU/ml) (1% v/v), in a total volume of 0.1 ml, for 30 minutes at 37°C.

(b) The enzyme is concentrated by precipitation with solid ammonium sulphate added to 85% saturation at 4°C, and centrifugation. It is redissolved to 40–60 U/ml in 0.025 M HEPES, 5 mM $MnCl_2$, 50% w/v glycerol, pH 7.3, and stored at −20°C. It is stable for at least 6 months under these conditions.

(c) The incubation mixture for radioactive labelling of peptidoglycan contains:

    Cell wall containing 20 µg peptidoglycan
    UDP-[U-^{14}C] galactose (1 µCi, >200 mCi/mmol)
    0.4 units autogalactosylated galactosyltransferase (40 U/ml)
    HEPES buffer, 0.05 M, pH 7.3, containing 0.01 M $MnCl_2$
    Total volume 0.15 ml

(d) Incubation is carried out at 37°C for 90 minutes, then wall is recovered by centrifugation and thoroughly washed with water to remove soluble radioactivity. Bound radioactivity is best measured following hydrolysis of the product in 2 M HCl at 115°C for 2.5 h.

C. LIPOPOLYSACCHARIDES

(i) Demonstration of heterogeneous chain length

Classically, lipopolysaccharide (LPS) was viewed as an amphipathic molecule consisting of three main regions: lipid A, core polysaccharide and *O*-antigen polysaccharide. However, it is now known to exist as a heterogeneous population of molecules varying in *O*-antigen chain length, substitution patterns, microheterogeneity in the core and, in some cases, overall structure.

Polyacrylamide gel electrophoresis (PAGE), with or without SDS, is the simplest method of examining the heterogeneity in *O*-antigen chain length. The technique developed by Tsai & Frasch (1982) for silver staining LPS on gels has been applied to an increasing range of Gram-negative bacteria to demonstrate heterogeneity of chain length in smooth LPS and to give an indication of chemotype of rough forms of LPS. This technique has been referred to briefly in the section on LPS preparation (see Chapter 4E).

(a) LPS is prepared by one of the methods described in Chapter 4E, and the amount of carbohydrate material present is measured by the method of Dubois *et al*. (1956) (see Appendix 1) in order to assess the amount to be applied to the gel.

(b) A 12% or 14% PAGE gel is prepared by the method of Laemmli (1970) (see Appendix 1) but omitting the SDS.

(c) The LPS is dissolved to a concentration of about 200 µg

carbohydrate/ml in the SDS-containing sample buffer of Laemmli (1970).

(d) 50 μl samples of the LPS solution (10 μg carbohydrate) are applied to the gel, and electrophoresis and silver staining are carried out as described in Appendix 1. An example of silver-stained LPS preparations from *Escherichia coli* 018, demonstrating the sensitivity of the technique, is shown in Figure 5.10.

It is possible to analyse the antigenic nature of the LPS by immunoblotting. Duplicate gels are prepared and the separated LPS is transferred to nitrocellulose membrane by the method described in Chapter 6D. SDS in the gel buffers inhibits the binding of LPS to nitrocellulose and should be omitted (Pyle & Schill, 1985).

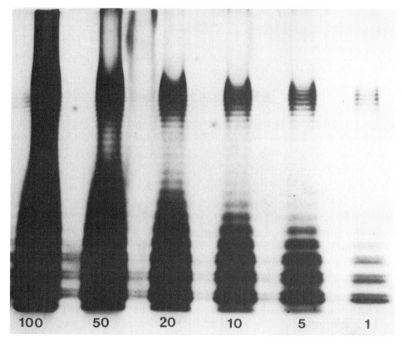

Figure 5.10 Heterogeneity of *O*-antigen chain length of LPS. LPS was extracted by the aqueous phenol method from *E. coli* serotype 086 and analysed on 14% polyacrylamide gels (without SDS) and stained with silver. The figure below each track refers to the number of micrograms applied. Each band of increasing molecular mass corresponds to the addition of one repeating unit of *O*-antigen.

(ii) Demonstration of polymeric O-antigen in some *Escherichia coli* strains

Certain serotypes of *E. coli*, such as 0111, will only agglutinate with O-serotyping antisera after heating to 100°C. This suggests that a capsule is present. However, K antigens are not detectable. Phenol extracts of such serotypes have been fractionated by isopycnic density gradient centrifugation into two fractions (Goldman *et al.*, 1982). The lower-molecular-mass fraction contained classical LPS, while the higher one consisted of polymerized O-antigen, without core polysaccharide or lipid A. This phenomenon has now been extensively studied by Peterson & McGroarty (1985). They have demonstrated that LPS from two *Salmonella* spp. and certain non-agglutinating *E. coli* strains can be fractionated by gel filtration chromatography in the presence of detergents into three major fractions. In the *Salmonella* spp. the high-, medium- and low-molecular-mass fractions were shown by SDS-PAGE to consist of LPS with average O-antigen chain lengths of 70 repeating units, 20–30 repeating units and one repeating unit respectively. In the *E. coli* strains the medium- and low-molecular-mass fractions were similar to those of *Salmonella* but the high-molecular-mass fraction could not be analysed by SDS-PAGE. It is made up of O-antigen polymers not attached to core or lipid A, and has no mobility in PAGE.

For details of procedures used, see Peterson & McGroarty (1985). Examples of their results are shown in Figures 5.11(a) and (b). These workers recommend increasing the concentration of SDS to 0.5% in the running gel to prevent the formation of LPS multimers which may appear as high-molecular-mass bands. Peterson & McGroarty used Tris–HCl–EDTA buffer at pH 8.0, containing 0.25% deoxycholate for dissolving and chromatographing LPS on Sephadex G-200. Other workers, such as Kasper *et al.* (1983) and Poxton & Brown (1986) have used glycine–EDTA buffer, pH 9.5, containing 3% deoxycholate with Sephacryl S-300 (Pharmacia) as the chromatographic medium. The LPS was dissolved in the latter buffer by raising the pH to 11.5 with NaOH, then back-titrating to

---

Figure 5.11   (a) Fractionation of LPS from *Salmonella minnesota* with Sephadex G-200. Fractions were analysed for KDO (●) and amino sugar (○). Silver-stained SDS polyacrylamide gels of column fraction are aligned under their appropriate fraction number. Note that the fractions in peak 1 are seen on the gel. (b) Fractionation of LPS from *E. coli* 0111:B4 with Sephadex G-200. Fractions were analysed for colitose (●) and neutral sugar (○). Silver-stained SDS polyacrylamide gels of column fraction are aligned under their appropriate fraction number. Note that the fractions in peak 1 are not seen on the gel. Both pictures were kindly supplied by Dr E. McGroarty and are printed by permission of the American Society for Microbiology.

(a)

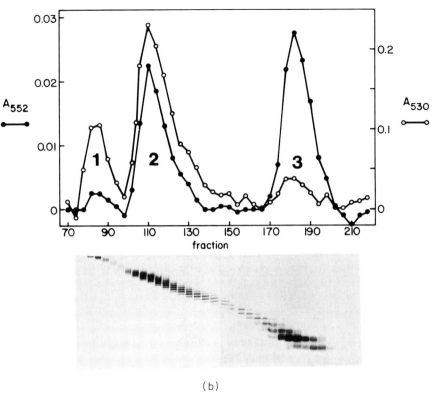

(b)

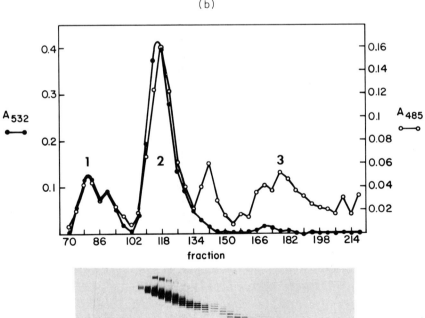

pH 9.5 again with HCl. Insoluble material was discarded after low-speed centrifugation. It is not known how these separation systems compare.

(iii) Separation of core and O-antigen

It is possible to hydrolyse the lipid A–polysaccharide linkage with dilute acid (see Chapter 4E). For smooth LPSs from enterobacteria and pseudomonads this procedure usually also cleaves the linkage between the core and O-antigen. In the polysaccharide fraction after removal of the lipid there remains a mixture of O-antigen chains of varying length and core oligosaccharides. Early work, before it was realized that the O-chains were of heterogeneous length, employed simple gel filtration on Sephadex G-50 to separate O-antigen and core (Schmidt *et al.*, 1969). A high-molecular-mass fraction that was excluded from the column was designated O-antigen, and the included fraction was regarded as core oligosaccharide. It is now known that this core fraction also contained short chains of O-antigen. As described in section C(ii) above, it is also possible that the high-molecular-mass fraction could be contaminated with polymeric O-antigen and also capsular polysaccharide, if present in the starting material.

The following procedure is recommended for the preparation of core and O-antigen:

(a) 2–10 mg of LPS is hydrolysed in 1 ml of 1% (by volume) acetic acid for 90 minutes in a sealed tube or vial in a boiling-water bath.
(b) Precipitated lipid A is removed by low-speed centrifugation.
(c) The polysaccharide solution is extracted with an equal volume of chloroform–methanol (2 : 1 by volume) and the organic extract is discarded.
(d) The aqueous phase is rotary evaporated to dryness and taken up in 1 ml of pyridinium acetate, pH 5.3 (4 ml pyridine, 10 ml acetic acid, made to 1 litre with water).
(e) The solution is applied to a column of Sephadex G-50 (250 × 20 mm) in the pyridinium acetate buffer, and eluted with the same buffer. Sixty 1 ml fractions are collected and alternate ones are assayed for carbohydrate by the method of Dubois *et al.* (1956) (see Appendix 1).
(f) The fractions corresponding to the two main peaks of carbohydrate are pooled. The peak eluting in the void volume consists of O-antigen of long chain length, while the included peak consists of core oligosaccharide contaminated with short chains of O-antigen.

For analytical purposes it is recommended that the two fractions are purified further by affinity methods. A suitable approach would be to use an immunoabsorbent column of monoclonal antibodies to the O-antigen.

The contaminating O-antigen molecules in the core oligosaccharide would be removed by this method. It can also be used to purify the O-antigen fraction if capsular contamination is suspected.

D. ACCESSORY POLYMERS OF GRAM-POSITIVE BACTERIA

(i) Introduction

Teichoic acids are major cell-surface components and antigens in a wide range of Gram-positive bacteria. They are readily detected in clean walls by the presence of large amounts of organic phosphorus (up to 5% by weight) and can be isolated from walls as described in Chapter 4C. They fall into four structural groups:

Type 1: polymers of polyol phosphates (usually glycerol or ribitol, rarely mannitol phosphates) substituted to various extents with sugars
Type 2: polymers in which sugars form, with polyol phosphates, part of the main polymer chain
Type 3: polymers with polyols or polyol phosphates, and sugar 1-phosphates in the chain
Type 4: polymers of sugar-, or oligosaccharide-1-phosphates alone.

The last two categories are typified by lability to dilute acid owing to the presence of the sensitive sugar 1-phosphate linkage. Phosphorus may make up between about 5 and 20 per cent of the weight of the purified polymer and may exist entirely in phosphodiester groups, or with a small proportion in terminal phosphomonoester groups, depending on the method of isolation. Related polymers, mainly of types 2 and 4, occur as capsular components in *Neisseria meningitidis*, *Escherichia coli*, *Pasteurella haemolytica*, *Haemophilus influenzae* and many pneumococci.

A preliminary, qualitative characterization of the polymer can be made by thin-layer or paper chromatography of the products of acid hydrolysis (see Table 5.2).

Specific stains for phosphates, polyols and reducing sugars give a considerable amount of preliminary information about the structure; identification can be confirmed, and quantitative analysis carried out, by gas–liquid chromatography (GLC) or high-performance liquid chromatography (HPLC).

Further quantitative information can be obtained by degradation with concentrated hydrofluoric acid (Anderson *et al.*, 1977). This reagent cleaves phosphate ester linkages without affecting glycosidic linkages or esterified acyl groups such as those found in lipoteichoic acids and lipopolysaccharides. It therefore yields only alditols and glycosyl alditols

Table 5.2 Products of acid hydrolysis of teichoic acids by 2 M HCl at 100°C for 3 h. Major products are shown in bold type.

Type of polymer	Products
Glycerol TAs	
Type 1	**Glycerol, glycerol monophosphates** and **diphosphates, sugars**
Type 2	Glycerol, **glycerol monophosphates, sugars,** sugar phosphates
Type 3	Glycerol monophosphates, **glycerol diphosphate, sugars,** sugar phosphates
Ribitol TAs	
Type 1	Ribitol, **anhydroribitol, ribitol monophosphates,** ribitol diphosphates, **sugars**
Type 2	**Anhydroribitol, ribitol monophosphates, sugars**
Type 4 TAs	**Sugar phosphates,** sugars

from type 1 polymers, glycosyl alditols from type 2 polymers, alditols and sugars from type 3 polymers and sugars or oligosaccharides from type 4. The products are conveniently analysed by GC (see section v, 2 below).

Alkaline hydrolysis can be valuable for distinguishing glycerolphosphate-containing polymers of type 1 and type 2. Alkaline hydrolysis of a phosphodiester linkage occurs solely by way of cyclization to an adjacent free hydroxyl group. Thus, for example, the phosphodiester link between two 2-glucosylglycerol residues in a type 1 glycerol teichoic acid is not susceptible to alkaline hydrolysis. On the other hand alkali causes the hydrolysis of the phosphodiester between the hydroxyls at C6 of the glucose and C3 of the glycerol in a type 2 polymer of 1-glucosylglycerol-3-phosphate repeating units, by cyclization of the phosphate to the free hydroxyl at C2 of the glycerol residue. Since the glucose residue has no hydroxyl group on C5, cyclization of the phosphate on to the glucose is not possible and the only major hydrolysis product is glucosylglycerophosphate. Complete alkaline degradation of type 1 ribitol teichoic acids is possible, even in a polymer carrying a glycosyl substituent on every ribitol residue, because the phosphodiester groups are still adjacent to free hydroxyl groups. In the latter case (see Figure 5.12) hydrolysis is unidirectional because the glycosyl substituent is present at, and blocks phosphate cyclization to, the D4 position so that the phosphate ester linkage to the D5 position must be cleaved with phosphate cyclization to the D2 position of the adjacent ribitol residue. The result of this is that the terminal glycosylribitol unit of the chain is released in unphosphorylated form, whereas all the other repeating units of the polymer yield glycosylribitol phosphates. The presence of esterified alanine does not interfere with this process as in alkali it is rapidly removed.

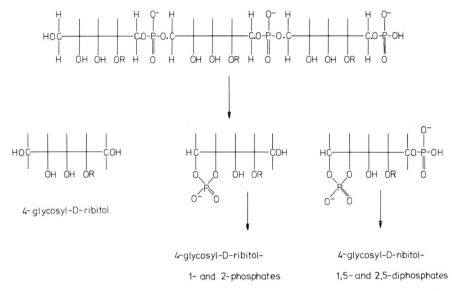

4-glycosyl-D-ribitol

4-glycosyl-D-ribitol-

1- and 2-phosphates

4-glycosyl-D-ribitol-

1,5- and 2,5-diphosphates

Figure 5.12 Alkaline hydrolysis of a ribitol teichoic acid. Alkaline cleavage of the phosphodiester linkages proceeds exclusively by way of five-membered cyclic phosphates. Where the C4 hydroxyl of the ribitol residue carries a glycosyl substituent (R) cyclization of the neighbouring phosphate on to that ribitol residue is impossible. Thus the left-hand terminal glycosyl ribitol unit is released entirely in the non-phosphorylated form, while the centre residue cannot give rise to a diphosphate. Subsequent hydrolysis of the cyclic phosphates results in mixtures of 1- and 2-monophosphate esters.

Hydrolysis with dilute hydrochloric acid can be used to cleave selectively the sugar-1-phosphate ester linkages in type 3 and type 4 polymers, yielding products with phosphomonoester groups and with reducing activity.

^{13}C NMR, as described in section A(vi) above, has been applied to the structural elucidation of both teichoic acids and teichuronic acids with considerable success. Examples of such analyses can be found in Bundle *et al.* (1974), Poxton *et al.* (1978), de Boer *et al.* (1978), Tarelli & Coley (1979), Lifely *et al.* (1980), Fournier *et al.* (1984) and Adlam *et al.* (1985).

(ii) Acid hydrolysis of teichoic acids

(a) Lyophilized polymer (up to 2 mg) is dissolved in 0.2 ml of 2 M HCl in a thick-walled reaction tube or vial and the container is evacuated or flushed with nitrogen and sealed. If the vial is provided with a Teflon/silicon rubber septum (e.g. 'Reactivial' from Pierce or 'Micro

V' vials from Wheaton), evacuation or gassing can be done through a hypodermic needle.

(b) The sealed vial is heated at 100°C for 3 h.

(c) The hydrolysate is frozen and taken to dryness over KOH pellets in a vacuum desiccator. Traces of residual acid are removed by adding water and drying again.

When N-acetylamino sugars are present in the polymer, hydrolysis under the above conditions is incomplete since, if de-N-acetylation precedes cleavage of the glycosidic bond of the N-acetylamino sugar, the resulting free amino group renders the glycosidic bond extremely resistant to acid hydrolysis. For complete hydrolysis under these circumstances, 4 M HCl for at least 6 h is required and the temperature must be raised to 100°C as rapidly as posible, by placing the vial in a preheated block heater or in a boiling-water bath immediately after sealing. Under these conditions, however, considerable destruction of neutral sugars occurs. Takemoto *et al.* (1985) have described a procedure that gives maximum recovery of hexosamines with only moderate destruction (about 20 per cent) of neutral sugars, permitting concomitant analysis of both groups of sugars by GLC or HPLC: lyophilized polymer is dissolved in a mixture of equal volumes of 4M trifluoroacetic acid and 4 M hydrochloric acid and heated in a sealed vial for 6 h at 100°C. Dilute acid hydrolysis of type 3 and type 4 polymers for analytical work is carried out in aqueous 66% (by volume) formic acid at 100°C for 15 minutes (0.5 ml acid, 10 mg polymer). The acid is removed by evacuation in a desiccator over KOH pellets, a few drops of aqueous ammonia are added to the residue to neutralize any remaining acid, and ammonia and ammonium formate are removed by evaporation *in vacuo*. Alternatively, selective hydrolysis can be carried out in 0.1 M HCl, at 100°C for 8 minutes, but the acid is more difficult to remove without causing further hydrolysis of the products.

(iii) Cleavage with hydrofluoric acid*

(a) Lyophilized polymer (up to 5 mg) is dissolved in 0.25 ml of HF (analytical grade, 48%) in a polythene tube fitted with a cap; a 1.5 ml 'microfuge' tube is satisfactory.

---

*CAUTION—When working with HF:

1. Do not use glass: only polythene vessels, pipettes and tubing are suitable. Note that polycarbonate, of which the lids of most plastic desiccators are made, is susceptible to HF. It is therefore vital that an efficient trap containing NaOH is included in the system and is refilled regularly.
2. Avoid contact with skin and inhalation of fumes.
3. Have special burn cream (calcium gluconate) available, and solid calcium gluconate for spillages. Seek immediate medical attention for splashes and burns.

(b) The solution is incubated at 0°C for 48 h.
(c) HF is removed by evaporation under vacuum in a *plastic* desiccator over NaOH pellets, with a sodium fluoride trap intervening between the desiccator and the pump.
(d) The dried material is redissolved in 1 ml of water, applied to a column (1 ml) of Dowex 2 (×8) in the bicarbonate form, and eluted in 2 ml of water.

Faster degradation can be obtained using 60% HF, but if only technical grade reagent is available the HF cannot be removed by evaporation as this leaves a strongly acidic residue that may lead to hydrolysis of glycosides or deacylation. In this case, when incubation is complete (after 16 h) the reaction mixture is transferred to a stirred suspension of Dowex 2 in the bicarbonate form (15 ml of resin in a total volume of 30 ml) and mixed until evolution of CO_2 is complete. The ion-exchange resin is filtered off and washed with 10 ml of water. The combined filtrate and washings are concentrated by rotary evaporation, then lyophilized.

(iv) Alkaline hydrolysis of teichoic acids

(a) Lyophilized teichoic acid (up to 10 mg) is dissolved in 1 ml of 1 M NaOH and heated for 3 h at 100°C.
(b) After cooling, the hydrolysate is desalted by application to a column (1 × 10 cm) of Dowex 50 in the ammonium form, washed off with 10 ml of water, and rotary evaporated to dryness.

Glycosylalditol phosphates formed by alkaline hydrolysis may be dephosphorylated for GC analysis by treatment with alkaline phosphatase. The dried residue following Dowex treatment is dissolved in 5 ml of 0.05 M $(NH_4)_2CO_3$ and incubated with 10 units of calf intestinal phosphatase for 24 h at 37°C. Ammonium carbonate is removed by repeated rotary evaporation.

(v) Chromatography of hydrolysis products

1. Thin-layer and paper chromatography

Thin-layer chromatography of polyol phosphates, polyols and sugars is best carried out on precoated cellulose TLC plates, with appropriate authentic markers. For phosphate derivatives the best eluting solvent is propan-1-ol/aqueous ammonia (sp. gr. 0.88)/water (6 : 3 : 1 by volume). Typical values of R_f are given in Table 5.3.

(a) Dissolve the hydrolysis products from about 2 mg of polymer in 20 μl of water.

Table 5.3 R_f values of hydrolysis products of teichoic acids after TLC in propanol/ammonia/ water (6 : 3 : 1 by volume).

Glycerol and anhydroribitol	0.61
Glycerol cyclic phosphate	0.56
Ribitol	0.46
Glycerol monophosphate	0.26
Ribitol- 1 or 2-phosphate	0.24
Ribitol-3-phosphate	0.27
Sugar phosphates	0.17
Glycerol diphosphate	0.08

(b) Apply samples containing about 10 μg of phosphorus as small spots in a line 1.5 cm from the bottom of duplicate 20 cm plates, with a glass capillary, drying in a stream of cold air.

(c) Develop the chromatograms, by upward flow, in a tank thoroughly pre-equilibrated with solvent, until the solvent front is close to the top of the plates (3–5 h).

(d) Allow the plates to dry at room temperature in a fume hood until the smell of ammonia can no longer be detected.

(e) Stain one plate with the Hanes–Isherwood reagent and the other with periodate–Schiff reagent, as described below.

Some batches of cellulose TLC plates are unsuitable for staining with the periodate–Schiff reagent owing to the development of high background coloration. Paper chromatography on Whatman no. 4 paper by the descending technique in the same developing solvent is a reliable alternative, though development takes longer (overnight). Compounds elute in the same order as in TLC.

Glycerol and anhydroribitol do not separate in this solvent, but can be distinguished by their reaction to the periodate–Schiff stain. Polyols, anhydroribitol and sugars are best separated in acetic acid/ethyl acetate/ pyridine/water (1 : 7 : 5 : 3 by volume) or on silica gel G TLC plates in propan-1-ol/ethyl acetate/water (7 : 2 : 1 by volume). The latter solvent mixture is also particularly effective for separating polyol glycosides. The periodate–Schiff stain can be used with silica gel plates.

Hydrolysis products separated by TLC can be detected by the use of sensitive, group-specific stains. The most useful are the Hanes & Isherwood reagent for the detection of phosphates, the alkaline silver stain for reducing sugars, ninhydrin for amino acids and amino sugars, and the periodate–Schiff reagent for polyols. The last stain gives different colour reactions with different polyols and is very valuable for distinguishing between polymers containing glycerol and ribitol.

The Hanes–Isherwood reagent for phosphates (Hanes & Isherwood, 1949). 5 ml of 60% perchloric acid, 10 ml of 1 M HCl and 25 ml of 4% ammonium molybdate are mixed and made up to 100 ml with water. The plate or paper is thoroughly sprayed with the reagent, then heated in an oven at 100°C until just dry (charring will occur, beginning at the edges, if heating is prolonged). The stain is developed by exposure to UV light for up to 1 h. Blue spots indicate phosphate-containing components. The reagent can be used on plates that have been stained previously with ninhydrin.

The silver reagent for reducing compounds (Trevelyan *et al.*, 1950). Three reagents are required and should be freshly prepared:

> Silver reagent: mix 0.1 ml of saturated silver nitrate solution with 20 ml of acetone and add water dropwise until the white precipitate just redissolves.
> Ethanolic sodium hydroxide: mix 1 ml of 40% w/v aqueous NaOH with 20 ml of ethanol.
> Thiosulphate destainer: 5% w/v aqueous sodium thiosulphate.

The plate or paper is immersed briefly in the silver reagent, then drained and allowed to dry in air in a fume cupboard. The dry plate is immersed in the ethanolic NaOH and allowed to drain. Reducing compounds produce black spots on a brownish background. As soon as the spots have developed the plate is rinsed gently in a bath of thiosulphate destainer to decolour the background. The plates can be kept longer if the thiosulphate is washed off with distilled water before they are allowed to dry.

The stain is less effective where TLC solvents containing ammonia have been used. It is important that in these cases the plates are very thoroughly air-dried before staining. The stain will also detect polyols and glycosides, with lower sensitivity; longer development is required, and heating to 100°C for example in steam, without drying, after treatment with the ethanolic NaOH.

The periodate–Schiff reagent for polyols (Baddiley *et al.*, 1956). Three reagents are required: 1% sodium metaperiodate solution; 1% *p*-rosaniline hydrochloride solution, decoloured by bubbling sulphur dioxide through it until a straw colour is obtained; and gaseous sulphur dioxide (small laboratory cylinders are available). Not all samples of *p*-rosaniline are suitable: that from Eastman–Kodak seems to give the most consistent results. The plate or paper is sprayed with sodium periodate solution and left at room temperature for 5 minutes. It is treated with a stream of sulphur dioxide in a fume hood until the brown colour initially formed just disappears again, then sprayed with the *p*-rosaniline reagent and left to develop. The colours of the spots depend on the products of periodate oxidation. Compounds giving rise to formaldehyde, such as glycerol, glycerol phosphate and ribitol, give an immediate purple coloration, while those forming malondialdehyde derivatives, such as ribitol phosphate, produce an intense yellow colour. Anhydroribitol and sugars give a blue colour that may take up to 1 h to develop.

Ninhydrin for amino compounds. The reagent consists of 0.2% w/v ninhydrin and 1% v/v acetic acid in water-saturated butan-1-ol. The paper or plate is sprayed with the reagent, then heated at 100°C for 5 minutes to develop purple or brown spots.

2. GC of degradation products from teichoic acids

Because of the difficulty of preparing sufficiently volatile derivatives, GC analysis of alditol phosphates and sugar phosphates has not found wide application. GC is more appropriate for analysis of free alditols and sugars produced by acid hydrolysis, and of alditols and alditol glycosides obtained by HF degradation. The analysis of sugars, as their alditol acetates, from surface carbohydrates generally, is discussed in section A above, and the methods described there are applicable to sugars and free alditols from teichoic acids. This section will deal with other methods for the analysis of alditols and alditol glycosides.

The following procedure is adapted from that of Endl *et al.* (1984) using trifluoroacetyl derivatives.

(a) The dry, acid-free products from acid hydrolysis or HF degradation are dissolved in 0.2 ml of trifluoroacetic acid and trifluoroacetic anhydride (1 : 50 by volume and heated at 100°C for 20 h in a Teflon-sealed vial. (Free polyols and sugars are derived in 1 h, but the longer incubation time is required for full derivation of glycosides.)
(b) The cooled reaction mixture is centrifuged if it is cloudy, and samples (up to 5 μl) are injected directly on to a column of 3% QF1 on Chromosorb W-HP 80/100 mesh (Hewlett Packard). The column temperature is programmed from 100 to 230°C at 5 degrees per minute and then maintained at 230°C for a further 15 minutes.

This system separates the alpha and beta glycosides of glycerol and ribitol, as well as glycerol and ribitol themselves. It is also suitable for sugar identification where only a small number of different species is present; but because sugars give multiple peaks by this method it is not very suitable for complex sugar mixtures.

3. Estimation of esterified-D-amino acids in teichoic acids

Ester-linked amino acids can usually be removed from teichoic acids by mild alkaline treatment and assayed spectrophotometrically using

D-amino acid oxidase. The method described below is a modification of that of McArthur & Reynolds (1980) and will detect amounts of D-amino acids in the range 10–80 nM. Thus amino acid from a maximum of 100 μg teichoic acid should be measurable.

(a) 1 mg of teichoic acid in 1 ml of 0.1 M sodium pyrophosphate, pH 8.3, is incubated at 60°C for 3 h in order to release all the esterified D-amino acid. It is advisable to determine the time-course of release for a new sample since the kinetics may vary depending on polymer structure and amino acid.

(b) The assay reagent is made up *freshly* from the following components:
   A: 0.1 M sodium pyrophosphate, pH 8.3
   B: flavin adenine dinucleotide (FAD) disodium salt in pyrophosphate buffer, 0.2 mg/ml
   C: horseradish peroxidase (c. 1000 U/mg), 10 μg/ml in water
   D: dianisidine sulphate, 5 mg/ml in water
   E: D-amino acid oxidase (from hog kidney, c. 15 U/mg), 5 mg/ml in water.
   Reagents A to E are mixed in the ratio 40 : 20 : 10 : 5 : 1 by volume just before use.

(c) Samples (75 μl) containing D-amino acids are mixed with 225 μl of the mixed reagent and incubated at 37°C for 15 minutes.

(d) The reaction is stopped by addition of 1.0 ml of 0.1% sodium dodecylsulphate and the absorbance is measured at 460 nm against a reagent blank incubated under the same conditions. A standard curve should be prepared with samples of the appropriate D-amino acid, plotting increase in A_{460} in 15 minutes, against weight of amino acid in the sample. 50 nmol of amino acid should give an increase in absorption of 0.3–0.4.

(vi) Determination of chain length of teichoic acids

Accurate chain-length determination depends on the quantitative measurement of a distinctive group from one end of the polymer chain. Since most techniques for the isolation of teichoic acid involve hydrolytic cleavage of the linkage to peptidoglycan, it must be borne in mind that chain cleavage may also occur, resulting in underestimation of the *in vivo* chain length—random cleavage of only one phosphodiester link per chain will halve the estimated mean chain length. The methods that can be employed for chain-length measurement depend on the type of teichoic acid and are too varied to be described in detail. They are summarized below—the techniques involved are all based on methods described in other parts of this chapter.

1. Measurement of terminal phosphomonoesters

Measurement of phosphomonoester groups as a proportion of total polymer phosphate, using phosphomonoesterase treatment of the polymer, was widely used in the past for acid-extracted teichoic acids. However, present knowledge of the chemistry of attachment of teichoic acids to peptidoglycan indicates that the acid-labile link in this attachment is an N-acetylglucosamine-1-phosphate so that undegraded acid-extracted teichoic acid would not bear a terminal phosphate group (see Chapter 4C). Terminal phosphates measured in earlier work must have originated by acid degradation of the main polymer chain, and it is not surprising that the method gave low values for chain length. However, this was a problem of extraction, not of the estimation technique—the method is valuable specifically for the identification and measurement of partially degraded polymers (see, for example, Lang & Archibald, 1982).

2. Acid or alkaline hydrolysis

As described above in section D(ii), dilute acid treatment selectively degrades teichoic acids containing sugar-1-phosphate linkages (types 3 and 4). The result of this is that internal repeating units are released as their phosphomonoesters whereas the product of the terminal group distal to the peptidoglycan attachment site does not carry a phosphomonoester group. Measurement of the ratio of the two types of product has formed the basis of a number of chain-length estimations (see, for example, Hussey *et al.*, 1969).

Analysis of the products of alkaline hydrolysis provides a useful technique for fully glycosylated polyribitol phosphate teichoic acids of type 1 because, as described in section D(ii) and Figure 5.12, only the terminal glycosylribitol residue distal to the peptidoglycan attachment site can yield an unphosphorylated product, which can be measured as a proportion of the total glycosylribitol of the sample (see, for example, Lang & Archibald, 1982).

3. Periodate oxidation

Periodate oxidation of a type 1 ribitol teichoic acid or an unglycosylated type 1 glycerol teichoic acid yields formaldehyde from the primary alcohol group of the terminal alditol residue, but from none of the others. The method is also applicable, in principle, to unglycosylated ribitol teichoic acids while still linked to peptidoglycan, but the possibility of the reaction of formaldehyde with amino groups and of the presence of other compounds yielding formaldehyde on periodate oxidation of the wall must be taken into account (Heckels *et al.*, 1975).

(vii) Teichuronic acids

There are no specific analysis procedures for individual uronic acids and aminouronic acids, and few authentic compounds are commercially available for use as standards in chromatography. The standard techniques of polysaccharide analysis (see section A above) can be applied, but where the presence of a uronic acid is suspected its identity can be confirmed by conversion to the corresponding aldose. This procedure can be carried out on the intact polysaccharide and is particularly valuable for polymers that contain aminouronic acids that are severely degraded during acid hydrolysis.

1. Carboxyl reduction of uronic acids to the corresponding aldoses

The procedure involves the formation of a derivative of the uronic acid carboxyl group, followed by its reduction to the aldose by treatment with borohydride. Derivation can be carried out either by treatment with methanolic HCl (Nasir-ud-Din *et al.*, 1985) or with 1-ethyl-3-(3-dimethyl-aminopropyl)-carbodiimide (EDC) (Taylor & Conrad, 1972). Both methods are given below:

Methylation with methanol–HCl. 10 mg polysaccharide are dissolved in 20 ml of 10 mM HCl in methanol and incubated for 48 h at 4°C. 50 ml of methanol are added, and the solvent removed by rotary evaporation under reduced pressure at 20°C. The methylated polymer is then dissolved in 5 ml of 10% aqueous methanol and mixed with 30 mg of sodium borohydride for 12 h at 4°C. Excess borohydride is destroyed by the addition of acetic acid until the pH remains at 5. The solution is dialysed against distilled water, then lyophilized.

Derivation with EDC. 10 mg of polysaccharide in water is converted to the free acid form by passing through a small column (1 g) of AG 50W × 8 ion-exchange resin in the hydrogen form. EDC (5 to 10-fold molar excess over uronic acid) is added while the solution is stirred at room temperature for 2 h. The pH is maintained at about 4.8 by the addition of 0.1 M HCl as necessary. 2 ml of 2M sodium borohydride (freshly prepared) is then added, and the mixture incubated at 37°C for 12 h. The product is treated as above under (a).

In both methods it is advisable to repeat the entire procedure at least once in order to modify a substantial proportion of the uronic acid residues. Complete conversion may require three or four treatments.

2. *Determination of polymer chain length*

Release of teichuronic acid from linkage to peptidoglycan using dilute acid produces a polymer chain terminating in a sugar with a free reducing group (see Chapter 4C). Quantitative reduction of this residue with [^{3}H]-borohydride permits the introduction of a radioactive label into the terminal sugar, which can then be measured as a proportion of the total sugar in the polymer. The same method of labelling has been applied to peptidoglycan and is described in detail in section B(ii) above.

6

Immunochemistry of Cell Surfaces

A. PREPARATION OF ANTISERA

Although monoclonal antibodies are replacing conventional antisera, there will remain many applications for polyclonal antisera—for example in preliminary investigations and in laboratories without access to tissue culture facilities.

Before attempting to raise antiserum the following points must be considered:

1. Is the antigen likely to be toxic to the animal?
2. How pure is the antigen?
3. Is the antigen likely to be immunogenic?
4. What is the physical nature of the antigen?
5. What is the natural immune status of the animal?
6. How is the antibody response to be monitored?
7. How much antiserum is needed?

When these questions have been answered it will be possible to know what type of animal to use, the route of inoculation, whether or not to detoxify or kill the antigen and if an adjuvant is necessary.

Only two general methods will be given here: one for raising antiserum to whole organisms and one for subcellular fractions. A useful generalization is that whole organisms can be injected intravenously without adjuvant, while subcellular or soluble components require adjuvant and are injected subcutaneously.

The New Zealand white rabbit is the ideal laboratory animal for the production of antisera.

(i) Whole bacteria

(a) Select rabbits weighing approximately 2 kg (ideally more than one rabbit should be used per antigen) and remove 5 ml of blood from the ear vein. This will act as a preinoculation control.

(b) If the organisms are likely to be lethal or toxic they must first be inactivated. Traditionally this was done by formalin treatment (0.5% formaldehyde in buffered saline for 18 h). Ultraviolet irradiation of a thin film of bacteria has proved successful and less damaging to protein antigens. When the antibody response required is to heat-stable carbohydrate antigens, it might be more convenient to heat the organisms for a few minutes in a boiling-water bath.

(c) For each rabbit prepare at least ten doses of antigen, each of between 10^7 and 10^9 bacteria in 1 ml volumes of buffered saline and store deep-frozen until required.

(d) Inject one dose per rabbit into the marginal ear vein with a 26-gauge needle on days 1, 2, 3, 8, 9, 10 and 22, and test bleed from the ear vein on day 29.

(e) Serum is obtained by allowing the blood to clot in a clean vial. After clotting it is shaken to separate the clot from the sides of the vial and then left overnight at 4°C. The serum is removed by pipette and centrifuged at low speed to clarify. If the antiserum is of sufficient titre the animal may be exsanguinated by cardiac puncture; or if smaller amounts of serum are required, up to 20 ml can readily be collected by nicking the ear vein.

(f) If the serum is not of sufficiently high titre it may be worth giving one or two further booster injections and testing a week after the last injection. From experience, however, this is unlikely to increase the titre more than twofold.

(g) If more than one rabbit per antigen was used the antisera should be pooled and divided into convenient amounts and stored deep-frozen. When required it should be gently thawed. Refreeze a minimum of times.

(ii) Subcellular fractions of soluble antigens

For this technique depot adjuvants are usually necessary, and the method detailed here has worked for a variety of antigens ranging from outer-membrane preparations to subfractions of lipopolysaccharides.

(a) See section A(i)(a) above.

(b) If antigen is toxic, detoxify by treatment with formalin (0.5% formaldehyde for 18 h). For each rabbit emulsify 0.1–2.0 mg of antigen (suspended/dissolved in 0.5 ml of water or buffered saline) in an equal volume of Freund's complete adjuvant. It is preferable that a water-in-oil emulsion is prepared and a small homogenizer is usually employed for this, although it is possible to prepare an emulsion by passing the mixture from one syringe to another through a fine-bore tube. The nature of the emulsion can be tested by placing a small drop on to water and observing that it does not immediately disperse. Further mixing is required if it does disperse rapidly.

(c) Inject each rabbit subcutaneously with a total of 1 ml of emulsion in four to six sites in the scapular region of the back. A wide-bore needle (21 gauge) is necessary and the back of the animal should be massaged firmly after the injections.

(d) After about four weeks the procedure should be repeated, but incomplete Freund's adjuvant should be used.

(e) From two weeks after the second series of injections the antibody response should be monitored (see section (i)(e) above.

(f) It might be necessary to boost further (see section (i)(f) above).

(g) See section A(i)(g) above.

(iii) Preparation of IgG from antiserum

In many applications it is preferable to extract the IgG from the antiserum. For example, in crossed immunoelectrophoresis a much cleaner background is obtained if pure IgG is used; and if an antibody is to be tagged with an enzyme or fluorescent label, a reagent of much higher specific activity can be produced if IgG is the starting material.

Several different methods exist for the isolation of IgG: they are based on ion-exchange chromatography, affinity chromatography, gel filtration or precipitation. A very simple method based on ammonium sulphate precipitation is recommended here, but see later in this section for purification of monoclonal antibodies.

(a) Place 1 ml of serum in an ice-bath and gently stir on a magnetic stirrer.

(b) Prepare saturated ammonium sulphate in water and add 0.67 ml dropwise at the rate of one drop per second.

(c) Continue stirring for 15 minutes in the ice-bath.

(d) Centrifuge at 3000g for 10 minutes at 4°C where the IgG forms a pellet.

(e) *Either* take up in 1 ml or less of 0.05 m/l Tris buffer (pH 8.0) and dialyse against 2 l of the same buffer for 24 h; *or* take up in a few millilitres of Tris buffer, place into an ultrafiltration cell fitted with a 10 000 MW cutoff membrane (e.g. Amicon PM10) and reduce to minimum-stirred volume several times, refilling with buffer each time, and reduce to a convenient final volume—this allows concentration of IgG.

(f) If no further treatment of the IgG is necessary it can conveniently be preserved by adding sodium azide to a final concentration of 0.1%, and 1% bovine serum albumin is often added as a protective agent. This can then be stored at 4°C.

B. PREPARATION OF MONOCLONAL ANTIBODIES

This operation should not be undertaken lightly. Although not technically difficult it requires a high degree of expertise in tissue culture and is labour-intensive. Now that many laboratories exist that specialize in this work it is recommended that monoclonal antibody production be carried out by experienced staff in such laboratories. Brief outlines only of the methodology will be given here (Figure 6.1); the reader should consult an expert or one of the following references before attempting such work: Kipps and Herzenberg (1986), Okumura and Habu (1986) or Hurrell (1982).

(i) Purification of IgG monoclonal antibodies from ascites fluid

Immobilized protein A is well known as a useful matrix for binding and purification of IgG molecules from a range of species. However, it varies greatly in its affinity for the IgGs of various species, especially the differential binding of the various subclasses of murine IgG. Pharmacia have recently developed a method for the purification of all subclasses of murine IgG (*Pharmacia Separation News*, Vol. 13.5). The technique is as follows.

(a) Pack a small column with 1 ml of protein A–Sepharose CL-4B and equilibrate with binding buffer (1.5 M glycine, 3 M NaCl adjusted to pH 8.9 with %M NaOH) and connect to a UV recorder to monitor the A_{280}.

(b) Apply 0.5 ml of ascites fluid diluted with an equal volume of binding buffer. Irrigate the column with binding buffer until the A_{280} reaches baseline.

(c) Elute stepwise with elution buffers of 100 mM citric acid adjusted to pH 6.0, 5.0 and 4.0 with 5 M NaOH. IgG1 elutes with the pH 6.0 buffer, IgG2a with the pH 5.0 buffer and IgG2b and IgG3 with the pH 4.0 buffer.

(d) Adjust the pH of the eluted IgG to Ph 8 as soon as possible.

(e) Regenerate the column with elution buffer adjusted to pH 3.0 with NaOH before equilibration with binding buffer.

(ii) Radiolabelling of monoclonal antibodies

(a) React purified antibody (100 μg) in PBS (200 μl) with 100 μCi ^{125}I labelled NaI in the presence of one Iodobead (Pierce Chemical Co.) for 10 minutes at room temperature (Fletcher *et al.*, 1986).

(b) Pass reaction mixture down a small column (10 ml) of Sephadex G25 eluting with PBS. The labelled antibody elutes in the void volume.

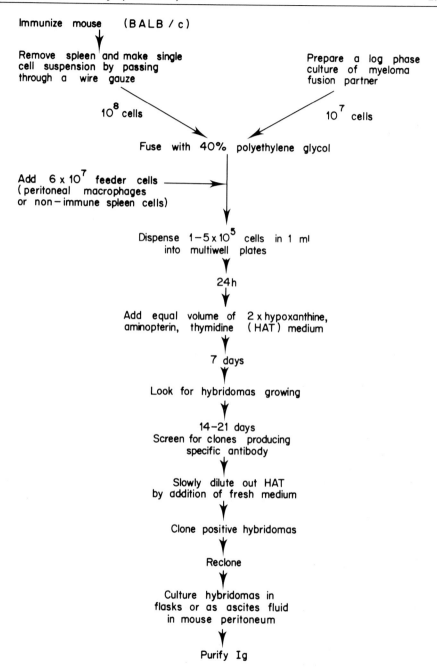

Figure 6.1 Procedure for the preparation of murine monoclonal antibodies.

C. PRECAUTIONS FOR WORKING WITH HUMAN SERA

In the following sections situations may arise when the antiserum or immunoglobulin used is of human origin. The potential risks involved cannot be stressed too strongly and great care must be taken.

For many years it has been recognized that human blood or blood products can harbour hepatitis B virus. More recently the virus (HIV) responsible for causing the acquired immunodeficiency syndrome (AIDS) has been shown to be transmitted through blood. It is quite feasible that other as yet unrecognized agents are present in human blood. It is thus important when work is done with products prepared from human blood, even when they are known to be free of hepatitis or AIDS virus, that sensible precautions be taken; for example, wear gloves, avoid creating aerosols, and if possible heat-treat serum at 60°C for 1 h.

D. IMMUNOCHEMICAL METHODS

(i) Gel diffusion

One of the simplest and most widely used immunological techniques for visualizing antigen/antibody reactions is the double diffusion in gel technique originally described by Ouchterlony (1948). Its major advantage is that it is extremely simple to perform in any laboratory. The disadvantages are that it is very insensitive, complex antigen mixtures cannot be resolved, it is difficult to make quantitative, it is slow, and it sometimes gives apparent false-positives if detergents or salts form precipitates. If nothing better is available it may have a use in early screening of antigen preparations or in monitoring antibody levels in animals. For most purposes it is not recommended.

(a) Prepare a 1% agarose solution in 0.85% NaCl by heating to 100°C in a boiling bath until dissolved, allow to cool to about 50°C and pour on to a clean glass slide to a depth of 3 mm (approx. 4 ml for a 50 × 50 mm glass slide).

(b) Cut circular wells 3–5 mm in diameter, 5–10 mm apart, either side-by-side, or more usually in a pattern with a central well surrounded by a circle of six or so wells (purpose-made cutters are available for these patterns).

(c) If several antigen preparations are to be tested against one antiserum, the antiserum is placed in the central well and the antigens in the outer wells. The slide is placed in a humid chamber (a sandwich box with damp tissues in it) and allowed to diffuse for up to 48 h at 4°C.

(d) Precipitin lines of antigen/antibody complex are viewed in oblique illumination against a dark background. They may also be stained with Coomassie blue by following the method described for crossed immunoelectrophoresis (see section D(ii)5).

(ii) Immunoelectrophoresis methods

The advantages that electrophoretic methods have over simple diffusion are that mixtures of antigens can be separated before reacting with the antibodies, thereby increasing the resolving power of the technique, and the reaction can be made faster by actively bringing together the antigen and antibody. Agarose gels are usually used in a pH 8.6–8.8 buffer. At this pH most biological molecules have a negative charge and will migrate towards the anode, and immunoglobulins at this pH are near to their isoelectric point and should be stationary. The electroendosmotic (EEO) properties of the agarose are important. In gels with high EEO agarose, the dissociable cations and their associated water molecules migrate towards the cathode, resulting in a flow of water through the gel in a direction opposite to the movement of the biological molecules. Immunoglobulins in such gels tend to be carried towards the cathode by EEO. This in useful in such techniques as counter-current immunoelectrophoresis, but for most other immunoelectrophoretic applications agarose of low EEO is recommended.

1. Recipes for buffers and agarose gels for immunoelectrophoresis

Electrophoresis buffer (Svendsen, 1973). This barbital/glycine buffer has high buffering capacity and low ionic strength with a final pH of 8.8:

Solution 1
Barbitone sodium	26 g
Barbitone	4.14 g
Distilled water	2 l

Solution 2
Glycine	112.5 g
Tris	90.4 g
Distilled water	2 l

Mix equal volumes of solutions 1 and 2 and check that the pH is 8.8. In electrophoresis it is used at full strength in the electrode reservoirs.

Agarose gel. Use agarose of low EEO for all applications except counter-current immunoelectrophoresis:

Agarose	1 g
Electrophoresis buffer (see above)	25 ml
Distilled water	75 ml

Mix together and heat in a boiling-water bath with constant stirring until the agarose is dissolved. Avoid prolonged boiling. At this stage it is sometimes necessary (e.g. if antigens are in membranous vesicles) to add Triton X-100 to a final concentration of 1% v/v. Either allow to cool to approximately 50°C before pouring plates, or dispense volumes into tubes for storage. For pouring on to 80 mm square plates, 15 ml amounts are suitable, or 4 ml for 50 mm square plates.

Glass plates are suitable for many simple applications, but for any method where washing, pressing and staining of gels is necessary it is advised that Gelbond™ (Marine Colloids) is used according to the manufacturers' instructions. Gels adhere much better to this medium than to glass. A flat levelling table fitted with a spirit level is essential for successfully pouring gels.

2. Simple immunoelectrophoresis

The simplest type of immunoelectrophoresis is a combination of electrophoresis of antigen in one dimension and diffusion of antibody in a second dimension, resulting in the formation of precipitin arcs. As this technique is of limited use, only brief details are given here. Antigen is placed in a central well cut in agarose on a microscope slide and electrophoresed in the direction of the long axis of the slide. Antiserum, which is placed in a trough cut parallel to the direction of separation of antigen, is allowed to diffuse towards the antigen and forms precipitation arcs. This is now seldom used, but Ørskov and co-workers derived much information on the immunochemistry of O and K antigens of enterobacteria with this technique (Ørskov *et al.*, 1977).

3. Countercurrent immunoelectrophoresis (CCIE)

This technique closely resembles gel diffusion and the sensitivity and resolution are similar. However, the antigen and antibody are actively forced together by electrophoresis in agarose of high EEO, thus considerably speeding up the precipitation reaction. It has gained a degree of popularity in recent years for rapid diagnostic procedures (see Chapter 7C).

(a) Prepare 1% v/v high-EEO agarose (e.g. BDH Electran™ Agarose 25) as above and pour on to the hydrophilic side of a sheet of Gelbond™.

(b) Cut pairs of wells 3–5 mm in diameter approximately 8–10 mm apart.

(c) Place antigen into wells at the cathodic side of the gel and antiserum into the anodic side.

(d) Place gel in an electrophoresis tank, fill buffer reservoirs with full-strength electrophoresis buffer, and connect the edges of the gel to the buffer by means of wicks of thick filter paper (e.g. Whatman 3MM).

(e) Apply a voltage of 25 V/cm for 1–3 h in a cold room.

(f) Precipitin lines are viewed either directly with dark-ground illumination or after staining as for crossed immunoelectrophoresis (see section D(ii)5 below).

4. Rocket immunoelectrophoresis (RIE)

This technique is recommended as a sensitive and quantitative method for detecting precipitating antigens. It is particularly suitable for screening large numbers of samples of antigens (e.g. for screening column fractions or steps in a purification procedure). Its main drawbacks are that it is difficult to resolve mixtures of antigens, and that it only works with precipitating antigens.

(a) For up to ten samples of antigen a 50 mm square sheet of Gelbond™ is required. Place it hydrophilic side up on to a supporting plate of glass and lay it on to a level surface.

(b) Blank off approximately two-thirds of the sheet with a glass plate or strip of Perspex and cast an agarose gel of low EEO on to the remaining third (use approximately 1.3 ml of agarose).

(c) Cut up to ten 3 mm diameter wells in a staggered pattern, as shown in Figure 6.2.

(d) If it is required to show identity between antigens, samples of antigen (10 μl) should be loaded into the wells at this stage and be permitted to diffuse into the agarose for 30 minutes in a humid atmosphere at 4°C. This variation to the technique is termed *fused rocket immunoelectrophoresis (FRIE)*. If identity between antigens is not to be demonstrated, then proceed to the next step and load samples of antigen after that.

(e) Prepare 2.2 ml of molten agarose and cool to 52–55°C in a water-bath. To this add up to 0.5 ml of antiserum—diluted if necessary in electrophoresis buffer (itself diluted 1 in 4 in distilled water); mix gently and pour immediately on to the remaining part of the plate.

(f) Place the gel into the electrophoresis tank with the antigen wells nearest to the cathodic reservoir and connect the edges of the gel to the buffer reservoirs with presoaked wicks of thick filter paper (Whatman 3MM).

(g) Apply a voltage of 12 V/cm for 16 h at 4°C.

(a) + **(b)**

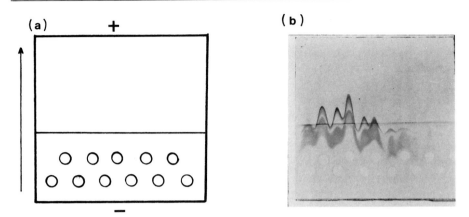

Figure 6.2 Rocket immunoelectrophoresis. (a) Template for positioning wells for fused rocket immunoelectrophoresis. (b) Example of FRIE. Samples (10 µl) of fractions of *Clostridium difficile* cell-lysate eluted from a Sepharose 6B column were placed in the wells and allowed to diffuse for 30 minutes. Agarose (3.5 ml) containing 125 µl of antiserum raised to whole cells of *C. difficile* was cast on to the upper part of the plate and electrophoresis was carried out at 12 V/cm for 16 h. After pressing and drying the gel was stained with Coomassie blue.

(h) Press, wash and stain gel as described for crossed immunoelec-
    trophoresis (see section D(ii)5 below).

An example of the use of rocket immunoelectrophoresis is shown in Figure 6.2(b).

5. *Crossed immunoelectrophoresis (CIE)*

This technique is a two-dimensional version of rocket immunoelectrophoresis. The first dimension involves simple electrophoresis in agarose, the pH of the buffer resulting in movement of antigens through the
agarose towards the anode as a function of charge. The second dimension
is electrophoresis of the separated antigen into antiserum-containing
agarose, resulting in precipitin lines. Its main advantage is that it allows
resolution of extremely complex mixtures of antigens and remains quantitative; the area enclosed by the precipitin arc being proportional to the
amount of antigen present. Routinely we incorporate Triton X-100 into the
agarose.

(a) Prepare the first-dimension gel by pouring 15 ml of molten agarose
    containing 1% Triton (see section D(ii)1 for recipe) on to a 80 mm
    square *glass* plate and allow to set.
(b) Place the plate on to a template (Figure 6.3) and cut out wells—

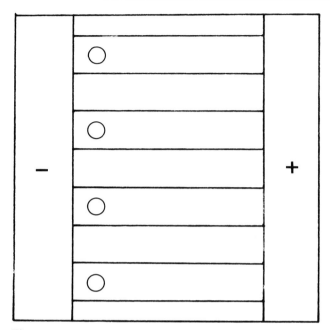

Figure 6.3 Crossed immunoelectrophoresis: template for first dimension. Wells should be large enough to hold 15 μl samples.

rectangular wells give improved resolution but circular ones can be used. Fill wells with 10–15 μl of antigen.

(c) Place the plate into an electrophoresis tank, with the wells nearer to the cathode, and connect the edges of the gel to the buffer reservoirs with presoaked thick filter paper.

(d) Apply a voltage of 100 V (12.5 V/cm) for 1–1.5 h at 4°C.

(e) Remove the gel from the tank, place on a template and cut out strips. Carefully transfer first-dimension strips to the edge of a 50 mm square sheet of gel bond, as in Figure 5.4.

(f) Prepare the second-dimension agarose by adding 0.5 ml of serum, suitably diluted in 1:4 diluted electrophoresis buffer, to 3 ml of molten agarose held at 52°C, mixing gently and casting this against the first-dimension strip.

(g) Arrange the gels in the tank with the first-dimension strips nearer to the cathode, connecting to the reservoirs as above. Up to eight second-dimension gels can be run simultaneously in most commercially available electrophoresis tanks. The gels are arranged in two rows of four, the rows being joined by a wick of filter paper but separated by a dialysis membrane.

(h) Apply a voltage of 12 V/cm for 16 h at 4°C.

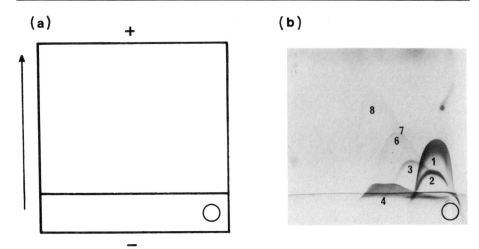

Figure 6.4 Crossed immunoelectrophoresis. (a) Template for second dimension. (b) Example of CIE. First dimension: 10 μl EDTA surface protein extract (15 μg protein) of *Clostridium difficile*. Second dimension run into 125 μl antiserum to whole cells of *C. difficile* in total of 3.5 ml of agarose. The gel is stained with Coomassie blue.

Staining of CIE gels. Place the gels on a flat surface and carefully cover each with a sheet of Whatman No. 1 filter paper, avoiding any air bubbles. Place several sheets of blotting paper over the gels, cover with a glass plate and apply pressure by means of a weight of about 1 kg (e.g. several books) for 15 minutes. Remove pressed gels, discard filter paper covers and wash in two changes of 0.01 M NaCl and one of distilled water, each for 15 minutes or more. The pressing is repeated as above and the thin gels are dried by means of a hot-air blower or incubator. When the gels are perfectly dry they are stained for 10 minutes in a solution containing 5% w/v Coomassie blue R250, 45% v/v ethanol, 10% v/v acetic acid and 45% distilled water. The backgrounds are destained in several changes of a solution as above from which the dye has been omitted. An example of CIE is shown in Figure 6.4(b).

Several more complex forms of CIE have been developed, such as tandem CIE and intermediate gel CIE. The reader is referred to the excellent *Manual of Quantitative Immunoelectrophoresis* by Axelson *et al.* (1973) for further details of these and much more information on crossed immunoelectrophoresis.

(iii) Immunoblotting (electroblotting/Western blotting)

The immunoelectrophoresis techniques described in section D(ii) have now been largely superseded by immunoblotting. For many years, polyacrylamide gel electrophoresis (PAGE) has been the method of choice for the analytical separation of proteins. Several immunochemical methods were developed for probing such separated proteins with antiserum, but all were technically difficult or results were unsatisfactory. It was not until the development of electrophoretic transfer from gels to nitrocellulose membranes by the method of Towbin *et al.* (1979) that satisfactory detection of PAGE-separated proteins became generally available to most laboratories. In the last few years the technique has become extremely popular.

The technique can now be applied to both proteins and lipopolysaccharide molecules. It is usually based on the Laemmli PAGE technique, with or without SDS, and after transfer and antibody treatment the antibody/antigen complexes are visualized with enzyme-linked or radiolabelled anti-first antibody immunoglobulins or radiolabelled protein A.

The method described below is for the separation of proteins and their subsequent probing and detection with an enzyme immunoassay and closely follows the Bio-Rad ImmunoBlot™. The variation developed for LPS is described in the section on analysis of LPS (see Chapter 4E).

1. Buffer recipes for electrotransfer and immunoassay

Transfer buffer
Tris	12 g
Glycine	57.68 g
Methanol	1 l
Distilled water	4 l

Dissolve Tris in methanol/water before adding glycine. If necessary adjust the pH to 8.3 with 1 M NaOH.

Tris-buffered saline (TBS)
Tris	4.84 g
NaCl	58.48 g
Distilled water	2 l

Dissolve Tris and NaCl in approximately 1.5 litres of water. Adjust the pH to 7.5 and make up to 2 l.

Tween Tris-buffered saline (TTBS)
To 2 l of TBS add 0.5 ml of Tween 20.

Blocking solution (3% gelatin)
Add 15 g of gelatin to 500 ml of TBS and dissolve by heating to 37°C. Dispense in 50 ml volumes. Autoclave 10 p.s.i. for 10 minutes and store at 4°C. Before use, melt in warm water.

Antibody/conjugate diluent (1% gelatin)
Add 5 g of gelatin to 500 ml of TBS and treat as above.

HRP colour reagent
Make up freshly immediately before use.
   HRP colour reagent (Bio-Rad)   30 mg
   Methanol (analytical grade)   10 ml
   Hydrogen peroxide (30% w/v)   30 μl
   TBS   50 ml
Dissolve HRP reagent in ice-cold methanol. Add ice-cold hydrogen per-
oxide to TBS. Mix both solutions and use immediately.

2. Procedure

(a) Prepare an SDS-polyacrylamide slab gel containing 10% w/v
     acrylamide with the Laemmli buffer system (see Appendix 1).
(b) Prepare samples of protein by solubilizing 50 μg of protein in 50 μl of
     sample buffer (see the Laemmli method) at 100°C for 3 minutes.
(c) Apply samples to gel and electrophorese until tracking dye has
     migrated a suitable distance (e.g. 10 cm).
(d) Remove the gel from the PAGE apparatus, cut away any tracks that are
     to be stained and place the remainder on the Scotchbrite™ pad which
     will be nearest to the cathode. Gently cover the gel with a sheet of
     nitrocellulose membrane that has been presoaked by floating on
     transfer buffer, carefully avoiding air bubbles. Assemble the cassette
     and place it into transfer buffer in the electroblotting apparatus (see
     Figure 6.5).
(e) Apply a voltage across the tank. We have found that 10–12 V and a
     current of approximately 40 mA for 16 h at 4°C is suitable for all
     transfers, including LPS. Faster transfer is possible, but a special
     power pack is required. It is often more convenient to transfer over-
     night and sensitivity is maximal.
(f) After transfer, remove the nitrocellulose membrane and wash in TBS
     for 10 minutes.
(g) Place the membrane into blocking solution by sliding it through the
     surface at an angle of 45° and leave it for 30–45 minutes.
(h) Remove it from the blocking solution and place it into the first
     antibody. Depending on the titre of the antiserum the dilution will
     range from 1 : 20 to 1 : 250. Treat for 3 h.
(i) Remove it from the first antibody, rinse it briefly in distilled water and
     then wash it for two 10 minute periods in TBS.
(j) Place it in peroxidase-labelled anti-first antibody conjugate which has
     been suitably diluted in TBS. For high-titre conjugates (e.g. Bio-Rad

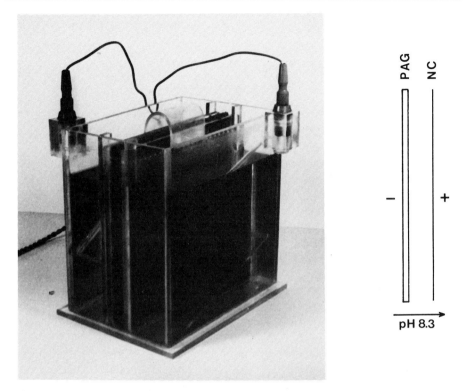

Figure 6.5 Immunoblotting apparatus: an example of the apparatus used in the author's laboratory, and a diagram showing the orientation of polyacrylamide gel (PAG) and nitrocellulose membrane (NC).

anti-rabbit IgG peroxidase conjugates) it is possible to use them at a dilution of 1:3000. Other conjugates may need to be diluted to 1:500. Treat for 1 h.

(k) Wash it as in (i).

(l) Place the membrane in freshly prepared HRP colour development reagent and allow it to develop for up to 30 minutes. Stop the reaction by placing it in several changes of distilled water.

(m) Dry it in the dark. The colour is permanent if stored in the dark.

Use gentle shaking at room temperature for stages (h)–(l). An example of an immunoblot is shown in Figure 6.6.

Immunoblotting has many advantages over other immunochemical assays. For example, it is extremely sensitive, quantitative, reproducible, it has high resolving power, it can be used to detect monovalent (non-precipitating) antigens and can easily be used on multiple samples. It has

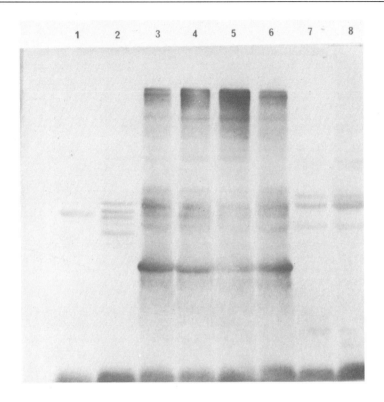

Figure 6.6 Example of immunoblotting used in immunochemical fingerprinting. EDTA surface protein extracts of eight clinical isolates of *Clostridium difficile* probed with a 1:200 dilution of antiserum raised against whole cells of *C. difficile*. For further details see Poxton *et al.* (1984).

several drawbacks, however. Perhaps the main one is the denaturation conditions that are necessary before the sample is run on PAGE. This may result in the destruction of epitopes. This is especially important when monoclonal antibodies (MAB) are being used as probes for protein antigens. By definition MABs recognize a specific epitope, and if this is lost no reaction will occur. To check on this possibility a dot of control, undenatured antigen should be applied to the side of the nitrocellulose membrane after stage (f) above. Another problem is that some antigens may not be transferred to nitrocellulose. They may pass through without sticking. If this is suspected it is worth attempting to remove the SDS from the gel buffers while retaining it in the electrode and sample buffers. This allows for more sensitive binding to nitrocellulose. Some antigens are too large to be analysed by PAGE: CIE is the best alternative.

3. Direct staining of proteins on nitrocellulose

A problem with immunoblotting was that there was not a satisfactory non-immunological stain that could be used to reveal the total number of protein bands transferred to nitrocellulose. Recently a commercially available Aurodye™ (Janssen Pharmaceuticals) has been introduced. This is a highly sensitive and simple to use gold dye which binds to proteins on nitrocellulose. Its sensitivity is equivalent to silver staining of gels.

(iv) Enzyme-linked immunosorbent assay (ELISA)

Since its introduction in the early seventies (Engvall and Perlmann, 1972), ELISA has developed into the most widely used of immunoassays. Its high sensitivity and adaptability to automation has resulted in many applications, ranging from routine diagnostic serology and clinical chemistry to a powerful research tool in all branches of modern biology, especially as a screening technique in monoclonal antibody production. As with all labelled antibody techniques it can be used directly and indirectly, and can be used in inhibition assays. The indirect assay is probably the most widely applicable and is described in most detail here.

1. Indirect ELISA

The principles of indirect ELISA are summarized in Figure 6.7. The antigen is immobilized on a solid support, usually polystyrene, by incubation in a high-pH buffer. After washing with Tween 20 solution, which both blocks any unbound sites and prevents subsequent non-specific

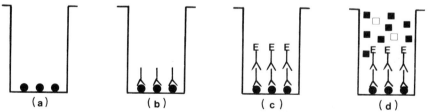

Figure 6.7 Diagrammatic representation of indirect ELISA. (a) Antigen is bound to a plate in pH 9.6 buffer. (b) After washing, antiserum is added and specific antibody binds to antigen. (c) An enzyme-labelled second antibody which is specific for the first antibody is added. (d) A substrate which produces a product that is coloured or a different colour is added and the colour change is measured. The colour change is proportional to the antigen/antibody complex formed at stage (b).

binding, the first antibody is added. After incubation and further washing in Tween, an enzyme-conjugated second antibody, which recognizes the first antibody as an antigen, is added. After further incubation and washing the enzyme substrate is added. This either changes colour or becomes coloured as the enzyme acts. The final colour change is proportional to the amount of antigen/first-antibody complex formed.

2. Buffers for ELISA

Antigen coating buffer
   0.05 M sodium carbonate buffer, pH 9.6
   0.2% sodium azide
Prepare 0.2 M carbonate/bicarbonate stock solutions. Solution A: 21.2 g of anhydrous sodium carbonate in 1 l of distilled water; solution B: 16.8 g of sodium bicarbonate in 1 l of distilled water. Mix 40 ml of A and 85 ml of B and make up to 500 ml. Check the pH and add 0.1 g of sodium azide.

Antibody/conjugate diluent
   0.05 M phosphate buffer, pH 7.4
   0.85% sodium chloride
   0.05% Tween 20
   0.02% sodium azide
Prepare 0.2 M phosphate buffer stock solutions. Solution A: 31.2 g of $NaH_2PO_4.2H_2O$ in 1 l of distilled water; solution B: 28.39 g of Na_2HPO_4 or 71.7 g of $Na_2HPO_4.12H_2O$ in 1 l of distilled water. Mix 23.75 ml of A and 101.25 ml of B and make up to 500 ml with distilled water. Check that the pH is 7.4. Add 4.25 g of sodium chloride and readjust the pH to 7.4 with 1 M NaOH. Add 0.25 ml of Tween 20 and 0.1 g of sodium azide.

Substrate solvent (for alkaline phosphatase)
   0.05 M sodium carbonate buffer, pH 9.8
   1 mM $MgCl_2$
Prepare 0.2 M carbonate buffer by mixing 55 ml of solution A and 70 ml of solution B (see antigen coating buffer above for recipes) and make up to 500 ml with distilled water. Add 102 mg of $MgCl_2.6H_2O$, check that the pH is 9.8 and adjust if necessary.

Washing solution
   0.9% NaCl
   0.05% Tween 20
Add 18 g of NaCl to 2 l of distilled water and add 1 ml of Tween 20.

3. General method for titration of antibody

The following is a general method that can be applied to the titration of antibody directed against a bacterial protein antigen. It assumes no specialized equipment for plate washing, etc.

Before beginning the titration of antibody, the procedure should be optimized so that the minimum amount of reagents can be used: antigens are often difficult and expensive to prepare in large amounts, antibodies are often only available in small quantities, especially when screening hybridoma culture supernates, and enzyme conjugates are expensive. It is usual to begin by performing a chequerboard titration of antigen versus conjugate, keeping antibody constant at the highest possible concentration. The conditions given below are those developed by Engvall and Perlmann (1972) for maximum sensitivity. It is possible, however, to speed up the procedure considerably by increasing the incubation temperatures and shortening the times, while recognizing that sensitivity might be sacrificed.

Flat-bottomed 96-well polystyrene microtitre plates are the usual solid phase for ELISA. There are, however, several purpose-designed plates made of a range of materials that are said to have greater binding properties and are less prone to background problems.

(a) Prepare dilutions of antigen in coating buffer and add to the microtitre wells. Protein concentrations are usually in the range 10–100 μg/ml and volumes of 50 or 100 μl are usual. Ensure that the whole of the bottom is covered and avoid air bubbles. Carefully seal the plate with plastic food covering or aluminium foil and incubate for 4 h at 37°C followed by 4°C for 16 h.

(b) Pour out unbound antigen and wash the plate three times by directing a jet of washing solution into the bottom of each well, shaking the plate dry between washes.

(c) Prepare dilutions of antibody in antibody/conjugate buffer. For the initial chequerboard titration dilute to a minimum—even undiluted may be required for some early screening of monoclonal antibodies, but for hyperimmune animals levels of 1:200 may be a useful first guess. Add volumes as above to the wells. Seal the plate and incubate at room temperature for 4 h.

(d) Wash as in (b).

(e) Prepare dilutions of conjugate in antibody-conjugate buffer. Alkaline phosphatase is commonly the enzyme. Most manufacturers give a guide to the working dilution of their conjugates, but it is worth while checking each new batch for activity. Most commercially available conjugates can be used at dilutions of at least 1:500. Add to wells as above, seal plates and incubate overnight at room temperature.

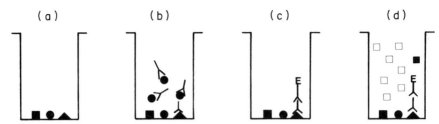

Figure 6.8 Inhibition of indirect ELISA. (a) The plate is coated with complex antigen mixture (e.g. whole bacteria or envelope preparation). (b) Antiserum that has been preincubated with a purified component of the complex antigen (e.g. LPS) is added. (c) Only antibodies that have not already bound to soluble antigen can bind to immobilized antigen on plate. (d) After substrate is added, the colour change is proportional to antibodies bound to the plate, thus giving an indication of proportions of antibodies to the specific antigen in the original antiserum.

(f) Wash as in (b).
(g) Add enzyme substrate. For alkaline phosphatase conjugates dissolve *p*-nitrophenol phosphate in substrate buffer to give a concentration of 1 mg/ml; Sigma produce 5 mg tablets for convenience. Incubate for 1 h at room temperature. If results cannot be read immediately, 10 μl of 1 M NaOH is added to stop the reaction. For most assays results can be read by eye and an end point easily determined. It is strongly recommended, however, to use an ELISA plate reader if one is available (e.g. the Titertek Multiskan from Organon Teknika).

It is important that on each plate there are control wells (i.e. without antigen and without antiserum) to observe any non-specific reactions. Most plate readers require negative controls on which to set the zero level of the assay.

4. Inhibition of indirect ELISA

This technique is a modification of the previous one and is useful for demonstrating cross-reactive or competing antigens, or for investigating the part played by purified antigens or components of antigens in a complex antigen/antibody reaction. The technique is summarized in Figure 6.8 and illustrated in Figure 6.9.

5. Special method for coating LPS on to plates

It is often difficult to coat plates with lipopolysaccharide (LPS), especially if it is rough-type, hydrophobic LPS. Recently a simple method has

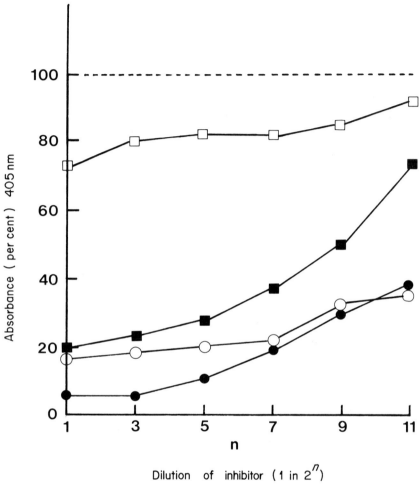

Dilution of inhibitor $(1 \text{ in } 2^n)$

Figure 6.9 Inhibition of ELISA, to determine the chemical nature of the major surface antigens of *Bacteroides fragilis*. A standard indirect ELISA was first performed to determine the 100% value (– – – –): outer membrane (OM) of *B. fragilis* was coated on to the plate and reacted with doubling dilutions of antiserum raised in rabbits to whole cells. The titre was taken as the highest dilution of serum that gave an absorbance reading at 405 nm of greater than 1.0. Inhibition of ELISA was by preincubating doubling dilutions of potential inhibitors with equal volumes of the serum diluted to twice the titre for 30 minutes at 37°C. Inhibitors were untreated OM (●) as positive control, OM heated to 121°C for 15 minutes (○), OM treated with 0.01 M sodium periodate (pH 5.0) for 16 h at 20°C (□), and purified lipopolysaccharide (LPS) (■). The results show that a heat-stable, periodate labile component of the OM was the major antigen and it was predominently LPS.

been developed which utilizes the great affinity that the antibiotic polymyxin has for LPS (Scott & Barclay, 1987).

(a) Mix 1 mg/ml of Polymyxin B sulphate (Sigma Chemicals) with 10 μg/ml of LPS together in pyrogen-free distilled water and stir for 30 minutes at room temperature.
(b) Place the mixture into a clean dialysis bag and dialyse overnight against distilled water to remove excess unbound polymyxin.
(c) Dilute the dialysed complex 1 : 16 in antigen coating buffer (see section D(iv)2 above) and add 100 μl volumes (equivalent to 62.5 ng of LPS) to microtitre plates. Incubate overnight at room temperature.
(d) Wash as in standard indirect ELISA.
(e) Post-coat with 100 μl volumes of 5% bovine serum albumen in PBS overnight at room temperature.
(f) Wash.
(g) At this stage plates can be wrapped in plastic film and stored for at least three months at 4°C. They are then used as in the standard indirect ELISA described earlier.

6. Direct ELISA

In this technique the specific antibody is labelled with the enzyme and can be added directly to the immobilized antigen. It suffers from the major drawback that a specific enzyme conjugate is required for each antigen. ELISA is therefore little used in this direct mode.

There is, however, a variation of the technique which can be used to detect antigen. This is *capture ELISA* or *sandwich ELISA* and is summarized in Figure 6.10. Specific antibody (unlabelled) is bound to the plates in the same manner as the antigen in indirect ELISA. Purified antigen or specimens suspected of containing antigen are added to the wells, as is the first antibody in indirect ELISA. Specific antibody/enzyme conjugate is then added, followed by substrate. A positive reaction signifies specific antigen capture.

Antibody-class capture ELISA. Capture ELISA techniques have been developed further to capture a specific class of antibody. For example, IgM can be assayed by coating a plate with anti-mu chain antibody. Serum is then added and all IgM should be captured. Specific antigen is then added which will only bind to those IgM molecules that recognize it. A specific enzyme-labelled antibody which recognizes the antigen is then added and the ELISA completed by addition of substrate. This technique is easily adapted to the biotin–avidin system described later in this section.

(a)   (b)

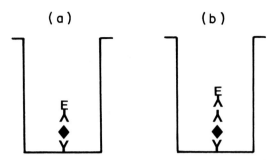

Figure 6.10 Sandwich or capture ELISA. Specific antibody (e.g. a monoclonal antibody) is bound to the plate. An antigen solution or a body fluid from an infected site (urine, blood, cerebrospinal fluid, etc.) is added and specific antigen will be captured by antibody. (a) Demonstration of the direct mode in which an enzyme-labelled antibody (which can be the same antibody as the capture antibody if the antigen has at least two binding sites) is added. (b) Demonstration of the indirect mode where a second antibody (which must not cross-react with the capture antibody, i.e. it is produced in a different species and is perhaps preabsorbed with the first species antibody) is added. The enzyme-conjugated antibody must be specific for the second antibody.

Preparation of enzyme conjugates. Whereas for indirect ELISA it is usual to buy quality-controlled commercial conjugates, for direct and sandwich ELISA it will be necessary to prepare conjugates. The method below is for the preparation of alkaline phosphatase conjugates and is based on the method of Voller *et al.* (1976).

(a) Add 1.4 mg of immunoglobulin (prepared by the method described in section A(iii) above) in 1.0 ml of PBS, pH 7.4, to 5 mg (5000 units) of alkaline phosphatase and mix at room temperature.

(b) Dialyse for 18 h at 4°C against two changes of 2 l of PBS. This step may be omitted if the alkaline phosphatase is supplied in NaCl solution (e.g. Sigma type VII-T).

(c) Add 25% glutaraldehyde solution to give a final concentration of 0.2% v/v. Incubate at room temperature for 2 h.

(d) Dialyse for 18 h at 4°C against two changes of 2 l of PBS.

(e) Transfer the dialysis sac to 2 l of 0.05 M Tris buffer, pH 8.0, containing 1 mM $MgCl_2$ and dialyse for a further 18 h with two changes of buffer.

(f)   After dialysis, add bovine serum albumin to 1% w/v and sodium azide
      to 0.02%. Store in the dark at 4°C.

7. Some recent developments

The ELISA methods described above are all based on the original
Engvall and Perlmann (1972) method with alkaline phosphatase conjugates. Recently a wide range of different enzyme conjugates has become
available, and variations of the basic principle, allowing amplification of
the reaction, have been developed.

Perhaps one of the most interesting developments, giving much greater
sensitivity, is the *biotin–avidin* system. Avidin, a glycoprotein, has an
extremly high affinity for biotin, and biotin-labelled immunoglobulins
allow the development of a flexible, sensitive and specific immunoassay
with avidin conjugates. Various enzyme–avidin conjugates, including
alkaline phosphatase and peroxidase, are commercially available.
Immunoglobulins can easily be conjugated to biotin. Activated biotin
reagents   are   commercially   available;   for   example,   biotin-*N*-
hydroxysuccinimide may be coupled to immunoglobulins through free
amino groups. Biotin anti-mouse IgG is commercially available and is of
great use for screening for mouse monoclonal antibodies. Figure 6.11
shows diagrammatically a sandwich ELISA based on biotin–avidin.

Another recent development is the exploition of enzymes which have
fluorescent products. A fluorimeter can be used to read the final product

Figure 6.11   Biotin–avidin ELISA. Antigen is coated to the plate and is followed
by first antibody which may be at an
extremely low concentration (e.g. a
monoclonal antibody hybridoma supernate). Anti-first antibody conjugated to
biotin (B) is added. Enzyme-avidin (E) is
added and many molecules of this bind
to the biotin, producing an amplification
of the reaction.

with much greater sensitivity than colour can be read in a conventional spectrophotometer. Fluorimeters for reading 96-well plates are now available.

(v) Dot blotting

A variation on indirect ELISA which is gaining popularity is the technique commonly known as dot blotting. This involves the binding of antigen to a nitrocellulose membrane and performing an indirect ELISA on the immobilized antigen. The major advantage it has over conventional ELISA in microtitre plates is that the binding is immediate because no time-consuming coating procedure is required. Also, it seems that a greater range of compounds can bind to nitrocellulose than to polystyrene, and the technique is extremely sensitive. The major disadvantage is that it is difficult to read results in an objectively quantitative way, unless a reflectance densitometer can be modified.

It is, however, gaining popularity as a means of screening column fractions, and in certain immunodiagnostic procedures. The technique is as follows:

(a) Mark a grid of 10 mm squares on a sheet of nitrocellulose.
(b) Wash the sheet for 10 minutes in TBS (see section D(iii)1 above) and allow it to dry on a sheet of filter paper in the air at room temperature. This takes 10–15 minutes.
(c) By means of a syringe or micropipette, place 1 μl samples of antigen solution on to marked squares and allow to dry.

For most purposes this will produce sufficient antigen, but if very dilute solutions or clinical specimens are being used it may be necessary to apply more dots of antigen, allowing drying between applications. Several manufacturers are marketing an apparatus based on the 96-well format for holding the nictocellulose membrane. Larger volumes of antigen solution can be drawn through the membrane by gentle suction from a vacuum pump. Various problems have been encountered in some apparatus: leaking occurs between wells. It is suggested that the apparatus is taken on approval before purchase.

(d) When antigen dots are dry the membrane is blocked with gelatin solution and treated exactly as for immunoblotting: see steps (g)–(m) in section D(iii)2.

(vi) Immunofluorescence

Antibody can be labelled with a fluorescent dye and used to visualize the antigen to which the antibody binds. In principle this test is very similar to ELISA and can be used directly or indirectly. It is necessary to use an ultraviolet microscope and a certain degree of skill is required in reading results. Interpretation of results is rather subjective.

As a diagnostic procedure this technique has gained many applications; and although with the advent of ELISA and immunogold labelling it is losing popularity, with the development of 'fluorescence activated cell sorters' (FACS) it will remain an extremely useful labelling technique.

1. Preparation of labelled antibody

(a) Prepare IgG fraction by the method given in section A(iii), but use 0.05 M phosphate buffer of pH 7.5 instead of Tris. Adjust the IgG concentration to approximately 2% w/v. Use 1 ml of this.
(b) Prepare 0.5 ml of 0.1 M Na_2HPO_4 containing 0.5 mg of fluorescein isothiocyanate (FITC).
(c) Add 0.25 ml of Na_2HPO_4 to 1 ml of IgG dropwise over 2–3 minutes.
(d) Add FITC solution in the same manner.
(e) Rapidly measure the pH and adjust to 9.5 with 0.1 M Na_3PO_4.
(f) Add 0.15 M NaCl to a final volume of 2 ml.
(g) Leave statically at 25°C for 30 minutes.
(h) Swirl in an ice-bath for a few seconds and remove any precipitate by slow-speed centrifugation.
(i) Prepare a small column (e.g. 30 × 1 cm) of Sephadex G-25 or equivalent in 0.05 M phosphate buffer, pH 7.5, containing 0.15 M NaCl (PBS).
(j) Apply FITC/IgG solution to the column and elute with PBS. Collect 2 ml fractions. The FITC IgG conjugate elutes in approximately 8 ml, while the unbound FITC elutes in 60–80 ml.
(k) The coloured fractions containing the conjugate are pooled, concentrated to 1–2 ml in an ultrafiltration cell fitted with a 10 kDa cutoff membrane, and bovine serum albumin is added to 1% w/v and sodium azide to 0.1%. The conjugate is stored at 4°C in the dark for short-term use or frozen in small volumes and stored at −20°C.

Rhodamine conjugates can be prepared in a similar manner with tetramethyl rhodamine isothiocyanate.

2. Procedure for direct and indirect immunofluorescence assay

(a) Prepare a weak suspension of bacteria or specimen in PBS and place drops (approx. 10 µl) on to a clean glass slide. Commercially available Teflon-coated slides with wells are recommended for this. Allow the drops to dry.

(b) Fix the bateria on to the slide by either heating briefly in a hot flame, or by placing in anhydrous acetone or methanol for 10 minutes.

(c) *For the direct assay*, apply 10 μl volumes of dilutions of the specific fluorescein-labelled conjugate. For most commercial conjugates, dilutions of undiluted to 1 : 16 or 1 : 32 are usual. Allow to incubate in a humid chamber for 30 minutes at a temperature between 20 and 37°C. *For the indirect assay*, dilutions of specific unlabelled antibody are added, and after incubation and washing in a jet of PBS for a few seconds the fluorescein-labelled anti-first species immunoglobulin is added and incubated as above.

(d) Wash the slide in a jet of PBS for a minute or so, followed by several changes of PBS over 5–10 minutes and a brief wash in distilled water. Allow to dry.

(e) Mount the slide under glass in glycerol mounting fluid (see below) and view in an ultraviolet microscope.

Reading results is somewhat subjective, but a bright fluorescent outline is usually accepted as maximum fluorescence. As the reaction becomes weaker the fluorescence becomes more diffuse over the whole of the cell.

Mounting fluid for UV microscopy

$NaHCO_3$	71.5 mg
Na_2CO_3	16.0 mg
Water	10 ml
Glycerol	90 ml

(vii) Immunoprecipitation

In its simplest sense this term means an antibody reacting with an antigen to produce an immune complex that can be recovered as a precipitate. The complex can then be dissociated and the antigen or antibody recovered in pure form.

In practice the term usually implies the detection or purification of a specific antigen by combining it with an antibody. The resultant complex is made insoluble by adding protein A linked to an insoluble support such as agarose beads, or to fixed *Staphylococcus aureus* cells which are naturally rich in protein A.

In theory it is a simple procedure, but in practice it can be extremely difficult, the main problem being that in a complex antigen mixture different antigens often tightly associate with one another. The resulting immunoprecipitate will therefore contain both the specific antigen and those that are bound to it. This necessitates the inclusion of detergents and other dissociating agents in the procedure. To allow ease of visualization, the antigen is usually radiolabelled and detected by autoradiography. The technique is then referred to as *radioimmunoprecipitation* (RIP).

Although sometimes a difficult technique, it is a necessary alternative to immunoblotting. In RIP the antibody binds to native antigen and the resultant complex is purified. In immunoblotting the antigen is denatured before being exposed to the antibody and the technique is dependent on denatured antigen retaining sufficient epitopes that can be recognized by the antibody. With conventional antiserum this is usually not a problem. However, with monoclonal antibodies, especially those to protein antigens, binding does not occur.

The RIP technique described below was developed by Zak *et al.* (1984) for investigating the surface protein antigens to which antibodies were produced in patients with gonorrhoea. The detergent system works for solubilizing antigens from *Neisseria gonorrhoea* but will not necessarily work for other bacteria. Conditions can only be determined by trial and error.

(a) Surface label cells with ^{125}I as described in Chapter 4A, section (i)6.

(b) Suspend in RIP buffer (0.3% v/v Empigen BB^{TM}, 0.1% w/v SDS, 0.05% w/v sodium azide and 0.1 M *p*-toluene sulphonyl fluoride in PBS). Incubate at 37°C for 1 h.

(c) Centrifuge at 100 000*g* for 2 h. Store supernate at 4°C until required. The specific activity of the protein should be in the range 0.1– 0.2 μCi/mg.

(d) For immunoprecipitation, add 0.12 μCi in 100 μl of RIP buffer to 40 μl of antiserum and 50 μl of a 100 μg/ml suspension of protein A Sepharose CL-4B in PBS in a 1.5 ml microcentrifuge tube. Mix for 1.5 h at 4°C.

(e) Centrifuge at 10 000*g* for 2 minutes and wash five times with RIP buffer.

(f) Add 50 μl of SDS-PAGE sample buffer to the pellet and heat in a boiling-water bath for 5 minutes.

(g) Remove the Sepharose beads by centrifugation at 10 000*g* for 2 minutes and apply the supernate to a linear gradient (10–25% w/v) acrylamide SDS-PAGE gel.

(h) After electrophoresis, shrink the gel in 50% v/v methanol, 10% v/v glycerol in water and investigate by autoradiography (Kodak X-Omat film at -70°C for 16–48 h in a metal cassette with a regular intensifying screen).

(viii) Co-agglutination

Agglutination used to be one of the most useful of immunological tests. In its simplest form it consists of mixing bacteria with dilutions of antiserum and observing macroscopic clumping of the cells if agglutinating antibodies are present in the serum. It is reasonably sensitive but can be non-specific. It probably still has a role in some diagnostic procedures but it is largely being superseded by ELISA. Because of its simplicity no more will be described here.

A related technique that is gaining popularity as a method for detecting antigen is termed 'co-agglutination'. The principle behind this technique is that specific antibodies are coated on to a particulate support and these are mixed with antigen: either soluble or particulate (whole cells). Agglutination results if the antigen is recognized by the antibody. The particulate support for the antigen is fixed *Staphylococcus aureus* cells, which are rich in protein A (latex particles are also often used as a support). Kits are commercially available for a range of bacterial species and are discussed in more detail in Chapter 7C.

1. Preparation of S. aureus cells for co-agglutination reagent

(a) Culture *S. aureus* strain NCTC 8530 on large plates of nutrient agar for 18 h at 37°C.
(b) Harvest into 0.01 M phosphate-buffered saline (PBS), pH 7.2. Wash three times in the same buffer. Weigh the pellet.
(c) Resuspend the pellet to a concentration of 10% w/v in PBS containing 0.5% v/v formaldehyde.
(d) Incubate the suspension for 3 h with gentle mixing at room temperature. Harvest and wash three times in PBS.
(e) Resuspend in PBS and heat to 80°C for 5 minutes. Wash twice in PBS and reweigh the pellet.
(f) Resuspend the pellet to a concentration of 10% w/v, divide into 1 ml volumes and store at −20°C until required.

2. Preparation of sensitized S. aureus suspensions

The antibody that is used for sensitization of the stabilized *S. aureus* cells can be whole antiserum or monoclonal antibody.

For high-titre polyclonal serum. Mix 1 ml of the above *S. aureus* suspension with 0.1 ml of serum for 15 minutes at room temperature. Centrifuge the suspension at low speed and wash twice in PBS. Finally resuspend in 10 ml of PBS containing 0.1% sodium azide to give a final cell suspension of 1 per cent.

For monoclonal antibodies. In hybridoma supernates, antibodies are usually present at low titre. To sensitize the *S. aureus* cells, 1 ml of the above suspension is first centrifuged and 1 ml hybridoma supernate is added to the pellet. After incubation at 37°C for 30 minutes the suspension is recentrifuged and a further 1 ml of supernate added and incubated for a further 30 minutes. This may be repeated several times. Wash the pellet three times and resuspend in 10 ml of PBS containing 0.1% sodium azide.

To detect unknown antigen, the specimen (which might be whole bacteria, culture supernate or body fluid) is mixed with the sensitized cell suspension on a clean glass slide or black tile, rocked backwards and forwards gently and agglutination observed. Alternatively the cell suspension can be prestained with methylene blue and the reaction performed on a white tile.

(ix) Immunoelectron microscopy

Since the advent of electron microscopy, immuno-methods have been attempted to locate antigens in both thin sections and negatively stained preparations. The main problems have been associated with the fixation procedure causing the denaturation of antigens, and the difficulty in preparing an antibody conjugate labelled with an electron-dense material that could be resolved in the EM.

The method of choice described here (personally communicated by Dr Sheila Patrick) uses a fixation procedure which seems to preserve the antigenicity of a wide range of molecules and an immunogold conjugate.

(a) After harvesting bacteria, wash twice in 0.01 M sodium cacodylate buffer, pH 7.2

(b) Fix in cacodylate buffer containing 2% w/v paraformaldehyde (prepared by heating the 2% solution to 60°C, allowing it to cool, adding 1 M NaOH dropwise until clearing occurs and then adjusting to pH 7.2 by addition of 1 M HCl) and 0.1% glutaraldehyde for 1 h at 4°C.

(c) Wash the cells in cacodylate buffer and dehydrate twice in graded alcohols (30, 50, 70, 95% ethanol) and finally twice in 100% ethanol that has been dried by filtering through sodium sulphate. Each step should take no longer than 15 minutes.

(d) Embed in LR (London Resin) white resin as follows. First mix in 50% resin, 50% ethanol for 1 h at room temperature. Next mix in 100% resin in open containers in a fume cupboard at room temperature overnight. After two further changes of resin at 2–3 h intervals, transfer the cells to gelatin capsules which have been dried at 60°C for 3 h, fill the remaining space with 100% resin, seal and polymerize the resin in a 60°C oven for 18 h.

(e) Cut thin sections in an ultramicrotome and place on nickel grids.

(f) Carry out immunoassay at room temperature by placing sequentially in (i) 1% bovine serum albumin (BSA) in PBS, pH 7.2, for 15 minutes to act as a blocking agent, (ii) antibody solution suitably diluted if necessary for 2 h, (iii) 0.1% BSA in 20 mM Tris–HCl saline, pH 8.2 (TBS), to act as a holding buffer prior to washing under a stream of drops of TBS coming from a burette (total washing volume of approximately 8 ml), (iv) 1% BSA–TBS for 15 minutes to act as a blocking buffer, (v) an appropriate dilution of gold conjugate (a 15 or 20 nm gold particle linked to the appropriate antispecies immunoglobulin, obtainable from Janssen Pharmaceuticals), (vi) 0.1% BSA–TBS holding buffer prior to washing in 0.1% BSA–TBS as above, and finally (vii) distilled water, washing as above.

(g) Stain with saturated aqueous uranyl acetate in the dark for 30 minutes.

(h) Wash once in distilled water, then in Reynold's lead citrate (see Appendix 1) and desiccate for several minutes before viewing in the EM.

For negatively stained preparations, bacteria are suspended in distilled water and applied to Formvar-coated nickel grids. The gold labelling is then performed as in (f) above and the specimens finally negatively stained with 1% w/v aqueous ammonium molybdate.

Examples of immunogold-stained thin sections and negatively stained preparations are shown in Figures 6.12 and 6.13.

(x) Immunoadsorbent chromatography

This affinity chromatographic technique allows the preparation of pure antigens or antibodies. Either the antibody or the antigen can be immobilized on to a chromatographic support and used to affinity-purify the other from complex mixtures. As the purpose of this section of the book is to give methods for purification of antigens, only methods for purifying antigens with immobilized antibodies will be described here. The technique described here is based on the method recommended by Pharmacia in *Affinity Chromatography: Principles and Methods*. It should be noted that elution of antigen is usually the most difficult step.

Factors that must be taken in to account include the affinity of the antibody and the stability of the antigen. The easiest methods for dissociating the complex are high and low pH (greater than 10 and less than 4). These pHs often denature antigens, so the use of chaotropic agents (e.g. Cl^-, I^-, CLO_4^-) and polarity-reducing agents (e.g. ethylene glycol) is recommended at pHs nearer to neutrality.

1. Procedure for coupling antibody to support

(a) Prepare IgG by the method given in section A(iii) above. About 5 mg of protein is required per millilitre of swollen gel (i.e. 50 mg for 3 g of unswollen activated Sepharose, see below). The required amount of

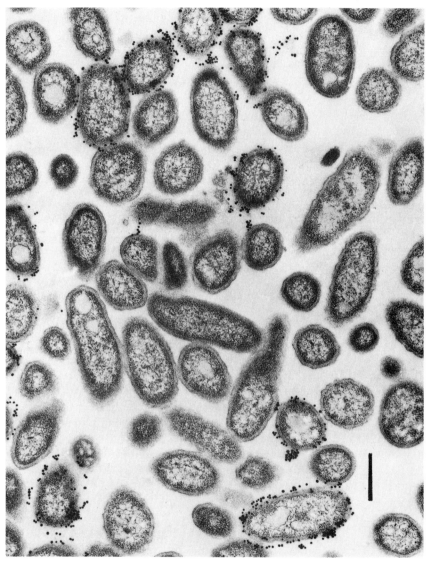

Figure 6.12 Example of a thin-section electron microscope preparation stained by immunogold: a monoclonal antibody specific for an undetermined cell surface polymer that is found in only a proportion of *Bacteroides fragilis* cells, used indirectly with anti-mouse IgG–gold conjugate. Bar marker: 0.5 μm. Reproduced by kind permission of Dr Sheila Patrick.

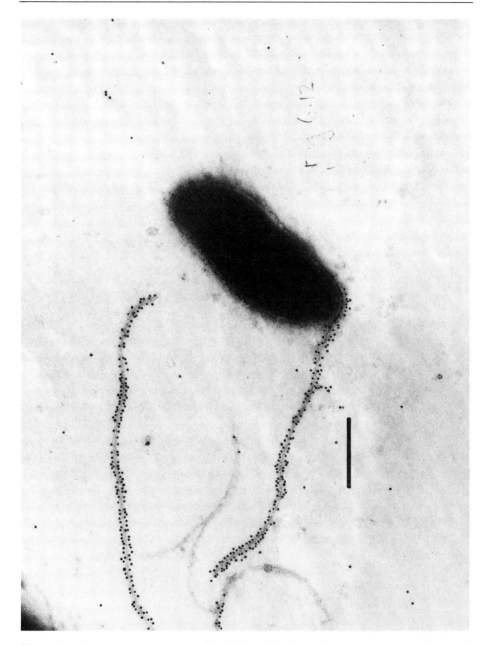

Figure 6.13 Example of negatively stained electron microscope preparation with immunogold: a monoclonal antibody specific for the flagellar protein of *Campylobacter jejuni* indirectly labelled with anti-mouse IgG–gold conjugate. Bar marker: 0.5 µm. Reproduced by kind permission of Dr Sheila Patrick.

IgG solution must be equilibrated in coupling buffer (NaHCO₃ buffer 0.1 mM, pH 8.3, containing 0.5 M NaCl) by dialysis or ultrafiltration and dissolved in coupling buffer to a final volume of 4–5 ml. Tris or other buffers containing amino groups must be avoided as these will couple to the gel.

(b) Weigh out 3 g of CNBr activated Sepharose 4B and swell for 15 minutes in 1 mM HCl. This will give sufficient support for a 10 ml column.

(c) Pour the swollen gel into a sintered glass funnel (porosity 3) and wash with 500 ml of 1 mM HCl.

(d) Wash the gel twice with 25 ml of coupling buffer in a funnel, sucking dry between each wash.

(e) Mix IgG solution (in coupling buffer) with the swollen gel, adding 6 ml of buffer to give a good slurry, and mix in an end-over-end mixer for 2 h at room temperature or overnight at 4°C. Do not use a magnetic stirrer.

(f) Transfer the mix back to the funnel and wash away excess IgG with five volumes of 10 ml, sucking dry between each volume.

(g) Transfer to 25 ml of blocking agent (0.2 M glycine).

(h) Transfer back to the funnel and wash with four cycles of 10 ml of coupling buffer followed by 10 ml of 0.1 M acetate buffer (pH 4), sucking dry between each addition, and finishing with coupling buffer. This high- and low-pH cycle removes ionically bound IgG.

(i) Store immunoadsorbent in 0.1 M phosphate buffer, pH 7.0, containing 0.5% Tween 80 and 0.01% Merthiolate.

2. Chromatography procedure

A small column (10–15 cm × 1 cm) fitted with end adaptors that can be adjusted to make certain that there is no dead-space in the column is required. Ideally the flow of the eluting buffer should be in the opposite direction to the flow of the sample-application buffer: this is so that if the column is not fully saturated with sample then the sample will not interact with unbound sites further along the column during elution.

(a) Pack the column with immunoadsorbent with a flow rate not exceeding 10 ml/cm² per hour. Slide up end adaptors to the liquid/gel interface. Equilibrate the column in 0.1 M phosphate buffer, pH 7.0.

(b) If protein antigens are being purified or if protein contamination is likely, connect the column to a UV recorder and set the baseline to zero.

(c) Apply sample (for 10 ml of immunoadsorbent several milligrams of antigen should be bound) dissolved in a few millilitres of phosphate buffer. Wash with buffer until the recorder has retuned to zero (40–50 ml is ample if a recorder is not being used).

(d) Reverse the flow and begin to elute with desorbing buffer.

[As described in the introduction to this section, dissociation of antigen can be a problem as the affinity of the antibody can vary greatly. A first attempt might be to use ethylene glycol/PBS (1:1). If this does not work consider using ethylene glycol in a 50 per cent mixture with a higher or lower pH buffer. For example, 0.2 M citric acid containing 0.5 M NaCl gives a pH of approximately 2.0, or 0.2 M carbonate buffer (see ELISA coating buffer, section D(iv) above) gives a pH of 9.6. Guanidine hydrochloride or urea might also be incorporated. If a non-neutral pH is used the eluted fractions should be neutralized immediately after collection. Also, as soon as the eluant has passed through, the column should be immediately re-equilibrated in the storage buffer (PBS with Tween and Merthiolate)].

(e) Monitor fractions for antigen by rocket immunoelectrophoresis or dot blotting. The UV recorder, although useful for some protein antigens, may not be sensitive enough for many antigens in low concentrations.

Bacterial Cell Surface Techniques
I. Hancock and I. Poxton
© 1988 John Wiley & Sons Ltd.

7

Present and Future Applications of Knowledge of Bacterial Surfaces

For this final chapter, three experts have been invited to review their areas of special interest and describe how a knowledge of the chemistry or immunochemistry of bacterial cell envelope structures is being applied, or might be applied in the future. These three reviews are of a medical nature because it is in this area that we are furthest forward. However, many other applications, particularly in biotechnology, can be envisaged. The physical properties of biomass, important during fermentation, down-stream processing and industrial utilization of free and immobilized bacteria, depend particularly on cell surface chemistry; a knowledge of cell wall chemistry and biochemistry is essential if we are to control or exploit wall autolysis during fermentation and product isolation; the use of biomass for recovery of toxic and precious metals depends on an understanding of cell wall chemistry and organization. Such applications are in their infancy, but show great promise.

A. BACTERIAL CELL ENVELOPES IN ADHESION

David C. Old

The recent surge of interest in the mechanisms whereby pathogenic bacteria attach to surfaces is understandable because, of course, it offers us the possibility of interrupting bacterial infections at their earliest stages. For most pathogenic bacteria and most infections, that earliest stage is the colonization of mucosal surfaces. The concept of *tissue tropism*, long preached by medical microbiologists, ensured at least an awareness of the specificity of these interactions between pathogen and host tissue. More recent has been the realization that for most pathogens colonization is mediated by specific *lock and key* interactions between surface structures or *adhesins* present on the pathogen that recognize specific *receptors* on the host cell. This is a review of our current knowledge of the structure and function of these components.

Table 7.1   Envelope fractions as adhesins of some important pathogens.

Species	Adhesin associated with:
Bordetella pertussis	Fimbriae
Coagulase-negative staphylococci	Slime
Corynebacterium renale	Fimbriae
Enterobacteriaceae	Fibrillae, fimbriae and outer-membrane proteins (OMP)
Neisseria gonorrhoeae	Pili, OMP II
Streptococcus pyogenes	Teichoic acid, M fibrillae
Streptococcus mutans	Polysaccharide
Streptococcus sanguis	Fibrillae, fimbriae
Vibrio cholerae	Flagella

As well as having an involvement with structural integrity and with the processes of transport and diffusion by the bacterial cell, the envelope fractions also influence the complex mechanism of adhesiveness by contributing to the surface charge and hydrophobicity of the cell surface. Furthermore, the macromolecules of the envelope may themselves function as the specific adhesins mediating receptor recognition. The diversity of the structures functioning as critical determinants of adhesion of some important Gram-positive and Gram-negative bacteria that are pathogenic for man and animals is clear from those listed in Table 7.1. Thus, the adhesin-bearing determinants are often the dominant envelope components themselves—flagella, slime, capsules, pili and outer-membrane proteins. But most intriguing perhaps is the realization that most of the adhesins thus far identified are associated with fimbriae or fibrillae. Recent studies have attempted to identify the molecular nature of adhesins and receptors in the hope that that knowledge might be used for prophylaxis, either by using analogues of receptors as competitive inhibitors at the site of infection or by the development of adhesin vaccines to produce appropriate anti-adhesin secretory immunoglobulins at the site of infection.

(i) Pili of *Neisseria gonorrhoeae*

Examination of the adhesive properties of the important human pathogen *Neisseria gonorrhoeae* shows only too clearly that the pathogen is far from being a passive participant in the disease process. Among its major surface antigens, two have been particularly well studied because of their likely role in colonization of host mucosal surfaces by gonococci. The first to be recognized as important were the filamentous, protein appendages called pili (or fimbriae) of which, it soon became clear, there was a great variety of types among clinical isolates of gonococci. Indeed, any one strain of gonococcus has the ability to produce at least seven antigenically distinct types of pili, though it usually expresses only one at any time (Sparling *et al.*, 1986). The antigenic changes, which can, indeed, even be seen to occur *in vivo* among serial isolates from a single patient, are sometimes accompanied by alterations in the adhesive properties of the gonococcus (Heckels, 1984).

Our understanding of the nature of the variability in antigenicity of gonococcal pili has benefited from sequencing analyses which have shown that the constituent protein subunits (or pilins) of different variant pili are structurally closely related. There is a highly conserved, immunorecessive region of around 50 amino acids at the N-terminal end of the molecule that tolerates minor changes only, presumably because of rigorous criteria for correct assembly of the pilus filament. The subsequent parts of the molecule show increasing variability towards the C-terminal end of the pilin. Indeed, there is little, or no, serological cross-reactivity among pili from different strains because of the hypervariability noted at the C-terminal end of the pilin (Hagblom *et al.*, 1985). As well as two loci involved with gene expression, many other gene segments contain 'silent' pilin sequences, and multiple-step recombinational events among these latter loci are thought to account for the observed spectrum of antigenic variability (Sparling *et al.*, 1986).

The other adhesin of gonococci resides with a class of structurally related outer-membrane proteins (OMPs) referred to as OMP II (Heckels, 1984). A strain of gonococcus may express different combinations of up to three of the six or seven different OMP IIs in its repertoire, and the changes in the expression of OMP II proteins, like those of pili, occur spontaneously and at high frequency. Again, antigenic diversity of OMP IIs is seen not only among different strains of gonococci but also during the epidemic spread of a single strain through a susceptible population (Schwalbe *et al.*, 1985).

It should also be noted that the expression of both kinds of gonococcal adhesins (i.e. the pili and the OMP IIs) is subject to phase variation by switch-on and -off mechanisms that function independently of each other. Thus, by means of antigenic variability of its major adhesins, the gonococcus presents a constantly changing front to the host's immune system and thereby facilitates its epidemic spread. The successful development of antiadhesin vaccines would seem likely to require the identification of a portion of the adhesin subunit from the immunodominant part of the molecule that is common to all fimbrial types; such portions have apparently been found (Hagblom *et al.*, 1985). Perhaps the lesson to be learned, however, from this story is that the gonococcus is unlikely to be exceptional, and we should not be too surprised if other bacteria are equally versatile.

(ii) Enterobacterial fimbriae

Since the early pioneering work of Duguid it has been customary to define the adhesins of Enterobacteriaceae into those that are inhibited by D-mannose and its analogues (the mannose-sensitive, MS, adhesins), and others (the mannose-resistant, MR, adhesins), the latter being a heterogeneous group whose common property is that D-mannose does not inhibit their activities (Duguid *et al.*, 1955). Ideally, of course, fimbriae should be classified according to their receptor-binding specificities, but despite intensive investigations over the past 30 years since Duguid's early work, remarkably few adhesin receptors have been defined at the molecular level. Accordingly, the above crude classification of adhesins as MS or MR must suffice in the meantime.

In Enterobacteriaceae the presence of MS adhesin almost always correlates with the presence of type-1 or common fimbriae which are widely distributed in most species of most clinically important Enterobacteriaceae. Type-1 fimbriae of enterobacteria have similar morphologies, being 7–8 nm wide with an axial hole, and have similar chemical properties. Their presence bestows on type-1 fimbriate bacteria certain properties, including the ability to agglutinate the red cells of a wide variety of animal species. Regardless of bacterial species, the qualitative patterns of their associated MS adhesins (the MSHAs) are essentially constant. Though red cells have been used commonly as substrates for detection of type-1 fimbriae, because they provide a useful tool for their classification, type-1 fimbriae bind to a wide variety of eukaryotic cells—animal, plant and fungal—to polymorphonuclear leucocytes, intestinal, buccal and urinary epithelial cells, mucus, and also to some tissue-culture lines (Duguid and Old, 1980). Thus, their receptors must be widely distributed, and recent work has confirmed that idea (Nesser *et al.*, 1986). Evidence accumulated over the last 15 years has implicated mannose residues on the surface of target cells as the sites of attachment of different mannose-specific enterobacteria. This was first studied by Old (1972) who showed that several activities associated with type-1 fimbriation, not only haemagglutination but also fimbrial-pellicle formation, were specifically inhibited by α-D-mannosides but not by β-D-mannosides. More detailed analysis showed that unmodified hydroxyl groups at the C2, 3, 4 and 6 positions of the D-mannopyranosyl molecule were necessary for inhibitory activity.

We have long known that type-1 fimbriae of Enterobacteriaceae have diverged considerably because of the evidence that they share little or no antigenic cross-reactivity, sometimes even within a species (Duguid, 1985). Nevertheless, it has been customary to group all type-1 fimbriae of Enterobacteriaceae into the same class on the basis of the presumed commonality of their mannoside-containing receptors. Recent evidence from several sources, however, suggests that the receptors of type-1 fimbriae of different Enterobacteriaceae are subtly different. Elegant studies from the Firon group have demonstrated that a series of mannose derivatives of different structures (see Firon *et al.*, 1983; 1984) inhibited the activities of type-1 fimbriate enterobacteria in a manner that was essentially species-specific. From their studies they concluded that the receptor site of type-1 fimbriae of *E. coli* and klebsiellae was one that best fitted a trisaccharide with a hydrophobic region, whereas that of *Enterobacter cloacae* had an extended region probably greater than a trisaccharide and was non-hydrophobic; that of salmonellae differed from both. Again, Old *et al.* (1986) demonstrated that a series of enterobacteria producing, as far as was known, type-1 fimbriae and MS adhesin only, presented a spectrum of activities of binding to HEp2 epithelial cells, and they explained these findings by suggesting that the type-1 fimbriae of different species recognized different receptors not all of which were equally abundant on the HEp2 cell surface. Thus, the grouping of the MS adhesins of type-1 fimbriae of Enterobacteriaceae into a single group has probably been too simplistic.

Current creed has tended to deny type-1 fimbriae any role as virulence determinants in infection because, for example, it was argued that in the urinary tract the trapping of type-1 fimbriate bacteria by Tamm–Horsfall protein (THP) or uromucoid represented an efficient host-defence mechanism for the elimination of type-1 fimbriate bacteria from the bladder (Ørskov *et al.*, 1980). Furthermore, by their propensity for attachment to PMN leucocytes, type-1 fimbriate enterobacteria were thought to trigger their own death by phagocytosis. However, Kuriyama and Silverblatt (1986) have now challenged that concept by their demonstration that type-1 fimbriate bacteria, when coated with THP which acts as a kind of pseudocapsule, were less susceptible to phagocytosis by PMN than type-1 non-fimbriate bacteria. The possession of type-1 fimbriae, therefore, might even considerably enhance the virulence of bacteria, especially in the bladder and kidney where serum activity is low and THP abundant (Kuriyama and Silverblatt, 1986).

What, then, is the role of type-1 fimbriae in urinary-tract infection? Studies with experimentally infected mice or rats have shown the following:

1. Type-1 fimbriate strains of *E. coli* and klebsiellae colonize bladder uroepithelium more effectively than type-1 non-fimbriate variants (Fader and Davis, 1982; Iwahi *et al.*, 1983; Hagberg *et al.*, 1983).
2. Receptor analogues (Aronson *et al.*, 1979) or type-1 fimbriate or receptor antibodies (Abraham *et al.*, 1985) protect against experimental urinary-tract infection.
3. Expression of the type-1 fimbriate phase is of critical importance

because type-1 fimbriate bacteria bind to the bladder uroepithelium whereas type-1 non-fimbriate bacteria are voided in the urine (Hultgren *et al.*, 1985).

4. Type-1 fimbriate bacteria lacking receptor-binding function are as ineffective at bladder colonization as type-1 non-fimbriate mutant strains (Keith *et al.*, 1986).

Thus, the evidence that type-1 fimbriae are important in the establishment of lower urinary-tract infection continues to accumulate.

Although there are reports that, when used as vaccine antigens, type-1 fimbriae protected piglets against colibacillosis (Jayappa *et al.*, 1985) and man against challenge with toxigenic *E. coli* (Levine *et al.*, 1982), the role of type-1 fimbriae in gastrointestinal infection is more controversial. That may be, however, because type-1 fimbriae are functionally heterogeneous, only some types contributing to enteroadherence *in vivo* (Sherman *et al.*, 1985). The study of type-1 fimbriae is further complicated by the finding that clinical isolates of *E. coli* may produce as many as three closely related variants of type-1 fimbriae, designated 1A, 1B and 1C (1A corresponding to the classical type with MS activity). The three types are heterogeneous with respect to both antigenicity and receptor recognition (Klemm *et al.*, 1982), although at this stage the receptors for types 1B and 1C have not been accurately defined nor have their roles in infection. Thus, with regard to antigenicity, receptor recognition and role in pathogenicity, the group of fimbriae known as type-1 is clearly complex even within a single species. In many ways, therefore, our understanding of the common fimbriae of Enterobacteriaceae is just about to begin, 30 years after their first definition by Duguid and co-workers.

(iii) Adhesins in urinary infection

In Enterobacteriaceae, *E. coli* is perhaps the most versatile of pathogens and one that has amassed a range of virulence determinants, each suited to particular ecological niches. Human uropathogenic strains of *E. coli*, perhaps the best studied of all enterobacteria, belong to a small number of clones, readily identified by serotype, and they exhibit critical virulence determinants including haemolysin, K antigen and adhesins (Väisänen-Rhen *et al.*, 1984; Ørskov and Ørskov, 1985). More than 90 per cent of strains involved in first episodes of acute, non-obstructive pyelonephritis in both young and adult females produce 'P' fimbriae, so called because of their association with pyelonephritis and P blood-group antigens (Domingue *et al.*, 1985; O'Hanley *et al.*, 1985). Because of their biological and clinical significance, they have been extensively studied in attempts to define the receptor-binding sites of P fimbriae. Significant breakthroughs came when it was shown that total glycolipid fractions extracted from human uroepithelial cells could inhibit the attachment of P-fimbriate strains of *E. coli* to uroepithelial cells, and that P-fimbriate *E. coli* agglutinated in MR fashion the red cells of most human bloods but not those of the rare $\bar{p}$ blood-group type. The P blood groups in man are represented by various combinations of the P blood-group antigens (P, P_1 and P^k) which are glycosphingolipids of known structure, each with a lipid moiety (ceramide) anchored to the cell membrane with carbohydrate portions exposed on the cell surface. The finding that red cells of the P_2^k type were agglutinated as strongly as those of the other types directed attention to the glycosphingolipid of the P_2^k antigen, which is present in all P blood-group types except $\bar{p}$, and suggested that α-D-Gal(1–4)-β-D-Gal, present in all P antigens, was the minimal receptor structure acceptable to P-fimbriate *E. coli* (Källenius *et al.*, 1980). Any glycolipid containing the Galα(1–4)Gal sequence, either as an internal or terminal part of its molecule, binds P-fimbriate bacteria, whereas glycolipids lacking these sequences do not. Conformational analysis of the structure of the binding glycolipids indicated the presence in the saccharides of a bend at the Gal–Gal region which was also highly hydrophobic (Bock *et al.*, 1985).

Although the identification of Gal–Gal as the minimal receptor structure was exciting, it did little to explain why more than 90 per cent of uropathogenic strains of *E. coli* should have P fimbriae or, indeed, why certain groups of patients had an increased susceptibility to recurrent urinary-tract infection. Lomberg *et al.* (1986) showed, however, that attachment of *E. coli* to human uroepithelial cells was determined not only by receptor recognition but also by distribution of the receptors on individual epithelial cells. The blood group determines the expression of carbohydrate-containing structures on the uroepithelial cells as well as on red cells, and the attachment of P-fimbriate *E. coli* to uroepithelial cells showed the same restriction as that found with red cells. They also showed that the receptors for P adhesins were equally well distributed on both transitional and squamous epithelial cells, whereas the distribution of receptors for adhesins other than P (i.e. non-P adhesins) were present at low density on transitional epithelial cells. Thus, non-P adhesins enhanced attachment mostly to squamous epithelial cells and did so independently of P blood group. In bacteria carrying both P and non-P adhesins, the attachment to squamous cells was independent and additive. P adhesins are the dominant adhesins among pyelonephritogenic strains of *E. coli*, and glycolipids extracted from the whole kidney or upper ureter are rich in Gal–Gal residues (Bock *et al.*, 1985). Non-P adhesins, on the other hand, are more common among cystitic than pyelonephritogenic strains, presumably because the receptors for non-P adhesins are absent from, or not abundant in, the upper urinary tract.

Adhesion varied, it was noted, among individuals otherwise of the same A, B, H or P blood groups in that their secretor status was found to be important. The exact function of the human secretor genes is unknown, but one function is the expression of a fucosyl transferase in the epithelial cells; the addition of A, B or H determinants by the transferase to the epithelial glycoconjugate blocks the receptors by modifying the surrounding molecules or the receptors themselves. Thus, it is of considerable interest that among women with recurrent urinary infections and scarring of the kidney, Lomberg *et al.* (1986) have shown that a higher frequency of patients are non-secretors than secretors. It may be, therefore, that among non-secretors, the receptors are simply more available, rather than, as previously thought, more dense. It must be clear that the detailed information about the chemistry of the red-cell membrane has proved crucial to our understanding of the specificity of P-fimbria recognition. Furthermore, as with the fimbria genes of *N. gonorrhoeae*, there are multiple copies of the P genes in strains of *E. coli* ensuring again a high degree of antigenic variability (Hull *et al.*, 1985).

About 10 per cent of urinary strains of *E. coli* carry adhesins recognizing receptors other than mannosides or Galα(1–4)Gal, and they are recognized by their MR haemagglutination of human red cells of the p̄ type. All have at various times been called 'X adhesins' which, apart from anything else, has produced considerable confusion in the literature. The binding specificity of one major group was designated 'S' and their associated adhesins 'S fimbriae' (Korhonen *et al.*, 1984). The observation that treatment of the erythrocytes with neuraminidase or trypsin abolished S-dependent haemagglutination was one finding that helped to identify the bound sialylgalactosides of glycophorin A, the MN red-cell sialoglycoprotein, as the receptor of S-fimbriate *E. coli*. Further detailed analyses identified the bacterial binding site on the red cell as O-linked acetylneuraminylα(2-3)galactose-β(1-3)galactosamine (Parkkinen *et al.*, 1986). However, although S fimbriae were found on about 5 per cent of pyelonephritogenic strains of *E. coli*, especially those of serogroups O2 and O4, their greatest association was with a few serotypes such as O18ac:K1:H7, commonly associated with neonatal septicaemia and meningitis (Korhonen *et al.*, 1985). It has been argued that S fimbriae, by binding to vascular endothelium, may facilitate spread to cerebrospinal fluid (Parkkinen *et al.*, 1986). *In vitro* tests, however, have failed to demonstrate the binding of S-fimbriate *E. coli* to the sialylgalactosides of brain gangliosides, perhaps because of inaccessibility of the receptors, and a direct role for S fimbriae in meningitis awaits proof; nor have the host-cell sialylgalacto-conjugates to which S fimbriae attach *in vivo* been established.

Yet other uropathogenic strains of *E. coli* carry M fimbriae which give blood-group-M specific haemagglutination. The antigens of the MN system are carried by glycophorin A and the M and N specificities are thought to reside in different N-terminal amino-acid sequences. Thus, M fimbriae, unlike other adhesins, seem to recognize amino acids rather than carbo-hydrates and their receptors are glycoproteins rather than glycolipids (Väisänen *et al.*, 1982). The specificity of M fimbriae, however, could be distinguished from that of S fimbriae, also associated with glycophorin A, because neuraminidase treatment of the red cells did not abolish M-dependent haemagglutination. Some urinary strains are haemag-glutinating only after treatment of the red cells with enzymes. One such novel receptor binding site is thought to involve N-acetyl-D-glucosamine; the associated N adhesin is rare, occurring on fewer than 1 per cent of urinary strains (Väisänen-Rhen *et al.*, 1983). The clinical importance of adhesins such as M and N has not been determined, and these and other adhesins, both fimbriate and non-fimbriate, still in the literature as X adhesins, are as yet inadequately characterized.

(iv) Adhesins in diarrhoeal disease

Having thus far discussed urinary adhesins, about which so much is known, it is correct to shift emphasis to those strains of *E. coli* that cause diarrhoea and which, for convenience, have been grouped into four classes according to their pathogenic mechanisms (Table 7.2). For each class, adhesion to gut receptors represents a critical step in the pathogenesis of their infections. Specific colonizing adhesins, fimbriae or fibrillae, were, of course, first discovered in enterotoxigenic strains of *E. coli*; recognition of receptors by these colonization factors allows specific cell-to-cell interaction and the delivery of enterotoxins to appropriate target cells in the small intestine of animals and man. Colonization factors, including the well-known K88, K99, P987, F41, CFA1, CFA2 and E8775, have been abundantly reviewed elsewhere both with regard to function and receptor recognition (see, for example, Gaastra, and de Graaf, 1982; Klemm, 1985), and will not be discussed further here.

Among the majority of EPEC strains, which were the first of the *E. coli* strains to be identified as agents of diarrhoea, a quite distinct kind of adhesion is seen. Of course, most studies have agreed that strains of *E. coli*

Table 7.2 Classes of *E. coli* causing diarrhoea.

ETEC	Enterotoxigenic	Infantile and travellers' diarrhoea
EIEC	Enteroinvasive (*Shigella*-like)	Dysentery
EPEC	Enteropathogenic classes I and II	Endemic and epidemic diarrhoea in infants
EHEC	Enterohaemorrhagic	Haemorrhagic colitis

of classical EPEC serotypes, if fimbriate, produce type-1 fimbriae only, and are generally negative for MR haemagglutination, though recent studies have shown that some EPEC strains produce MR haemagglutinating activity when grown in appropriately buffered media. Two classes of EPEC strains have been defined. Class I contains the major clones of EPEC strains associated with more severe forms of infantile diarrhoea and which usually occur in epidemic form. In tests with tissue-culture lines such as HEp2 or HeLa, strains of class I show a phenomenon known as localized adherence (LA) in the presence of D-mannose (Scaletsky *et al.*, 1984; Nataro *et al.*, 1985). This is an interesting observation because most strains of *E. coli*, certainly those with type-1 fimbriae or MRE fimbriae, adhere poorly in this particular *in vitro* model of adherence. The presence of plasmids of 50–70 MDa confers on these strains the ability to adhere to HEp2 cells and, when fed to volunteers, to cause diarrhoea (Levine *et al.*, 1985). Because EPEC strains produce type-1 fimbriae only, OMP profiles were examined and the plasmid was found to determine a 94 KDa outer-membrane protein. Volunteers who are ill from EPEC-induced diarrhoea mount serum IgA and IgG responses to OMP. The genes for this particular adherence factor (EAF), though generally plasmid-determined, apparently have a chromosomal location in a few EPEC strains of class I (Nataro *et al.*, 1985). Class II, on the other hand, generally contains minor clones of EPEC strains associated with sporadic diarrhoea and with a less severe clinical presentation. These strains, too, adhere to HEp2 cells, but in a diffuse fashion, and that diffuse adherence is also associated with plasmids. When, however, class II strains were probed with EAF-gene probes, genes for EAF were not found. Thus, although class II strains adhere to HEp2 cells, they must, it is clear, do so in a manner independent of EAF (Nataro *et al.*, 1985). Though the molecular nature of the adhesins of EPEC strains is unknown, they are probably non-fimbrial OMPs rather than fimbriae or fibrillae.

Dysentery in man is the illness characteristically produced by EIEC strains. The pathogenic potential of EIEC strains and the taxonomically related shigellae correlates with their ability to invade and multiply within cells of the colonic epithelium. Initiation of cell invasion involves cell-to-cell interaction, and this attribute is associated with large 120–140 MDa plasmids which encode between five and nine different OMPs. More recently, a 38-kilobase protein essential for invasiveness has been identified and shown to be conserved in the plasmids of all shigellae and EIEC strains tested (Maurelli *et al.*, 1985; Watanabe and Nakamura, 1986). The OMPs may be constituents of bacterial receptors that sequentially bind to unidentified host-cell determinants, though the mechanism whereby the OMPs induce uptake or invasion is unclear. Yet other studies with shigellae indicate the unusual but interesting possibility that carbohydrate-containing proteins present on the host-cell surface act by recognizing receptors on the bacterial surface (Izhar *et al.*, 1982). These interactions of EIEC and shigellae are little understood, except that some of them are inhibited by glucose and fucose residues. Similar OMP-dependent interactions with host epithelium are recognized among the intracellular yersiniae, and possibly also among salmonellae causing enteric fever.

In view of the findings that the adhesion of EPEC and EIEC strains are probaby mediated by non-fimbrial materials of OMP origin, it is interesting that as far back as 1955 Duguid *et al.* described strains of *E. coli* that produced mannose-resistant and eluting haemagglutinins (MREHAs) with most of the properties of classical MREHAs, and yet were non-fimbrial. These latter MREHAs, patterns 6–10 of Duguid *et al.* (1979), were associated with narrow-spectrum haemagglutinating activities; they agglutinated, for example, only one or two red-cell species of the 14 that were routinely tested. Again, like the MREHAs sometimes produced by EPEC strains, these narrow-spectrum MREHAs were best produced in phosphate-buffered liquid media (Old, 1985). Exhaustive searches with the electron microscope have only occasionally revealed filamentous structures on a minority of the strains; on the whole, present observations tend to substantiate Duguid's original description of them as non-fimbrial. However, we shall examine a few of the difficulties encountered in this area.

E. coli strains that produce narrow-spectrum MREHAs of pattern 9 of Duguid *et al.* (1979), colloquially referred to in the literature as the 'man only' haemagglutinin, have attracted much interest of late. For example, Ørskov *et al.* (1985) have recently demonstrated that this kind of adhesin aids the attachment of *E. coli* to urinary and buccal epithelial cells. Our own work (Tavendale and Old, 1985) and that of Knutton and Williams have further established the role of the 'man only' adhesin in attachment to HEp2 cells. There is general agreement that the adhesin is associated with a hydrophobic protein of subunit molecular mass of 14.5 kDa, but little agreement about the nature of the associated surface structure which has been described, variously, as follows:

1. When present on pyelonephritogenic strains of *E. coli* of serogroup O2 (one of the X adhesins in the literature), it was described as non-fimbrial by Labigne-Roussel *et al.* (1984).
2. When present on strains of serogroups O21 and O25 recovered from patients with diarrhoea or UTI, it was called the Z antigen by Ørskov *et al.* (1985), who managed with great difficulty to demonstrate, on just a very few bacteria and with uranyl acetate only, fine fibrils of about 2 nm diameter. Their best published electron micrographs illustrate well the problems of characterization of the Z antigen (Ørskov *et al.*, 1985).

3. Others working with non-serotypable strains recovered from infants
   with a dysentery-like illness (Williams *et al.*, 1984) postulated the
   involvement of a polysaccharide glycocalyx, and their excellent electron micrographs seem to afford an excellent demonstration of a
   mesh of fine fibrils of 2 nm diameter.

Were these different groups dealing, therefore, with one and the same
adhesin or three different 'man only' adhesins? With the degree of
technical difficulty presently associated with EM characterization of these
structures—fibrillae, protein capsule, polysaccharide glycocalyx or non-
fimbrial OMPs—it will be appreciated that the molecular nature of their
receptors is still pending. Furthermore, similar difficulties have been
encountered with the characterization of the adhesins of the CFA2 complex in ETEC strains of *E. coli* (Smyth, 1984).

Among the more intriguing recent observations has been the demonstration that the ability to form fimbrial fibres is genetically separable
from the ability to form the usually associated adhesin, a situation now
confirmed for type-1 fimbriae and the major antigenic classes of P fimbriae
(see, for example, Maurer and Orndorff, 1985; van Die *et al.*, 1986). Yet the
two properties remain functionally associated in most wild-type clinical
isolates. The idea that fimbriae are no more than arms whereby the
adhesins can be projected beyond the limits of the O antigens is a
fascinating one. Equally interesting, however, is the possibility that some
of the adhesins presently classified as non-fimbrial are, in fact, representa-
tives of adhesins that can be expressed independently of the presence of
fimbriae. Present EM techniques, however, may simply be too insensitive
to establish the existence of structures that are fibrillar, or even sub-
fibrillar, appendages. These latter, therefore, may be more common than
is yet appreciated. There is without doubt much more effort required to
resolve the many current difficulties associated with this aspect of adhesion.

In concluding this section on Enterobacteriaceae, we should return and
deal again with some of the more general aspects of adhesion. For
example, among the Enterobacteriaceae causing most current problems to
the clinical microbiologists are the constituent species and genera of the
tribes Klebsielleae and Proteeae which are important opportunistic
pathogens involved in hospital-acquired infections of many clinical sites,
especially of the urinary tract. Two broad classes of MRHAs are present,
along with MSHA, in these other Enterobacteriaceae (Duguid and Old,
1980; Old, 1985). First, and most common, is the mannose-resistant and
klebsiella-like haemagglutinin (MR/K-HA) first described in klebsiellas
but now known to be present in many genera. In all genera MR/K-HA is
associated with type-3 fimbriae which are morphologically identical,
being 3–4 nm in diameter and non-channelled, but they are antigenically

diverse. Type-3 fimbriae mediate attachment to a wide array of substrates, including plant root hairs, fungal and yeast cells and immune cells, but not to red cells. However, red cells can be made agglutinable after treatment with tannic acid. It is not known, however, whether tannic acid acts by modifying acid-labile carbohydrate residues, such as neuraminic acid, furanoses or deoxy-sugars, or by stripping off exposed residues to reveal underlying receptors. These type-3 fimbriae are almost as widely distributed throughout Enterobacteriaceae as type-1 fimbriae. The second class, the mannose-resistant and proteus-like haemagglutinins (MR/P-HAs), are more heterogeneous, giving both narrow- and broad-spectrum haemagglutination patterns. Again, some of the fimbriae in genera like *Providencia* are non-haemagglutinating. These latter MRHAs differ from the narrow and broad MRHAs of *E. coli* in lacking the important property of elution. Although we originally described them as MR/P haemagglutinins, that term has subsequently been used to describe the P-antigen recognizing activities of P fimbriae. But gene probing confirms that P fimbriae are absent from other important urinary pathogens, the enterobacteria and Gram-positive cocci alike. Neither the receptors nor the role in infection have been established for either the MR/K- or MR/P-HAs.

(v) Concluding remarks

In this brief overview, it has been convenient to approach the question of the role of envelopes as adhesins in a narrow, selective manner. Accordingly, by reference to *N. gonorrhoeae* and *E. coli* of the Enterobacteriaceae—perhaps the two pathogens best suited in terms of adhesion—I have attempted to highlight some of the current interest in their adhesive processes. It would have been equally valid to have discussed similarly the role of other envelope fractions such as the flagellum of *Vibrio cholerae*, the teichoic acids and M-proteinaceous fibrillae of *Streptococcus pyogenes*, the complex polysaccharides and pili of other streptococci, or even the slime of coagulase-negative staphylococci. However approached, the message is clear, namely that adhesins are diverse in terms of their chemical, physical and structural characters. Since the classical study of Duguid *et al.* (1955) first illustrated that diversity, adhesins and their role in infectious diseases have proved popular and productive research areas, as evidenced by the continuing emergence of large numbers of reviews and books on the subject. As might be expected, much is known about the adhesins and receptors of some pathogens, much less about others. Indeed, in certain areas, for example, Enterobacteriaceae other than *E. coli*, our knowledge is thin and even basic. Many questions have been answered, many others posed and in some ways it seems that our understanding of adhesins and their function in infectious diseases is just about to begin. As has been discussed in other chapters of this book, the application of modern techniques to the study of the bacterial envelope has significantly enhanced our knowledge of the complexities of its structure, chemistry and function. The resultant progress in our understanding of the bacterial envelope has gone hand in hand with a greater understanding of the role of the envelope in adhesion. The study of bacterial adhesins and adhesion and the role of the envelope fractions in these processes seems likely to prove attractive and popular for many years ahead.

B. DEVELOPMENT OF DEFINED BACTERIAL VACCINES

Gordon Dougan

Other chapters in this book describe in detail structures located on the bacterial cell surface which play a key role in regulating the interaction between bacteria and their environment. In the case of pathogenic bacteria, that environment includes the surfaces and tissues of a potential host organism. Many bacterial surface structures have evolved to interact in a specialized manner with the host. An obvious example are structures which enable bacteria to attach to the epithelial surfaces of a potential host. The specific adhesion pili found on enteric bacteria, which interact with receptors located on the membranes of intestinal brush-border cells, come to mind here. Perhaps the key factor which enables pathogenic bacteria to survive on or within a mammalian host is their ability to evade the host immune system. Thus many bacterial surface structures have a function to protect the organism from either direct or indirect killing and clearance by the immune system.

For the immune system to succeed in controlling the growth of an invading pathogen, it must be able to recognize and respond to key surface structures present on the pathogen. During the course of a natural infection an immune response is normally provoked in the host to a variety of bacterial antigens. However, in order for the response to be protective the correct antigens must be recognized by relevant arms of the immune system. The host organism often shows some form of immunity to a pathogen if it recovers from an infection. Immunity can be induced artificially by vaccination. Vaccines can be of several forms but all usually contain crude or purified preparations of surface antigens from a pathogen. This section discusses how knowledge of the structure of the bacterial cell surface can aid the development of novel vaccines.

(i) Requirements of modern vaccines

In the past, vaccines consisted of either live attenuated or inactivated whole cells. Although some of the vaccines showed reasonable efficacy, the use of genetically ill-defined attenuated strains or crude whole-cell preparations for vaccination was fraught with problems. Many were highly reactogenic in the host and the live strains often reverted to virulence. The modern trend is to move towards safer and more efficient vaccines. These can be of two main forms: genetically defined live attenuated strains or subcellular vaccines (Winther & Dougan, 1984). One potential aim is to synthesize chemically completely artificial vaccines such as synthetic peptides. These two main approaches can be considered separately for convenience.

(ii) Live attenuated bacterial vaccines

People recovering from a natural bacterial infection often show strong immunity against reinfection. Thus live infections can induce the correct immune response for protection. This was recognized by Jenner in his pioneering work on the use of vaccinia in smallpox vaccination. If it is possible to create artificially or isolate attenuated bacterial strains these can be of potential use in vaccination. An example of a currently employed live attenuated vaccine is the BCG vaccine against tuberculosis. Live vaccines have proved especially useful where it is necessary to stimulate a cell-mediated or mucosal immune response as this is often difficult to achieve with dead vaccines. Also, with many pathogens specific protective antigens have not yet been identified. Thus, in those cases, at present it is impossible to develop subcellular defined vaccines. Much of the recent activity in the field of live attenuated bacterial vaccines has been with enteric pathogens for which there are few effective vaccines currently available.

Two main approaches have been employed to isolate attenuated bacteria suitable for use as live vaccines. Serologically related micro-organisms which induce cross-protection in the absence of a full-scale clinical infection can be employed. Vaccinia serves as a classical example here. Secondly, a fully pathogenic micro-organism can be attenuated by selective procedures. In the past this was often achieved by passaging a pathogenic strain many times *in vitro* or in an unnatural animal host until virulence was reduced. Such vaccines can be hazardous because of the ill-defined nature of the attenuating lesions, which can result in reversion to virulence or genetic drift leading to difficulties in quality control of vaccines. Genetic manipulation can be used to overcome some of these problems as it is now possible to introduce stable non-reverting attenuating lesions into many pathogens. This can be in a manner which conserves the major immunogens of the bacterial cell surface. Since much of the work on live attenuated bacteria has been carried out with *enteric pathogens* we can use this group as an example of modern approaches.

The current crop of commercially available inactivated whole-cell vaccines against enteric pathogens are, on the whole, unsatisfactory. They do not induce strong local responses at the gut surface and fail to induce a good cell-mediated response required for protection against invasive pathogens such as *Salmonella typhi*. Live vaccines can induce these responses. The development of subunit vaccines against enteric bacteria has been hampered by lack of progress in identifying key protective antigens. Success has been achieved in controlling infections by enterotoxigenic *Escherichia coli* in domestic animals following the identification of surface-associated adhesion pili, but with the main human bacterial enteric pathogens progress has to date been slow (Gaastra & De Graaf, 1982; Morrissey & Dougan, 1985).

1. Cholera vaccines

The current cholera vaccine consists of inactivated whole cells of *Vibrio cholerae*. The efficacy of the vaccine is low and the duration of immunity is short-lived. Cholera is caused when viable *V. cholerae* organisms establish a non-invasive infection in the small bowel of the host. Since cholera is primarily a toxigenic infection a great deal of interest has centred on the potent cholera exo-enterotoxin. However, it is likely, but not proven, that both antitoxic and antibacterial immunity are required for good protection. Since people recovering from cholera often show long-lasting immunity to subsequent infection, attempts have been made to create live attenuated oral cholera vaccines which stimulate a local gut immune response. Genetic techniques have been used to construct stable non-toxigenic mutants of *V. cholerae*. Honda & Finkelstein (1979) constructed a *V. cholerae* strain, which they named Texas-Star, which produced the non-toxic but highly immunogenic cholera toxin B subunit but not the A subunit required for full toxicity. This work was followed up by Kaper and Mekalanos working independently who used genetic engineering techniques to isolate strains similar to Texas-Star but which contained more defined lesions (Kaper *et al.*, 1984; Mekalanos *et al.*, 1983). These attenuated strains were tested under controlled conditions in human volunteers. The genetically engineered strain of Kaper proved highly efficacious in these trials, but unfortunately some of the volunteers developed mild diarrhoea shortly after taking the vaccine and thus more work is needed to perfect a practical oral cholera vaccine.

2. Typhoid vaccines

Typhoid is an invasive infection caused by *Salmonella typhi*. *S. typhi* persists in the body for an extended period of time and it is probably located mainly as an intracellular parasite within cells of the reticuloendothelial system. Although the current inactivated whole-cell vaccines show some efficacy, especially in areas where typhoid is endemic, they are highly reactogenic. Since cell-mediated responses are believed to play a key role in protection against *S. typhi*, much work has been undertaken on the development of live oral typhoid vaccines. Germanier and co-workers used *S. typhimurium* infections of mice as a model for developing candidate vaccine strains. They concentrated on the effects of mutations in the lipopolysaccharide biosynthetic pathways. Strains of *S. typhimurium* harbouring *gal*E mutations, which render the cells sensitive to killing by exogenous galactose, were shown to be avirulent and capable of inducing strong immunity in mice following vaccination with live cells. A stable *gal*E mutant of *S. typhi*, named Ty21a, was isolated by the same workers and put forward as a candidate human typhoid vaccine (Germanier & Furer, 1975). The strain performed well when tested in controlled human volunteer studies and in a large-scale field trial in Egypt. The strain is now being more thoroughly evaluated in an extended trial in Chile.

3. Metabolic blocks

Other mutations in specific genes encoding enzymes involved in biosynthesis pathways are known to affect bacterial virulence. The idea of using a metabolic block to attenuate a pathogen is an attractive one since it should be possible to cripple a pathogen *in vivo* whilst preserving all the major surface antigens and the ability of the bacteria to grow *in vitro*. An example of this approach is the work of Hosieth & Stocker (1981). These workers used genetic manipulation to introduce stable deletion mutations into the *aro*A gene of *S. typhimurium*. Mutations in *aro*A render bacteria dependent for growth on aromatic compounds including para-amino benzoic acid. These aromatic mutants of *S. typhimurium* grow poorly *in vivo* and are attenuated, but they are able to induce a protective immune response in experimental animals. Work is going on to create candidate human vaccines based on these metabolic mutants. Candidate genes include those encoding enzymes of the aromatic and purine pathways.

(iii) Multivalent live vaccines

Genetic manipulation techniques can be used to move genes encoding surface antigen between bacterial species. This makes possible the development of multivalent vaccines. Live attenuated bacterial vaccines can in theory be used to carry heterologous antigens to the immune system. Much of this work has been carried out with *Salmonella* carriers. Since *Salmonella* is closely related to *E. coli* it is relatively easy to introduce cloned genes into *Salmonella* vaccine strains to create experimental vaccines. Genes from any major pathogen could be expressed in *Salmonella* if manipulated in the correct manner.

Several different bacterial antigens have been expressed in *S. typhimurium* and *S. typhi gal*E mutants; examples include *Shigella sonnei* O-antigen, the B subunit of the heat-labile enterotoxin of enterotoxigenic *E. coli* (LT-B), and the K88 adhesin (see Dougan *et al.*, 1987). These strains can present the cloned antigen to the immune system to stimulate secretory and systemic immune responses. *Aro*A *S. typhimurium* strains have also been used as carriers of LT-B and K88 (Fig. 7.1). Antigens do not have to be expressed at the surface of the carrier in order to be recognized by the immune system. *Aro*A *S. typhimurium* can be used to deliver the intracellular enzyme β-galactosidase to the mouse system and induce antibodies and cell-mediated responses (Brown *et al.*, 1987). Currently a variety of bacterial, viral and parasite surface antigens are being tested in this model system.

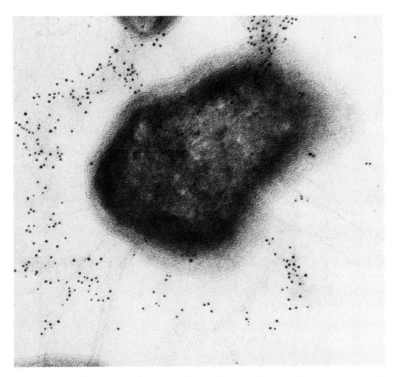

Figure 7.1 Electron micrographs of *Salmonella typhimurium Aro*. A vaccine strain expressing the *Escherichia coli*-derived K88 adhesion fimbriae. The fimbriae are labelled with anti-K88 antisera coupled to gold particles.

(iv) Subunit vaccines

Most currently available inactivated vaccines consist of whole bacterial cells. In some cases these empirically designed vaccines are reasonably effective, but serious side-effects can be caused by toxigenic components present on whole cells such as LPS. Since we now have a solid understanding of the structure of the bacterial cell surface, it is possible to undertake a more rational approach to the development of subunit vaccines which are less antigenically complex and safer. In order to achieve this aim it is important that candidate protective antigens should also be identified, characterized and isolated in large quantities.

Some antigens are relatively easy to identify. For example, the toxins responsible for the clinical symptoms of tetanus and diphtheria have been known for many years, as have the capsular polysaccharides of the pneumococci. Some important antigens, especially those components of fastidious bacterial pathogens such as treponemes, present more difficulties, especially if they are integral components of cell membranes. Using a combination of modern techniques, including monoclonal antibodies and gene cloning, it is possible to dissect the antigenic architecture of pathogens and identify individual candidate antigens. These antigens can then often be cloned and expressed in heterologous hosts such as *E. coli* K12. Gene cloning can also be used to modify the toxicity of potentially dangerous antigens and facilitate, along with monoclonal antibodies, the purification of the antigens on a large scale (Winther & Dougan, 1984). Once the amino acid sequence of a protein has been elucidated, peptides corresponding to specific sequences of the protein can be synthesized chemically or, *in vivo*, attached genetically to larger carrier proteins. Peptides have proved useful in defining protective epitopes within protein antigens. Such epitopes can be only a few amino acids in length (Lerner, 1982).

Although peptides have proved to be very valuable tools for identifying protective regions on proteins, they have yet to be developed into effective practical vaccines. They stimulate an extremely specific immune response in experimental animals which is often weak in comparison with that stimulated by whole proteins. Nevertheless there is great potential in this area. We can now consider examples of bacterial subunit vaccines.

1. Antiadhesin vaccines

The attachment of the bacterial cell to the host epithelial surface is an important primary step in the establishment of infection by bacterial pathogens. If this step can be controlled by the immune system then it is possible to prevent disease.

Attachment to surfaces is often mediated by so-called adhesion factors present on the bacterial cell surface. The interaction between adhesion factors and the epithelial cell surface is often very specific, involving receptor/ligand interactions. Antibodies raised against the adhesion factor can often inhibit attachment. However, antibodies—usually IgA class— have to be transported to the local epithelial surface in order for active protection to occur *in vivo*. Passive antiadhesion immunity can be mediated by maternally derived antibodies. Many candidate adhesive factors have been described in the literature, although few of these have been shown to have real roles *in vivo*. Classic examples of adhesion factors are the adhesion pili located on the surface of the enterotoxigenic *E. coli*. They are proteinaceous in nature although not all form piliated structures as defined by electron microscopy. Adhesion pili can often be purified using simple procedures from the bacterial cell surface (see Chapter 4A) and the pure piliated fractions can form the basis of subunit vaccines. An example is the K88 vaccine used for protecting neonatal piglets. Immunological heterogeneity and the inability to identify adhesive factors have prevented more antiadhesive vaccines being developed (Morrissey & Dougan, 1985).

2. Toxoid vaccines

Vaccines based on chemically inactivated tetanus and diphtheria toxins have been in general use with few problems for many years. Recent genetic studies of the diphtheria and tetanus toxin genes will make possible the development of new vaccines against these toxins if there is a demand in the future. For example, tetanus toxin is synthesized as a 150 kDa polypeptide protoxin which is normally split by protease activity to generate two polypeptides of molecular masses 100 and 50 kDa, which are held together by disulphide bridges. A 50 kDa polypeptide, known as the C toxin, can be released from the carboxy-terminal of the protoxin by papain digestion (Figure 7.2). Purified C fragment is non-toxic and can be used to vaccinate animals and protect them against tetanus. Polypeptide sequences corresponding to C fragment have recently been expressed in *E. coli*, and the *E. coli* derived material has been used to protect experimental animals against tetanus toxin (Fairweather *et al.*, 1986). Eventually it may be possible to develop novel oral and peptide tetanus vaccines for use in humans.

Unfortunately the role of many toxins in diseases and/or protection is less clearly defined than the examples used above. For example, so far vaccinations with cholera enterotoxin toxoids have had limited success. However, there may be a role for such toxoids in a new defined cholera vaccine. Since cholera enterotoxin is composed of two distinct polypeptide

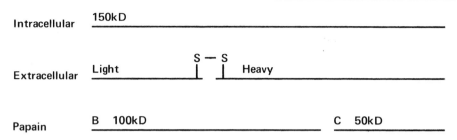

Figure 7.2 Structure of tetanus toxin showing the main polypeptide domains.

subunits, it was relatively straightforward to express the immunogenic but non-toxin B subunit free of the A subunit required for full toxicity. Isolated B subunit is an effective toxoid since antibodies against B neutralize the cholera holotoxin. The *E. coli* heat-labile toxin has a similar structure to cholera enterotoxin. Gene cloning can be used to construct *E. coli* or *Salmonella* strains expressing LT-B type toxoid (Figure 7.3).

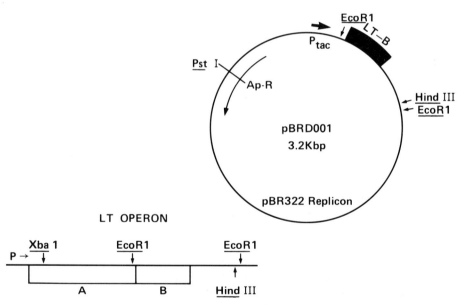

Figure 7.3 Map of the recombinant plasmid which directs the expression of LT-B from the tac promoter. The plasmid is a pBR322-based replicon. Ap-R is the ampicillin resistance gene. The arrow designates the direction of transcription from the tac promoter. The closed box designates the LT-B cistron. In the bottom left of the diagram is a representation of the organization of the LT toxin operon. P shows the position of the natural LT promoter in front of the A subunit cistron. *E. coli* or *Salmonella* cells harbouring this plasmid synthesize high levels of LT-B protein.

58
MET−VAL−ILE−ILE−THR−PHE−MET−SER−GLY−GLU−THR−

PHE−GLN−VAL−GLU−VAL−PRO−GLY−SER−GLN−HIS−ILE−

           83   1
ASP−SER−GLN−LYS−ASN−THY−PHE−TYR−CYS−CYS−GLU−

                                                    18
LEU−CYS−CYS−TYR−PRO−ALA−CYS−ALA−GLY−CYS−ASN−

Figure 7.4 Structure of the LT-B/ST peptide used by Klipstein, Houghton and co-workers to raise neutralizing antibodies against both ST and LT toxins.

Enterotoxigenic *E. coli* strains secrete an enterotoxin known as the heat-stable toxin (ST). ST can be isolated as a poorly immunogenic peptide about 18 amino acids in length. Although ST itself is poorly immunogenic, neutralizing antibodies against ST can be raised if the ST peptide is coupled chemically to the LT-B subunit of enterotoxigenic *E. coli* (Klipstein *et al.*, 1986). Animals vaccinated with chemically coupled LT-B and ST generate neutralizing antibodies against both toxins. Recently a completely synthetic LT-B/ST based peptide has been synthesized which stimulates antibodies to both toxins (Figure 7.4). This approach could be applied to toxins produced by a variety of different pathogens.

3. Combining adhesin and toxoid vaccines

More potent vaccines may be obtained by combining the protective potential of antiadhesive and antitoxic immunity. This approach has been employed to combat the enterotoxigenic strains of *E. coli* causing diarrhoea in pigs where both K88 antigen and LT-B have been incorporated into vaccines. An example of this approach for human vaccination is the new acellular pertussis vaccine preparation currently being tested around the world. *Bordetella pertussis* is the causative agent of whooping cough. The current vaccine contains inactivated whole cells, but although the vaccine is very efficacious there have been reports of serious side-effects in young children. *B. pertussis* produces a potent toxin often referred to as pertussis toxin. The toxin can be isolated together with a surface structure known as fimbrial haemagglutinin or FHA. FHA could be an adhesin. The acellular pertussis vaccine is made by chemically inactivating a relatively pure pertussis toxin/FHA complex. Since the genetic determinant for pertusis toxin has recently been cloned it may not be long before a genetically engineered pertussis toxoid is available (Locht & Keith, 1986).

(v) Polysaccharide vaccines

The role of polysaccharides present on the surfaces of bacteria has been well covered in other chapters of this book. The importance of poly-saccharides as components of vaccines has been recognized for many years. Capsular polysaccharide vaccines have been in use since the 1920s. Since there is often quite a lot of serotypic variation between capsular material produced by a single species of bacteria, capsule vaccines are often composed of mixtures of several different capsule serotypes. An example of such a vaccine is the *Streptococcus pneumoniae* capsule based vaccine where immunologically distinct capsules from several different serotypes of the organism are incorporated into the vaccine. Not all polysaccharide structures are good immunogens. Many polysaccharides are T-independent antigens and as a consequence do not effectively induce T-independent memory. Vaccines based on the capsules of *Neisseria meningitidis* have been developed. Those containing group A or C capsular material are effective against *N. meningitidis* strains of the homologous serotype in people over 18 months of age. However, attempts to develop vaccines against group B *N. meningitidis* have been hampered by the poor immunogenicity of the B capsular material. The immunogen-icity of the B polysaccharide can be increased if it is covalently linked to protein carriers. For example, experimental vaccines have been prepared which consist of the B polysaccharide in complex with outer membrane proteins of the *N. meningitidis* cell envelope. Such complexes have been used to raise anti-B capsule antibodies in experimental animals, although they have yet to be tested in humans. One problem with these vaccines is that they induce mainly IgM class antibodies and any protection they give tends to be short-lived (Gotschlich, 1984).

Although LPS can be highly toxic to mammals, attempts have been made to prepare LPS-based vaccines. LPS is often recognized early on by the host during infection and it is usually a good immunogen. Trials have been carried out in humans using *Pseudomonas aeruginosa* extracts in which the protective component appears to be LPS. Up to 16 different LPS serotypes of *P. aeruginosa* are incorporated into the vaccine to cover the main clinically relevant serotypes of this species (MacIntyre *et al.*, 1986). LPS has also attracted interest as a component of *Salmonella* vaccines. The toxicity of LPS resides almost entirely within the lipid A portion of the molecule. This is the most immunologically conserved part of the LPS and work has been carried out to develop polyvalent and monoclonal anti-bodies for passive protection against endotoxic shock. A deep-rough mutant of *E. coli* known as J5 has been used in this work. Another approach has been to use synthetic *O* side-chain molecules attached to protein carriers. Much of this work has been carried out on *Salmonella* LPS (Saxen *et al.*, 1986).

(vi) Bacterial outer-membrane proteins as vaccines

There are a number of examples of work suggesting integral outer-membrane proteins as potential vaccine candidates. However, to date there are no commercially available vaccines based on purified outer-membrane proteins. After failure of the group B meningococcal capsules and *Haemophilus influenzae* type B capsule vaccines because of poor immunogenicity, much recent work has concentrated on the outer-membrane proteins as alternatives. In the case of *N. meningitidis*, work has focused mainly on the serotype 2 antigen since it is frequently associated with strains causing invasive diseases and antibodies to it are bacterioci-dal. Purified preparations of this serotype 2 antigen have been tested in humans (Gotschlich, 1984). The immunogenicity of outer-membrane vaccine may be enhanced when they are coupled to capsular material. Work on *H. influenzae* outer-membrane proteins has led to the identification of several candidate proteins. In particular, a protein of 44 kDa has been identified by monoclonal antibodies. This antigen is protective in experimental animals when administered parenterally with complete Freund's adjuvant (Robbins *et al.*, 1984). Other vaccine development work has centred on the OMP A related proteins since they are relatively immunologically conserved between some Gram-negative organisms and they are exposed at the cell surface; but this work is still at an early stage.

More recently outer-membrane proteins have been used to deliver epitopes from heterologous pathogens to the surface of the *E. coli* K12 cell envelope (Charbit *et al.*, 1986). The regions of the lamB protein which are exposed at the bacterial cell surface have been identified. Since the complete DNA sequence of lamB is known, it has been possible to insert DNA sequences encoding foreign epitopes into sequences within the lamB gene which encode surface exposed regions of the protein. For example, a DNA sequence encoding the C3 epitope of polio virus has been fused internally into the lamB gene. *E. coli* K12 strains harbouring the recombinant lamB gene express on their surface a lamB/C3 polio fusion protein. This approach could be a very powerful method of stimulating antipeptide immunity, especially if the approach is combined with the live attenuated vaccines described above.

(vii) Conclusions

The ability to move genetic determinants between bacterial species, coupled with our increasing knowledge of the structure of bacterial cell surfaces, has opened up new approaches to vaccine development. However, in order to design new vaccines rationally we need to fully understand protective, as against non-protective, immune responses. Thus vaccine design is a multidisciplinary task. In this review I have described some of the current work in the field. It should be emphasized that to date few subunit vaccines are available for general practical use and much of the work is still being carried out on an experimental level. Nevertheless, significant progress is being made and it is not unrealistic to expect to see a crop of new bacterial vaccines to appear within the next 20 years. Workers have put forward almost every type of bacterial surface antigen as vaccine candidates, and it will be interesting to see how many of these find a place as a practical vaccine.

C. THE USE OF BACTERIAL CELL WALL COMPONENTS IN DIAGNOSTIC ASSAYS

David Parratt

(i) Introduction

Medical microbiology has for long used the principle of measuring antibody as a means of diagnosis. The concept is basically simple in that one looks for antibody in a patient's serum produced in response to an infecting micro-organism. It follows, of course, that a suspicion of the presence of a particular organism must exist before it is possible to search for the antibody, although it is often difficult to refine this suspicion to a single or a few possibilities and it is usual to adopt the approach of 'serology screening', whereby the patient's serum is tested for antibody to a range of micro-organisms or their antigens. The recognition of the need for screening is important in designing the serological tests to be used, for it determines that the assays must be sufficiently convenient to allow frequent and easy use.

The range of infections for which serological diagnosis is used is wide, but it generally includes both those produced by organisms which are difficult or impossible to grow and those which are dangerous to grow in the laboratory. In these situations the serological approach has a prime place in diagnosis. However, it is not unusual to see a patient for the first time several days or weeks after the acute episode of infection has taken place and at a time when culture of the responsible organism is unlikely to be successful. Brucellosis is a good example of this. Here, a 'retrospective' diagnosis by the measurement of antibody to *Brucella* organisms is valuable. An extension of this approach is epidemiology, where one is not concerned with diagnosis in the individual but where it is important to establish the prevalence of past infection in the community, or in certain groups in the community. Serum hepatitis B and HIV infections are topical examples of this need.

These examples collectively produce a large amount of serological testing for the average microbiology laboratory. In general terms it is likely that serology will account for about 15–20 per cent of the work of a diagnostic laboratory, emphasizing the need for simple, accurate and rapid assay methods. This review will summarize the types of assay in common use and will indicate where improved bacterial antigens in the form of isolated cell wall components have a place.

However, there is now more to 'serology' than the measurement of antibody. With improved technology it is possible to assay microbial antigen in serum and tissue fluids and to measure immune complexes which contain a mixture of the antigen and the patient's antibody. These new approaches require some discussion for they promise to provide rapid and precise diagnosis in many infections. Diagnosis by the demonstration of antibody to a micro-organism is always diagnosis by inference; that is, one 'infers' that the antibody is present in the patient's serum because the patient has had or is suffering from a particular infection. However, the demonstration of microbial antigens in the patient's fluids, or of immune complexes containing microbial antigens in these fluids, provides *direct* and unequivocal diagnosis. It is likely that, in the future, serology will consist of a combination of tests using antibody measurement, antigen measurement and immune complex measurement at the same time. This is already the case in some infections (e.g. hepatitis B), and it is therefore useful to consider the relationships between these different components. With an understanding of the nature of serological response, it is possible to consider the usefulness or otherwise of different immunoassays, and to see where improvements can be made.

(ii) The pattern of the serological response to infection

The serological events of an infection are represented in Figure 7.5. This simplified scheme is used only to demonstrate the sequence of events and is not intended to be an exact description.

From the time of infection, micro-organisms multiply within the body and liberate antigens. The latter may be enzymes, capsular material or cell wall components liberated by the breaking up of the organism. In the case of viruses, capsid or 'core' antigens are likely to be liberated. The release of these antigens stimulates antibody formation, which begins within a few hours of contact being made with the antigen. It is popularly believed that antibody is not produced for 7–14 days after an infection, but this erroneous belief is based on a misunderstanding of the events that are

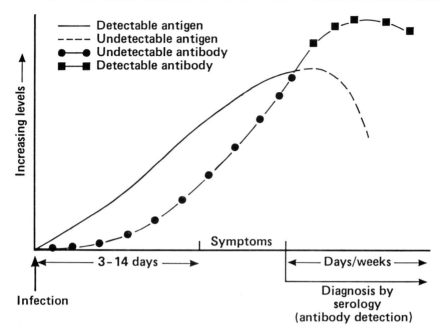

Figure 7.5 Antigen and antibody dynamics in an infection.

occurring. As shown in Figure 7.5, *detectable* antibody, measurable in serum, is present 7–14 days after the infection has begun. But this antibody is only detectable because sufficient antibody has been generated at this time to 'saturate' the microbial antigen in the body. The figure shows that a large amount of antibody is generated in the first 14 days, but all of it is used up by combining with antigen and generating immune complexes. Clearly, it may take a variable period for the 'saturation' point to be reached, and in some infections it takes longer than 14 days before one finds antibody in the patient's serum. An example is legionnaires' disease (although it must be added that here, as in other instances, patient to patient variation will affect the time taken to produce an excess of antibody in serum: some patients will have vigorous antibody responses whilst others have sluggish responses, and it is an important skill of the serologist to understand these variabilities and be able to interpret them correctly).

When antibody becomes detectable in serum, it usually increases rapidly in amount, and the general rule for the serologist is to look for a 'rising titre' (level) of antibody which gives the most definitive evidence of recent infection. Generally, at the time of presentation, an infected patient has little or no serum antibody to the causative organism, but 7–10 days later the level should be higher. Therefore it is the custom to take 'acute' and 'convalescent' specimens of serum for comparison and for demonstration of the rising titre. As shown in Figure 7.5, the excess of antibody increases during this convalescent period and therefore becomes more easily measured. However, the disadvantage is that diagnosis will be retrospective and will not be helpful at the time the patient is acutely ill. This situation can be improved by using very sensitive assays for antibody which can detect significant changes in antibody level over shorter periods. For example, specimens could be taken on the first three days of illness and be shown to have increasing levels of antibody. The modern assays, particularly ELISA and radio-immunoassay methods, have this capability and therefore have an advantage over the more traditional tests for antibody. This aspect of refinement of antibody measurement is discussed in more detail below.

As indicated in the introduction, serology is now concerned with antigen detection and measurement, and immune complex analysis. The reasons for this are clear from Figure 7.5. During the early part of an infection, antigen from the micro-organism is present in excess and is measurable by suitable antigen assays. As the infection proceeds more of the antigen becomes complexed with antibody in the form of immune complexes. Therefore, early diagnosis is best achieved by assaying (free) antigen or antigen complexed with antibody. This is clearly of most relevance when considering the period during which the patient is symptomatic, at which time either free antigen, complexed antigen or small amounts of free antibody are present. Antigen and complexed antigen methods are likely to be most effective when they use sensitive techniques, so that small changes in levels, over short intervals, can be defined.

With this background it is now possible to look at different assays and describe their advantages and disadvantages.

(iii) Traditional tests for antibody

The 'traditional' tests for antibody include agglutination tests, complement fixation tests and precipitation tests. In addition, neutralization tests are commonly used for detecting antibody to viruses.

1. Agglutination tests

These tests often use a suspension of bacteria, which is mixed with dilutions of the patient's serum. If the latter contains antibody, and particularly IgM antibody, the particles of bacteria will be agglutinated. The 'titre' of the serum is the highest dilution which will cause agglutination. These tests are notoriously difficult to standardize and they give only a semi-quantitative result. Agglutination tests are being phased out and replaced with more modern assays such as those described below.

There is, however, one important category of agglutination test which is relevant to studies of cell wall components: this is passive agglutination, using latex or similar particles, or passive haemagglutination using erythrocytes. In these tests soluble fractions from bacterial cell envelopes or cytoplasm are used to sensitize the particles. The addition of antibody in the patient's serum causes agglutination. This adaptation is more economical of reagents than the straight agglutination test and is particularly useful where it is difficult to obtain the microbial antigen. Passive agglutination tests have a high degree of specificity and sensitivity if the correct microbial antigens are chosen. A good example is the 'Treponema pallidum haemagglutination assay' (TPHA) which uses treponemal antigens absorbed to preserved sheep red cells. The test has a sensitivity and specificity approaching that of the FTAabs (fluorescent treponemal antibody absorbed) test, but is very much simpler to carry out. Ordinary agglutination tests with T. *pallidum* would be unthinkable for the routine diagnosis of syphilis, but the adaptation of the TPHA produces a valuable screening test.

Passive agglutination is used for several microbial antibody tests, and it is likely that their use will become more widespread in coming years, especially where it is possible to identify and isolate cell envelope components which have an important part to play in the pathogenesis of infection or in the response to infection. On a cautionary note, these tests can only be quantitated by carrying out the procedure with a series of dilutions of the patient's serum, and this is notoriously inaccurate. For screening purposes the passive agglutination tests are useful and valuable, but for quantitation of antibody, which is so important in diagnosis, there are preferred assays available.

In the above description of passive agglutination reference has been made only to testing for antibody. These tests are of no practical value in immune complex determination and of limited use in assays of 'free' antigen. For the latter, the *co-agglutination* principle, which is discussed below, is more appropriate.

2. Complement fixation tests

Complement fixation tests can be difficult to perform. In principle the patient's serum is reacted with antigen in the presence of a known amount of complement. If antibody is present in the serum, it combines with the antigen and complement is used up. In a second stage, sensitized erythrocytes are added as an indicator. If complement has been utilized in the first stage, no haemolysis of the erythrocytes results. However, if there is no antibody in the patient's serum, there cannot be any antibody–antigen combination and complement is therefore not used. In this case the indicator system is attacked by the remaining complement and the erythrocytes are lysed.

The problems of complement fixation tests revolve mainly around the fact that complement activation by antibody–antigen complexes varies with the nature and proportions of each component within the complex. Accordingly, pretest calibrations are required to determine the optimal amounts of each component in the test, and the measurable range of antibody level is usually narrow. A further problem is that even within the measurable range the accuracy of the test is poor. The antigen used in complement fixation tests is usually a crude extract of the micro-organism of interest. Refined antigens would be preferable, but the techniques of antigenic analysis have probably arrived too late to stimulate a resurgence in the use of complement fixation tests for diagnosis. A more modern direct approach to measurement of antibody is usually preferable.

Complement fixation tests cannot be used for immune complex measurement and in practice are not used for antigen detection.

The main objection to the use of agglutination and complement fixation tests is that they measure the *activity* of the antibody and not the antibody binding. In diagnostic terms this can be disastrous because a patient may produce large amounts of non-agglutinating antibody, which clearly signals a response to infection, but records a negative serological test simply because the antibody does not perform in the test. Direct binding assays of the ELISA and RIA (see below) type are not subject to this problem and are therefore preferred. The precipitation assays discussed below are also subject to the same limitation in that they will only detect antibodies if these have the ability to precipitate antigen, which is not always the case. However, their simplicity and their adaptability to antigen detection justifies their retention as useful serological procedures at present.

3. Precipitation tests

Precipitation tests (or assays) come in many forms. Basically precipitation may occur when antibody combines with *soluble* antigens. Classical examples of the latter are bacterial toxins, enzymes and capsular materials. If the antibody is of a precipitating type (e.g. IgG), an insoluble complex may form after combination with the antigen. However, as with complement fixation, the ability to precipitate is dependent on the ratio of antigen to antibody in the complex. A large excess of either component prevents precipitation and the complex remains insoluble. To test a patient's serum for precipitating antibody to even a single antigen can therefore be laborious because a range of concentrations of the serum and the antigen have to be reacted together to give the best chance for precipitation to occur. A short-cut is the use of the Ouchterlony principle of double gel diffusion. In this, two circular wells spaced 1–2 cm apart are cut in agar contained in a plate. One well is filled with the antigen and the other with the patient's serum. With time the fluid from each well diffuses and establishes a concentration gradient in the gel. The antigen gradient will be in the opposite direction to the serum gradient and at some point in the gel the interacting gradients will produce the proportions of antigen and antibody which will precipitate. At this point a 'band' of precipitated material appears. This simple and useful test has been used for the diagnosis of many infections and still has fairly wide applications. Its main drawbacks are its insensitivity and its long reaction time.

The lack of sensitivity is probably due to many factors, such as the need for a large 'critical mass' of reactants to produce visible precipitation, the limitation of the amount of reactants which can be added to the wells in the plate, and the nature of the antigen mixture used. The long reaction time is due to the fact that diffusion of the reactants is not complete for 2–10 days.

Attempts have been made to overcome these problems whilst retaining the basic simplicity of the test. In counter-current immunoelectrophoresis, the diffusion of antigen and antibody towards each other is speeded up by applying an electric current (Figure 7.6). Fortunately, at pH 8.3, most microbial antigens are positively charged whilst antibody molecules are negatively charged, and therefore the two components can be made to move towards each other in the electric field. By this means the diffusion time can be reduced to a few hours. Precipitin lines which are too weak to

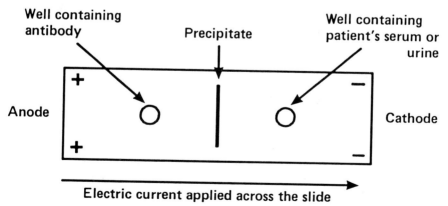

Figure 7.6 Diagrammatic representation of the principles of counter-current immunoelectrophoresis for the detection of bacterial antigen.

be seen by the naked eye can be intensified by protein staining, and if necessary the gel can then be scanned by a densitometer to locate areas of antigen–antibody reaction. These simple modifications make the test more rapid and more sensitive.

One would think that improvement in the sensitivity of this type of test might be obtained by carefully selecting the antigen or antigens used. Usually, the antigen preparation has tended to be a crude extract of the micro-organism obtained by sonication or freeze–thaw disruption of the cells. Little further purification of the extract is used, except to remove cell debris and concentrate the extract. It could be reasonably assumed that the antigen to which the patient has responded constitutes only a small proportion of the total material present in the extract and therefore the sensitivity of the test will be limited. If the antigen can be identified and extracted in large amounts in pure form the test could theoretically be made more sensitive. Attempts have been made to do this. For example, in 'farmer's lung', which is a hypersensitivity response to the antigens of the thermophilic actinomycetes *Micropolyspora faeni* and *Micromonospora vulgaris*, specific cell wall components have been identified as antigens. The patterns of patient response are variable and the selection of an 'antigen mixture' must therefore be sufficiently broad to avoid false-negative tests owing to the failure to include some important antigens. Characterization of the antigens may help, but only if numerous antigens are prepared and these are tested against a large number of patients' serum specimens. The corollary to this is that fractionation and purification of bacterial antigens in the hope of finding a 'diagnostic' antigen is often unrewarding, because of the variability in the way that individuals respond to a micro-organism. The concept of identifying and thereafter purifying and defining antigens is valid, but it should be a task approached with caution and the results must be carefully monitored.

The techniques for bacterial cell wall analysis described elsewhere in this book will make this approach both reasonable and practicable, and, provided that care is taken in the final selection of the antigens, improved sensitivity and specificity of diagnostic tests can be expected.

The precipitin tests described above can all be used for antigen detection by simply reversing the procedure and testing the patient's specimen against a known antibody. Thus, if one well is filled with anti-meningococcal antiserum, meningococcal antigen can be detected in the patient's serum. Apart from meningococcal infection the technique is frequently used in the diagnosis of pneumococcal infections and systemic candidosis. The success of these diagnostic techniques is probably due to the fact that it is relatively easy to produce an antibody with broad specificity for the microbial antigens present in these infections. Indeed, in this situation one is looking for the broadest specificity possible, and the sensitivity is not seriously disturbed by this because high-titre anti-body can be prepared in experimental animals. It is notable that, in these tests, monoclonal antibodies which have a defined and narrow specificity have not to date been satisfactorily employed. Although precipitin tests are able to detect antigen in the examples indicated above, they may not be sufficiently sensitive for all infections. In legionnaires' disease, for example, antigen detection is usually carried out by the more sensitive ELISA.

It is perhaps worth stressing at this point that the demonstration of significant amounts of bacterial antigen in the patient's serum or urine specimen often has greater diagnostic significance than the demonstration of antibody. In meningococcal, pneumococcal and *Candida* infections, it is likely that many individuals will have serum antibodies because the organisms frequently exist as commensals and the antibody may simply reflect this. The finding of antigen in the serum, or in the urine, is unusual in commensal states and is invariably associated with infection. It is this clarity of diagnosis that makes antigen detection tests so appealing. A technique which has been developed for this, co-agglutination, is described in the following section.

For further information on traditional antibody tests, see Cruickshank *et al.* (1975), Weir (1978) and Wilson & Dick (1983).

(iv) Modern serological assays

1. Co-agglutination tests

Co-agglutination is a special form of passive agglutination. It depends on the ability of certain strains of *Staphylococcus aureus* to bind immuno-globulin molecules, particularly IgG (Figure 7.7). The binding of the IgG occurs in such a way that it does not interfere with the antigen-binding activity of the immunoglobulin, and the *S. aureus* can therefore be used as highly efficient carrier particles for the antibody.

The procedure for a co-agglutination test is to prepare an antibody against the microbial antigen of interest, attach it to *S. aureus* in suspension (see Chapter 6D(viii)) and react the latter with the patient's sample which might contain the microbial antigen. Antigen in the specimen combines with the antibody and the staphylococcal particles are aggluti-nated. Monoclonal antibodies, or mixtures of monoclonal antibodies, are frequently used, and systems are available for the detection of the antigens of pneumococci, meningococci, *Haemophilus* sp. and *Streptococcus pyo-genes*. These examples are in common use in diagnostic laboratories, but other experimental assays have been reported.

The co-agglutination test is usually more sensitive than other passive

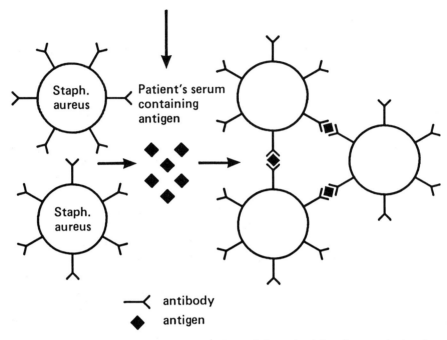

Figure 7.7   Diagrammatic representation of the principle of co-agglutination.

agglutination tests because the absorbed antibodies are efficiently used, in contrast to latex particles and erythrocytes, where much of the antibody is attached 'face-down' so that the antigen binding sites are not able to function. Although simple to perform and read, the co-agglutination test is difficult to quantitate; usually the only measurement is semi-quantitative, by determining the highest dilution of the patient's specimen which gives a positive test.

There is considerable scope for improvement of co-agglutination tests. If the relevant antigens of bacteria can be identified and separated, monoclonal antibodies of high specificity can be developed. This will probably improve the scope and the sensitivity of these tests. The problem here is of choosing the most appropriate antigens, because of the difficulties indicated earlier of variability in the response of the patient. For example, monoclonal antibodies have been produced against a large number of the surface antigens of *Legionella pneumophila*, but these antibodies do not combine with the antigens liberated from the organism *in vivo*. Clearly, in such instances more analytical work is required on the organisms to determine the appropriate targets for antibody production.

Finally, there is little scope for co-agglutination tests for antibody or immune complex measurement.

2. Enzyme-linked immunosorbent assays

As indicated earlier, ELISA is probably the procedure of the future as far as antigen and antibody measurement are concerned. There are many variations of ELISA available, but differences essentially involve the choice of solid phase, the type of enzyme and substrate used and the means of measuring the generated signal. Space does not allow these aspects to be discussed in detail.

The general principle is simple and is shown diagrammatically in Figure 6.7. In an assay of antibody, the microbial antigens of interest are attached to a solid-phase. The most popular solid phase is plastic, to which the antigen is non-covalently bound under appropriate conditions of pH and molarity. The plastic can be in the form of tubes, beads, microtitre wells or plates, or in specially formulated designs. After coating, the plastic is brought into contact with the patient's specimen (serum) and incubated to allow any antibody to attach to the antigen on the solid-phase. Excess serum proteins are removed by washing and an enzyme-labelled anti-immunoglobulin is added to the plastic. The enzyme is attached to the anti-immunoglobulin in a separate procedure and the labelled reagent is usually used in very low concentration. Many anti-immunoglobulins, labelled with different enzymes, are commercially available, although the labelling procedure itself is simple enough for most laboratories to perform themselves. If the patient's antibody has attached in the first stage of the test, the antiglobulin will bind to the antibody. Excess anti-immunoglobulin is removed by washing, and a substrate, appropriate for the enzyme used, is added to the solid phase. Enzyme on the solid phase can only be there because there is patient's antibody present, and the amount of enzyme bound is proportional to the amount of this antibody. When the enzyme comes into contact with the substrate, a change in colour of the latter will occur, the intensity of which will represent the amount of enzyme bound (and therefore the amount of anti-immunoglobulin, and in turn the amount of patient's antibody). Measurement of the colour change by spectrophotometry quantitates the amount of patient's antibody present. The labelled anti-immunoglobulin can have wide specificity (i.e. a polyvalent antiglobulin) which will detect all types of patient's antibody, or it can be specific to an individual immunoglobulin class. The test is simple to perform, rapid and economical, and prepared reagents have a reasonable storage time. However, the major advantage of ELISA is that it provides a format which can be used for almost any infection, thereby simplifying the procedures required in the laboratory. Furthermore, no dilution of the sample is required because quantitation is achieved by direct measurement rather than by establishing a 'titre', and the test relies only on the ability of the patient's antibody to bind to microbial antigen and not on other biological activities of the antibody. Freedom from the above constraints makes ELISA methods of antibody measurement very sensitive and specific.

The ELISA method can be 'reversed' for measurement of antigen. Thus an antibody, often monoclonal, prepared against a suitable microbial antigen is attached to the solid phase and after addition of a patient's sample any antigen in the latter is *captured* by the solid-phase antibody. A second antibody reactive with microbial antigen and labelled with enzyme is added in a second step and this will bind to antigen if it has previously bound to the solid phase. Addition of substrate produces a colour change which is a measure of the amount of antigen present. The two antibodies which react with the antigen should be different so that they will attach to different parts of the antigen molecule.

The attraction of ELISA for antigen capture assays is its sensitivity. In general terms, ELISA is about 40 times more sensitive than counter-immunoelectrophoresis, and 20 times more sensitive than co-agglutination. The limitation on the sensitivity of ELISA for antigen is usually the strength and specificity of the antibody, and it is advisable to identify the most appropriate microbial antigen or antigens and purify these before producing antibody.

The range of ELISA design for antibody assays is wide. Some investigators use whole bacteria, whilst others use lipopolysaccharide, polysaccharide or cell-membrane proteins as the solid-phase antigens. The argument in favour of whole bacteria is that, when the important antigens are not known, the use of the whole organism should ensure that no important ones are omitted. Whilst this is true, the sensitivity or measuring range can usually be improved by using a higher concentration of refined antigenic material. The methods of cell wall analysis described elsewhere in this book are relevant to this.

For further information on ELISA, see Voller *et al.* (1981) and Bergmeyer (1986).

3. *Radio-immunoassay*

Radio-immunoassay of antibody and antigen follows the same procedures as described for ELISA above. The sensitivity of radio-immunoassay is as good as, or better than, ELISA, and it is a simpler assay in theoretical terms. The only difference between the two is that radio-immunoassay uses a radio-isotope as the label in place of an enzyme. Isotopes are small molecules and therefore their attachment to an antibody does not cause problems of steric hindrance which can result with the attachment of large enzyme molecules. The isotope activity can be directly estimated without the need for substrate addition, and interfering substances in the patient's sample or the bacterial antigen do not occur, as they do sometimes for ELISA. Against this is the difficulty of handling and containing radio-isotopes.

Thus for the majority of assays ELISA is preferable, but for precise work, particularly in a research context, radio-immunoassay should be considered. The isotopes most commonly used are gamma emitters, such as ^{125}I and ^{131}I, which can be easily linked to antibodies by simple chemical procedures.

For further details of RIA in antibody tests, and clinical applications, see Parratt *et al.* (1982).

4. *Fluorescent antibody assays*

Fluorescent antibody tests have been used successfully in the diagnosis of many infections. The principle is the same as for ELISA and RIA, but the label attached to the detecting antibody is a dye which is activated by ultraviolet radiation to produce fluorescence. The dyes most commonly used are fluorescein derivatives and the resulting fluorescence is apple-green in colour.

The antigen is usually fixed to a microscope slide and the resulting fluorescence assessed by eye. This is tedious and difficult, requiring skilled observers and placing a severe limitation on the method for routine use. Photometers are available to measure the emitted fluorescence, but their performance is not very good and they do not make the test as efficient as ELISA or RIA for routine use.

Nonetheless, there are situations where the fluorescence method has advantages. An example is in the diagnosis of syphilis where it is difficult to obtain reasonable amounts of *Treponema pallidum* as the antigen. The fluorescence test is economical in the use of this reagent, although recently ELISA methods have been described which appear to be as efficient and much easier to use.

Fluorescence has also been used for antigen detection, and here it has had a particularly useful record as, for example, in the rapid diagnosis of respiratory infections by demonstration of the causative viruses or bacteria in secretions. The limitations noted above, however, still apply, and many investigators are adapting the ELISA principle to this type of test.

A new generation of fluorescence assays, using the principle of time-resolved immunofluorescence, is appearing in which the fluorescent label is a lanthanide chelate. These labels, which are readily attached to proteins, emit powerful but short-lived fluorescence. Sophisticated measuring equipment can accurately quantitate the emission over a period of about 50 microseconds. The results obtained with these assays are often better than for ELISA and are very rapid. The main disadvantage of the system is the considerable equipment expense. The same comments in respect of choice of antigen and antibodies in the assay apply as for ELISA. In time-resolved fluorescence, the measurement is faster and probably more accurate, but the basic design of the assay system is the same.

(v) Assays of immune complexes

It was indicated earlier that, for much of the course of an infection, immune complexes exist which are formed from antigen and the antibody produced against the antigen. There are many methods available for the measurement of immune complexes and it is beyond the scope of the current discussion to consider them all. What is required for the diagnosis of microbial disease are methods which can detect the antigen contained within immune complexes—so-called antigen-specific immune complex assays. The alternatives, of antigen non-specific immune complex assays, are useful for monitoring the progress of infections; but because they do not discriminate between complexes formed from many different stimuli, including cancer and autoimmune diseases, they have limited diagnostic application compared with antigen-specific assays.

Antigen-specific immune complex assays, however, are highly diagnostic. As can be seen from Figure 7.5, the only way in which antigen can be incorporated into an immune complex is if it is stimulating the formation of antibody, and the only circumstance in which this will occur is a genuine infection. To detect immune complexed antigen is therefore an ideal way of approaching diagnosis, particularly in the early stages of an infection. Unfortunately, methods for achieving this type of testing are at present primitive and require further investigation.

This is an area where careful analysis of bacterial cell wall antigens may produce benefit. Clearly, the antigens incorporated into complexes will tend to be small and soluble and the identification of such antigens and their characterization will do much to aid their detection in immune complexes. This approach, coupled with refinement of immune complex testing technology, should make antigen-specific immune complex assays important diagnostic tests in the future.

1. *Immune complex assays based on physical separation*

Because immune complexes are larger than uncombined antigen or uncombined antibody, they can be separated on the basis of their size. Ultracentrifugation is an example of a suitable method in which the complexes can be removed from serum. An easier approach is to use polyethylene glycol which will differentially precipitate complexes. Once the complexes have been separated they can be dissociated and suitable immunochemical techniques used to identify the antigens. In theory this sounds straightforward, and in research studies both these approaches have been successful. For regular diagnosis, however, both are laborious and insensitive and have not proved valuable.

2. Immune complex assays based on structural change of antibody

When antibody combines with antigen its structure is modified and parts of the molecule are exposed which are not normally accessible. Reagents which recognize the newly exposed receptors have the property of binding to immune complexes. Several reagents are known to do this, but the most commonly used example is rheumatoid factor, an IgM antibody which is present in the serum of many patients with auto-immune disease, and which reacts with altered IgG molecules. Rheumatoid factor can be either separated from the serum of such patients or prepared in animals, and can be attached to a solid phase as a 'capture' reagent for immune complexes. Once the complexes have been immobilized on the solid phase, the antigen component can be detected and measured by using a labelled antibody. The limitations of this type of assay are lack of sensitivity and cross-reactions with other autoantibodies. However, in its defence it has to be stated that the assay has been used for measuring immune complexes in autoimmune diseases where cross-reacting antibodies are common, and that in infections the same limitations might not apply.

3. Immune complex assays based on complement binding

When immune complexes form they often activate complement, a series of proteins present in serum which interact in a sequential fashion and produce numerous biologically active components. Some of the complement components are incorporated into the complexes and they can be used as receptors for binding complexes in immune complex assays.

Ciq binding assays. Ciq is the first component of the complement sequence, and it binds to receptors on antibody molecules which are exposed when the antibodies combine with antigen. If Ciq is separated from normal serum and attached to a solid phase, complexes which have not activated complement, but are capable of doing so, will bind to the solid phase. The presence of the complexes can then be established by adding a labelled anti-immunoglobulin, and in theory the use of labelled antiantigen antisera will recognize the antigenic component of the complex. In practice few investigators have reported the use of this type of test for antigen-specific immune complex tests.

Conglutinin assays. Conglutinin is a protein found in the serum of cattle which binds to a subunit of C3, the third component of complement. Immune complexes which have activated complement contain the C3 subunit and hence can be bound by conglutinin. If conglutinin is prepared and attached to a solid phase, it will separate immune complexes which can then be analysed either for their antibody or antigen constituents. Thus adding a labelled anti-immunoglobulin will identify the presence of antibody, whilst the antigen can be demonstrated by adding a labelled antibody to the antigen. This forms the basis of a useful and practical procedure which has been used to determine the nature of complexed antigens in several infections. In addition, solid-phased conglutinin can be used to separate immune complexes for purification of the components of the complex. Immune complexes can be readily eluted from conglutinin by washing with a chelating agent such as EDTA. The separated complexes can then be dissociated and the components analysed.

Antigen-specific immune complex assays depend largely on the ability of the detecting antibody to recognize the antigen in the complex.

Raji-cell assays. Raji cells are a lymphoblastoid cell line which can be maintained in tissue culture and which exhibit on their surface numerous receptors. One of these receptors is specific for the same C3 subunit that conglutinin uses, and therefore the Raji cell can be used as a substitute for conglutinin in immune complex assays. In practice, conglutinin is simpler to use and the performance of the assay is little different; but it is notable that the Raji-cell assay has been successfully employed as an antigen-specific immune complex assay in studies of hepatitis B.

Anti-C3d assays. The subunit of complement which binds to conglutinin and Raji cells is called C3d. Recently, assays have been described in which an antibody (usually monoclonal) has been developed to C3d and the antibody has been used, on a solid phase, to bind immune complexes. The antibody is a substitute for conglutinin or Raji cells, and the performance of the assay is the same as for the other assays. Antigen-specific immune complex assays are therefore possible with the anti-C3d tests.

4. Other immune complex assays

Many other techniques for the measurement of immune complexes have been described, most of which exploit the normal physiological responses to immune complexes, such as platelet aggregation, macrophage and neutrophil ingestion. Although useful as antigen non-specific assays, these other tests do not have potential as antigen-specific immune complex assays.

Further details of assay of immune complexes are described in Espinoza & Osterland (1983).

(vi) Conclusions

Diagnostic serology requires a sophisticated, precise approach if it is to be clinically useful. It is clear that the measurement of antibody needs to be sensitive and accurate in order to detect small amounts of antibody and small changes in the amount of antibody over short periods. Antigen assays must also be sensitive because the amount of free microbial antigen to be detected is small and may be difficult to differentiate from background 'noise' in any assay. Immune complex assays must be sensitive for the same reasons, and the added complication here is the need to isolate and then characterize the complexes.

For all of these assays the modern approach to immunoassay is the clear way forward, particularly using the ELISA principle; the 'traditional' assays will probably disappear within the next few years. The ELISA principle has many advantages, some of which have been discussed, but an important point can be made here, in this concluding section. It is that accurate serological diagnosis and an ability to understand the progress of an infection will increasingly depend on antigen measurement, antibody measurement and antigen-specific immune complex measurement coincidentally. ELISA is one of the few assay systems which can encompass all of these activities. The advantage to the diagnostician of having a single format for many assays, and of having the sensitivity, specificity and convenience, is obvious.

However, all assays are only as good as the components they incorporate. When assaying antibody, well-selected and well-characterized antigens are essential. In contrast, for antigen assays and antigen-specific immune complex assays, potent and well-defined antibodies are needed. For the study of bacterial infections, the production of all of these can be assisted by a clear knowledge of the structural components of the relevant organisms.

Appendix 1
General Methods

1. ESTIMATION OF PENTOSES AND HEXOSES

This method will quantitate hexoses, pentoses, uronic acids and deoxyaldoses and glycosides containing these sugars, including polysaccharides. Prior degradation of polysaccharides is *not* necessary. The method will not detect 2-deoxy 2-amino sugars. It follows the procedure of Dubois *et al.* (1956).

Reagents
A. Concentrated sulphuric acid (AR grade)
B. Phenol (AR grade—5% w/v aqueous solution)

Method
(a) Mix 2 ml of sample, containing between 10 and 100 μg of sugar, with 1 ml of 5% phenol in a 1.5 cm diameter heat-resistant tube.
(b) Rapidly add 5 ml of sulphuric acid and mix thoroughly (danger—the tube contents become very hot).
(c) Leave the mixture at room temperature for 10 minutes, then incubate at 30°C for 20 minutes. Measure the absorbance at 480 nm for pentoses, uronic acids and deoxyaldoses, or at 490 nm for hexoses, against a reagent blank. Construct calibration curves using authentic free sugars. The colours produced are stable for several hours.

2. ESTIMATION OF URONIC ACIDS

This procedure can be used to measure uronic acids in polysaccharides without prior degradation, as well as free sugars. It is not satisfactory for aminouronic acids (Blumenkrantz & Asboe-Hansen, 1973).

Reagents
A. 0.0125 M disodium tetraborate in concentrated sulphuric acid (analytical grade)

B. 0.15% w/v *m*-phenylphenol (*β*-hydroxydiphenyl) in 0.5% aqueous
sodium hydroxide

The reagents can be stored at 4°C in the dark for several weeks.

Method
(a) Vigorously mix aqueous samples (0.6 ml) containing up to 50 μg of
uronic acid with 3.6 ml of reagent A and incubate in a boiling-water
bath for 5 minutes. Cool in ice.
(b) Add 0.06 ml of reagent B and measure absorbance at 520 nm against a
reagent blank exactly 5 minutes after addition.

Use standards of glucuronic acid or glucuronolactone. 10 μg of glucuronic
acid give an absorbance of approximately 0.25.

3. DETECTION OF HEXOSAMINES IN GLYCANS USING MBTH

This method detects both free and glycosidically-linked hexosamines that
are not *N*-acetylated, and can be used to detect *N*-acetylated amino sugars
following de-*N*-acetylation with acid. Quantitation may depend on the
nature of the glycan and on the extent of de-*N*-acetylation achieved by
acid hydrolysis. The method has been used successfully to quantitate
chondroitin sulphate and cartilage glycosaminoglycans (Lane-Smith &
Gilkerson, 1979), peptidoglycan and hexosamine-containing teichoic acids
(unpublished).

Reagents
A. For de-*N*-acetylation, 0.5 M HCl
B. 2.5% w/v sodium nitrite
C. 12.5% w/v ammonium sulphamate
D. 0.25% w/v 3-methyl-2-benzothiazolone hydrazone hydrochloride
(MBTH)
E. Ferric chloride solution (0.83 g of $FeCl_3.6H_2O$ in 100 ml of water)

All the reagents are best prepared freshly, but MBTH solution can be
stored for approximately one week.

Method
(a) De-*N*-acetylation, if required: dissolve samples in 0.5 ml of 0.5 M HCl
(reagent A) in reaction vials with Teflon septa and heat at 110°C for 2 h
in an oven or block heater.
(b) Mix samples, containing up to 50 nmol hexosamine in 0.2 ml of 0.5 M
HCl, with 0.4 ml of sodium nitrite solution (B) and incubate until no
more gas is evolved (about 15 minutes).
(c) Add 0.2 ml of ammonium sulphamate solution (C) and incubate at
room temperature for 5 minutes.

(d) Add 0.2 ml of MBTH solution (D) and incubate at 37°C for 30 minutes.
(e) Add 0.2 ml of ferric chloride solution (E) and incubate at 37°C for 5 minutes.
(f) Measure absorbance at 650 nm immediately, avoiding vigorous agitation of the mixture during transfer to spectrophotometer cuvettes.

4. ESTIMATION OF HEXOSAMINES

Although quantitative estimation of sugars in acid hydrolysates of polysaccharides and related compounds is, in general, best done by GLC of their alditol acetates (see Chapter 5A), analysis of amino sugars by this technique can be difficult and a chemical assay procedure is valuable. The most widely used techniques make use of the Morgan–Elson reaction between N-acetylhexosamines and dimethylaminobenzaldehyde in borate buffer. Since acid hydrolysis leads to complete or partial de-N-acetylation of N-acetylamino sugars, an N-acetylation step precedes the assay. The sensitivity of the assay and its relative responses to different amino sugars depend critically on the assay conditions and have been the subject of many studies. The method described here is based on the work of Oguchi & Oguchi (1979).

Reagents
A. 1.5% by volume acetic anhydride in dry acetone (prepared fresh)
B. 0.7 M potassium tetraborate. Dissolve 21.4 g of potassium tetraborate in 75 ml of water and adjust the pH to between 9.1 and 9.3 with conc. HCl (3–4 ml). Make up to 100 ml with H_2O
C. 10% w/v dimethylaminobenzaldehyde (DMAB) (analytical grade) in glacial acetic acid containing 5% by volume conc. sulphuric acid
D. Glacial acetic acid

Method
(a) Acetylation: Mix samples (0.4 ml) of hexosamine hydrochloride solutions with 0.05 ml of acetic anhydride in acetone and incubate for 5 minutes at room temperature.
(b) Assay of N-acetylhexosamines: add to the reaction mixtures produced above 0.15 ml of 0.7 M tetraborate solution, cover the tubes and heat in a boiling-water bath for 3 minutes.
(c) Cool to room temperature and add 0.3 ml of DMAB reagent, followed by 2.7 ml of acetic acid. Mix well and incubate at 37°C for 20 minutes.
(d) Cool for 5 minutes, then measure the absorbance at 585 nm. The colour fades at approximately 5% per hour. 100 nmol N-acetylglucosamine gives an absorbance of about 0.5.

In addition to N-acetylhexosamines the method will detect N-acetylhexosamine-6-phosphates and N-acetylhexosaminuronic acids, though with lower sensitivity. Substitution of the sugars at C1 or C4 prevents detection.

5. GENERAL COLORIMETRIC ASSAY FOR PERIODATE-OXIDIZABLE POLYSACCHARIDES AND RELATED COMPOUNDS
(Mantle & Allen, 1978)

This is a valuable method for the determination of carbohydrate-containing material and has found particular application in monitoring fractions from column chromatography. Any compound susceptible to periodate oxidation will react, but the intensity of the response will, of course, depend on the number of periodate-oxidizable sites within the molecule. The response for glycogen, in which all sugar residues except those at branch-points are susceptible to periodate, is A_{555} 0.25 for 10 μg of polysaccharide.

Reagents
A. Periodic acid. Add 0.01 ml of periodic acid (50% solution) to 10 ml of 7% acetic acid, immediately before use
B. Schiff reagent. Dissolve 1 g of basic fuchsin (Eastman Kodak Co.) in 100 ml of boiling water, cool the solution to 50°C and add 20 ml of 1 M HCl. Treat the mixture twice, by mixing vigorously for 5 minutes and then filtering, with 300 mg of activated charcoal, and store the solution in the dark at room temperature. Immediately before use, this solution is decoloured by the addition of solid sodium metabisulphite to 1.67% w/v. Incubate the mixture at 37°C until the red colour changes to pale yellow or colourless (up to 2 h).

Procedure
(a) Mix 2 ml of aqueous sample, containing up to 100 μg of poly-saccharide, with 0.2 ml of periodic acid (reagent A), and incubate at 37°C for 2 h.
(b) Add 0.2 ml of reagent B, stand for 30 minutes at room temperature, then measure the absorbance at 555 nm, against a reagent blank.

The response is linear to an A_{555} value of at least 1.3

6. ESTIMATION OF 3-DEOXY-D-*MANNO*-OCTULOSONIC ACID (KDO) IN LIPOPOLYSACCHARIDE

Measurement of KDO is often used for the detection and quantitation of lipopolysaccharide in preparations of bacterial envelopes, but it must be emphasized that neither of the spectrophotometric methods available are specific for KDO (sialic acids and 2-deoxy sugars both react, and the semicarbazide method yields chromophores with many keto acids). Two procedures are described here, both of which depend on the prior release of KDO from the lipopolysaccharide by hydrolysis in dilute acid. NMR spectroscopy has shown that this process leads to degradation of the KDO and that some of the degradation products are reactive in the assay. Control of the hydrolysis conditions is therefore vitally important. This process and the assays themselves are fully discussed by Batley *et al.* (1985) and Brade *et al.* (1983). Carof *et al.* (1987) have shown that where KDO is phosphorylated at C4, as in lipopolysaccharides from *Bordetella pertussis* and *Vibrio cholerae*, dilute acid treatment leads to elimination of phosphate with the formation of an unsaturated derivative that gives no response in the thiobarbiturate reaction. In this case, prior removal of the phosphate substituent by treatment with HF (see Chapter 5D) permits subsequent assay of the KDO.

Release of KDO from LPS
Mix samples containing from 1 to 20 μg of KDO with 1 ml of 0.125 M sulphuric acid and heat in a boiling-water bath for 8 minutes. If insoluble material is present, remove it by centrifugation.

Reagents for periodic acid/thiobarbituric acid method
(based on Skoza & Mohos, 1976; Kharkanis *et al.*, 1978)
A. 25 mmol periodic acid in 62.5 mmol H_2SO_4
B. 2% w/v sodium arsenite in 0.5 M HCl
C. 0.6% w/v thiobarbituric acid adjusted to pH 9 with sodium hydroxide
D. Dimethylsulphoxide (analytical grade, stored frozen)

Method
(a) Mix 0.5 ml samples of the acid hydrolysate with 0.25 ml of periodic acid reagent and incubate at 37°C for 30 minutes.
(b) Cool, and add 0.25 ml of sodium arsenite solution, mixing until the brown colour of iodine disappears.
(c) Add 0.25 ml of thiobarbituric acid solution and heat at 100°C for 7.5 minutes. While the reaction mixture is still hot, add 1 ml of dimethyl-sulphoxide.
(d) Cool, and measure absorption against a reagent blank at 548 nm. 10 μg of KDO gives an A_{548} of about 0.75.

Since the method measures acid breakdown products of KDO it is essential that standard KDO samples are subjected to the same hydrolysis procedure as the LPS before construction of a calibration curve.

Reagents for semicarbazide method
(Batley *et al.*, 1985)
A. 1% w/v semicarbazide hydrochloride in 1.5% w/v aqueous sodium acetate trihydrate

B. 2.5 M NaOH
C. Chloroform

Method

(a) Neutralize 0.5 ml samples of LPS hydrolysate with 0.05 ml of 2.5 M NaOH (B). Add 0.5 ml of semicarbazide reagent (A) and heat at 60°C for 30 minutes. Cool and add 1.5 ml of water.

(b) Add 0.5 ml of chloroform and mix vigorously. Centrifuge to separate phases and retain the upper aqueous layer.

(c) Measure the absorbance at 250 nm against a reagent blank.

Note the comment about KDO standards in the previous method.

7. ESTIMATION OF ORGANIC PHOSPHORUS
(Chen *et al.*, 1956)

This procedure requires the prior digestion of organic material with a strong oxidizing agent to release inorganic phosphate. Two methods are given for digestion; the sulphuric/perchloric acid method is somewhat more consistent, but requires special fume-hood facilities because of the danger of accumulation of explosive metal perchlorates. Both methods involve vigorous heating which is best achieved for multiple samples through the use of an electrically heated Micro-Kjeldahl digestion rack or dry-block heater (capable of heating to at least 350°C) and infrared heating lamps.

Reagents

Digestion mixture: *either* (A) concentrated sulphuric acid (AR grade) and 60% perchloric acid, mixed 3:2 v/v, *or* (B) 10% w/v magnesium nitrate in absolute ethanol (Ames, B. N., 1966, *Meths. Enzymol.* **8**: 115–18).

Phosphate reagent: 2% w/v ascorbic acid in 0.6 M sulphuric acid/0.5% w/v ammonium molybdate (analytical grade) (C). The reagent is unstable and can be prepared from stock solutions of 3 M sulphuric acid, 2.5% molybdate and freshly prepared 3.3% w/v ascorbic acid in water. The reagent should be stored cold, and used within a few hours.

Method

(a) Digestion: Solutions or suspensions containing up to 10 μg of phosphorus are reduced to dryness in borosilicate glass tubes, by heating on a rack or block. An infrared heating lamp focused on the upper portion of the tubes is used to prevent condensation of steam. Add to the dry samples precisely 0.1 ml of digestion mixture A, heat to boiling on the same heater, without the infrared lamp, and reflux for

20 minutes. A clear solution should be obtained and no charred material should remain.

OR

Digestion: Samples, 0.1 ml in water, containing up to 10 μg of phosphorus, are mixed with 0.1 ml of magnesium nitrate reagent B in borosilicate glass tubes and heated until the samples are dry and no more brown nitrogen dioxide is evolved. Add 0.3 ml of 0.05 M HCl and heat to dryness again.

(b) Assay: Allow the digested samples to cool. Add 3.9 ml of water to each tube and mix thoroughly.

(c) Add 4 ml of phosphate reagent C, mix, and incubate at 37°C for 90 minutes.

(d) Measure absorbance at 820 nm, against a reagent blank. 1 μg of phosphorus gives an absorbance of approximately 0.1.

Standards should be prepared using a potassium orthophosphate solution of known concentration, and should be subjected to the same digestion procedure as the samples. Samples containing very large amounts of non-phosphate organic material may require digestion with more digestion mixture. All vessels and glassware used for reagents or assay should be thoroughly cleaned before use by boiling in acid or a phosphate-free detergent.

8. MEASUREMENT OF BOUND SULPHATE IN POLYSACCHARIDES
(Silvestri *et al.*, 1982)

Sulphate is not a common constituent of microbial polysaccharides, but has been found in a polysaccharide from *Arthrobacter*, in coat glycoprotein from halophilic bacteria, and in the form of taurine in some staphylococcal capsular polysaccharides. The following method is suitable for measurement of sulphate in partially purified samples, but it gives a significant response (up to 20 per cent of that with sulphate) with inorganic phosphate; phosphate-containing polymers also interfere to varying extents, depending on the extent of their degradation during the initial ashing procedure.

Reagents

A. Barium buffer. Mix 5 ml of 2 M acetic acid, 1 ml of 0.01 M barium chloride or barium acetate, 4 ml of 0.2 M sodium hydrogen carbonate, and adjust the volume to 150 ml with ethanol.

B. Rhodizonate. Dissolve 5 mg of potassium rhodizonate and 100 mg of ascorbic acid in 20 ml of water and adjust the volume to 100 ml with

ethanol. Leave at room temperature for 15 minutes, then store in a dark
bottle at $-20°C$.

Method

(a) Sample preparation: Mix samples of polysaccharide solution (up to
    3 ml, containing up to 200 nmol sulphate) with 0.02 ml of 0.1 M NaOH
    in 14 mm (i.d.) borosilicate glass tubes and evaporate the mixture to
    dryness as described above for phosphate analysis.
(b) Ash the dry samples by heating the dry tubes evenly to dull red heat in
    a gas flame, and immediately cooling, or alternatively by heating in a
    furnace at 600°C for 10–15 minutes.
(c) Assay: When the ashed samples are cool, redissolve in 0.5 ml of water
    on a vortex mixer, ensuring that all the surfaces of the tubes that have
    been in contact with the sample are washed by the liquid.
(d) Add 3 ml of barium buffer solution A, then 1.5 ml of rhodizonate
    reagent B. Mix, incubate for 10 minutes at room temperature and
    measure the absorbance at 520 nm against a reagent blank.

9. QUANTITATIVE MEASUREMENT OF PROTEIN

The dye-binding assay of Bradford (1976) as modified by Read & Northcote (1981) is convenient and suitable for measuring soluble protein in the
range 0.1 to 5 μg in 0.05 ml samples. There is no significant interference
from common biochemical reagents except from detergents at concentrations above about 1%.

Reagents

A. A stock dye solution, prepared by dissolving Serva Blue G (Serva
   GMBH, Heidelberg) 0.333% w/v in a mixture (2:1 by volume) of
   88% w/w (16 M) phosphoric acid and 93.5% v/v ethanol. This solution
   is stable indefinitely at room temperature in a dark bottle.
B. The working reagent, made when required by mixing 30 ml of stock
   dye solution, 80 ml of phosphoric acid (88%, 16 M) and 37.5 ml of
   absolute alcohol and adjusting the volume to 1 l with water. The
   solution is filtered through Whatman no. 1 filter paper before use and
   stored in a dark bottle. It is stable for a few weeks, but may need
   refiltering and calibration each time it is used.

Method

(a) 0.05 ml of sample solution containing between 0.1 and 5 μg of protein
    is mixed with 0.95 ml of reagent in a 1 ml plastic spectrophotometer
    cuvette.
(b) After 4 minutes the absorbance at 595 nm is measured against a
    reagent blank. A standard curve is constructed using ovalbumin.

Note that, like all protein assay techniques, the sensitivity of this method differs from protein to protein, though it is optimized for minimum variation. It cannot, therefore, be relied upon to give absolute quantities for a given protein or mixture of proteins, and it is essential that all data should be obtained using the same standard protein.

Because of interference by SDS, the method cannot be applied directly to samples dissolved in sample buffer for SDS-polyacrylamide gel electrophoresis. For this purpose, SDS must be removed from the sample first (Zaman & Verwilghen, 1979). Samples of 10–100 μg of protein in 0.02 ml of SDS-PAGE sample buffer are mixed with 0.05 ml of water and 0.45 ml of potassium phosphate buffer (100 mM in phosphate, pH 7.4–7.5) in a microfuge tube, then left standing at room temperature for 10 minutes. Precipitated SDS is removed by centrifugation in a microfuge for 10 minutes and the supernatant is assayed for protein as above.

10. POLYACRYLAMIDE GEL ELECTROPHORESIS

The buffer system developed by Laemmli (1970) is in widespread use with slab gels in a variety of commercially available apparatus.

Electrode buffer, pH 8.3
  0.025 M Tris
  0.192 M glycine
  0.1% sodium dodecylsulphate (SDS)
Weigh out 6.06 g of Tris (analytical grade) and 28.3 g of glycine (chromatographically homogeneous). Dissolve in about 1.5 l of distilled water and adjust the pH to 8.3 with 1 M NaOH. Add 2 g of SDS (specially pure for electrophoresis) and make up to 2 l with distilled water. Store in screw-capped bottles at room temperature.

Stacking gel buffer, pH 6.8 (double-strength)
  0.25 M Tris–HCl
  0.02% SDS
Weigh out 15.14 g of Tris (analytical grade) and dissolve in about 300 ml of distilled water. Adjust the pH to 6.8 with 1 M HCl. Add 1.0 g of SDS and make up to 500 ml with distilled water. Store as for electrode buffer.

Separating gel buffer, pH 8.8 (double-strength)
  0.75 M Tris–HCl, pH 8.8
  0.2% SDS
Weigh out 90.86 g of Tris (analytical grade) and dissolve in about 700 ml of distilled water. Adjust the pH to 8.8 with 1 M HCl. Add 2.0 g of SDS and make up to 1 l with distilled water. Store as for electrode buffer.

The stacking gel and separating gel buffers can be made up without SDS for certain applications e.g. immunoblotting LPS. These buffers must be stored at 4°C.

Sample buffer, pH 6.8 (double-strength)
    0.125 M Tris–HCl, pH 6.8
    4% SDS
    20% glycerol
    2% 2-mercaptoethanol
    0.002% bromophenol blue
Weigh out 1.51 mg of Tris, dissolve in about 50 ml of distilled water and adjust the pH to 6.8 with 1 M HCl. Add 4 g of SDS (specially pure). After dissolving, add the solution to a 100 ml volumetric flask. Add 20 ml of glycerol (analytical grade), 2 ml of 2-mercaptoethanol and 4 ml of a 0.05% aqueous solution of bromophenol blue (water-soluble). Make up to 100 ml with distilled water. Store in the dark at 4°C, but remember to bring to room temperature before use as the SDS precipitates in the cold.

Acrylamide stock solution (40%)
    100 g acrylamide (electrophoresis grade)
    2.7 g methylene bis acrylamide (electrophoresis grade)
Weigh out in separate containers and dissolve each separately in about 50–100 ml of water. Mix in a 250 ml volumetric flask, washing out the containers several times into the flask. Make up to 250 ml. Store at 4°C in the dark and label as toxic.

Preparation of polyacrylamide gels of various acrylamide concentrations
Mix reagents as in Table A.1.

Table A.1 Preparation of polyacrylamide gels.

Reagent	Volume (ml) to give acrylamide concentration of:			
	10%	12%	14%	4% (stacking gel)
Distilled water	6.95	5.2	3.45	3.5
Double-strength separating buffer	17.5	17.5	17.5	–
Double-strength stacking buffer	–	–	–	5.0
Acrylamide stock solution (40%)	8.75	10.5	12.25	1.0
TEMED	0.05	0.05	0.05	0.02
Ammonium persulphate (15 mg/ml)	1.75	1.75	1.75	0.5

11. STAINING OF POLYACRYLAMIDE GELS

Coomassie blue R was for many years the dye choice for staining proteins separated by gel electrophoresis. However, the progressive simplification of silver-staining procedures, originally based on histological stains, now means that these are little more complex or time-consuming than the Coomassie blue method. Moreover, the silver stain is approximately 100-fold more sensitive than the blue dye stain. A simple and reliable procedure for each type of stain is given here. For a detailed consideration of these and variant methods, see Ragan & Cherry (1986).

Coomassie blue R-250 staining procedure

Reagents
A. Fixer: 25% ethanol and 10% acetic acid by volume in water
B. Stain: 0.02% w/v Coomassie blue R in 25% ethanol, 10% acetic acid as above
C. Destainer: 10% methanol and 10% acetic acid by volume in water

Method
(a) Immerse the gel in fixer and shake gently at 37°C for at least 2 h.
(b) Transfer the gel to the stain and incubate for 90 minutes at 37°C.
(c) Pour off the stain and rinse the gel in destainer, then shake gently in destainer until the background is acceptably clear. A change of the destainer after about 2 h may be advantageous.

Alternative Coomassie blue stain

This sequential staining procedure takes slightly longer than the previous method but gives excellent results.

Reagents
Solution 1
  500 mg Coomassie brilliant blue R-250
  650 ml distilled water
  250 ml propan-2-ol
  100 ml acetic acid
Solution 2
   50 mg Coomassie blue
  800 ml distilled water
  100 ml propan-2-ol
  100 ml acetic acid

Solution 3
   25 mg Coomassie blue
   900 ml distilled water
   100 ml acetic acid
Solution 4
   500 ml distilled water
   400 ml methanol
   100 ml acetic acid
Solution 5
   900 ml distilled water
   100 ml acetic acid

Method
Place gel in solution 1 overnight, then take sequentially through 2–5 for 30–60 minutes at room temperature with gentle shaking throughout.

Silver stain (Thompson, S., unpublished)

Reagents
A. Dithiothreitol, 5 μg/ml
B. 0.1% w/v silver nitrate
C. Formaldehyde solution: 0.1 ml of 40% formaldehyde (analytical grade) in 200 ml
D. 3% w/v sodium carbonate
E. 2.3 M citric acid

Method
(a) For SDS-PAGE gels, fix the gel in ethanol–acetic acid fixer described above for the Coomassie blue method. For isoelectric focusing gels containing ampholytes, fix in 11.5% trichloroacetic acid/3.5% sulpho-sulphanilic acid for 1 h.
(b) Wash the gel in distilled water for at least 1 h.
(c) Soak in dithiothreitol solution (A) for 2 h.
(d) Pour off the dithiothreitol and add the silver nitrate solution (B) without rinsing. Soak for at least 2 h. Gels with small amounts of protein will require longer treatment.
(e) Rinse quickly with two changes of distilled water.
(f) Develop with formaldehyde (C) (0.1 ml) in sodium carbonate solution (D) (200 ml). Development may take 10 minutes or more.
(g) Stop the reaction by addition of 10 ml of citric acid (E), and soak for a further 30 minutes. The gels may subsequently be stored in distilled water.

Silver-stained gels can be successfully treated with standard photographic intensifiers, to increase the density of staining (proportional intensifiers) or increase contrast between the image and background (super-proportional intensifiers). These reagents also change the colour of the image to some extent and may be useful if the gels are to be photographed.

When stained gels are also to be subject to autoradiography it is valuable to maintain fixed dimensions during drying. This can be accomplished by casting the gel on a transparent plastic backing, such as 'Gel-Bond PAG' (FMC BioProducts), to which it adheres firmly during all subsequent procedures, including drying. If a hydrophilic cross-linker such as 'AcrylAide' (FMC BioProducts) is used in place of bis-acrylamide in preparation of the gel, and the gel is soaked in 1% aqueous glycerol before drying, drying can be accomplished in the air at room temperature or in a ventilated oven, without the application of a vacuum, and with minimal change in dimensions, when it has a plastic backing as described above. FMC BioProducts provide a valuable booklet of techniques using these two proprietary materials.

Silver stain for LPS
(Modified from Tsai & Frasch (1982), Oakley *et al*. (1980), and Hitchcock & Brown (1983))

Use absolutely clean containers and shake all steps.

(a) Fix in 200 ml of 25% propan-2-ol, 7% acetic acid overnight.
(b) Oxidize in freshly prepared periodic acid (1.05 g) in 150 ml of distilled water containing 4 ml of the above propan-2-ol fixative for 5 minutes.
(c) Wash in at least four changes of 200 ml of distilled water over 4 h.
(d) Drain, add *fresh* ammoniacal silver nitrate solution [1.4 ml of ammonia solution (SG 0.88) + 21 ml of 0.36% NaOH: with vigorous agitation add slowly 4 ml of 19.4% $AgNO_3$ solution (20 g $AgNO_3$ in 100 ml water); a brown precipitate may appear and after it has disappeared add distilled water to 100 ml]. Stain for 15 minutes. Rinse away silver solution, because it is explosive when dry.
(e) Remove the gel from the silver solution and wash it at least four changes of distilled water (200 ml) over 40 minutes.
(g) Transfer to fresh 0.005% citric acid in 200 ml of 0.019% formaldehyde (dilute 38% formaldehyde 1 in 2000) at 25°C.
(h) When the desired staining intensity is reached (5–15 minutes), wash repeatedly in large volumes of distilled water.

12. ELECTRON MICROSCOPY OF BACTERIA

Transmission electron microscopy (TEM)

As for all biological material, bacteria and their subcellular fractions must be treated with electron dense stains before they can be viewed in the electron microscope. Techniques vary considerably but they can be divided into four main types: (i) simple negative staining of whole cells; (ii) heavy metal shadowing; (iii) staining of thin sections; and (iv) affinity staining (immuno- or lectin-staining).

Simple negative staining

This technique involves mixing whole bacterial with a heavy-metal salt solution. When viewed in the TEM the only details that are visible are the shape and size of the organism, and any surface appendages (viz. flagella and fimbriae/pili). Very little fine structural detail can be discerned. When subcellular fractions (e.g. isolated wall components) are stained by this method, it is sometimes possible to see a degree of fine detail such as regular protein arrays.

(a) Harvest bacteria from broth or solid medium and wash once in phosphate buffered saline. Resuspend to a milky turbidity in 1% w/v ammonium acetate.
(b) Mix equal volumes of suspension and 2% v/v phosphotungstic acid (neutralized to pH 7 with sodium hydroxide) on the surface of a glass slide.
(c) Transfer a drop of the specimen to a Formvar-coated copper grid. Leave for 30 seconds and remove excess fluid by touching drop with filter paper.
(d) Place grid in a desiccator for several minutes before viewing.

An example of this technique can be seen in Fig. 3.1b.

Heavy-metal shadowing

In this technique a heavy metal, usually a mixture of gold and palladium, is vapourized in an evacuated chamber containing the specimen. Depending on the relative position of the specimen and vapourizing metal, a 'shadow' is cast of the specimen. The finely divided state of the metal gives a much more detailed result than simple negative staining. The technique can be applied to whole organisms to define surface appendages in some detail, and also to freeze-fractured preparations to give fine detail of intra-envelope structures.

Specialized equipment is required for this technique and the manufacturer's recommendations should be followed. Only the procedure for specimen preparation is given here.

(a) Suspend the sample in 1% ammonium acetate and transfer to a Formvar-coated copper grid by means of a platinum loop.
(b) Remove excess liquid with filter paper and place in a desiccator for at least 5 minutes before shadowing.

(c) Shadow with gold/palladium mixture according to the manufacturer's instructions.

Examples of shadowing are shown in Figs. 1.11 and 4.1.

Staining of thin sections

Preparation of thin sections of bacteria allows investigation of the internal structures of the cell, especially the details of the cell envelope. Briefly the technique consists of fixing the cells, preliminary staining, dehydration, embedding in resin, thin sectioning, mounting and finally one or more secondary staining procedures. Depending on the overall method and the skill of the electron microscopist, a great deal of detail can be discerned. The possible artifacts introduced by dehydration must be recognized, and if highly hydrated surface structures such as capsules are present an attempt must be made to preserve them. The following procedure includes a Ruthenium red step for capsule staining. It is recommended that this stain be included in all stains for thin sections of bacteria as the rigid layer (peptidoglycan) appears more clearly if the dye is present.

(a) Harvest bacteria from a plate or liquid culture and suspend in a mixture of 1 ml of Ruthenium red (1.5 mg/ml) in water, 1 ml of 3.6% w/v glutaraldehyde and 1 ml of 0.2 M cacodylate buffer, pH 6.5, and hold at 0°C for 1 h.
(b) Wash three times in 0.07 M cacodylate buffer.
(c) Resuspend in 1 ml of Ruthenium red (1.5 mg/ml), 1 ml of 4% w/v osmium tetroxide, and 1 ml of 0.2 M cacodylate buffer *in a fume cupboard* and incubate for 3 h at 27°C.
(d) Wash once in 0.07 M cacodylate buffer.
(e) Dehydrate in successive 10 minute steps in 25%, 50%, 75% and 90% ethanol, and in absolute ethanol for two periods of 1 h.
(f) Add sufficient propylene oxide to cover the pellet for two 10 minute periods, and then add propylene oxide/Epon Araldite (1 : 1, v/v), mixture.
(g) Embed the pellet in fresh Epon Araldite in disposable capsules and maintain at 45°C for 24 h and at 60°C for 48 h.
(h) Cut sections (40 nm, or up to 70 nm if capsules are to be seen) in an ultramicrotome and place on rhodium-coated copper grids.
(i) Stain with saturated uranyl acetate in 75% ethanol in the dark for 30 minutes.
(j) Wash once in 75% ethanol and once in distilled water.
(k) Stain for 2 minutes with Reynolds' lead citrate. [Mix 1.33 g of lead nitrate and 1.76 g of sodium citrate in 30 ml of distilled water in a 50 ml volumetric flask. Shake vigorously for 1 minute and then intermittently for 30 minutes to convert all the lead nitrate to citrate.

(a)

Figure A.1 Scanning electron micrographs of bacteria in the mouse colon: various types of bacteria associated with (a) the mucous blanket and (b) the microvilli. Bar marker: 2 μm.

(b)

Add 8.0 ml of 1 M NaOH and dilute to 50 ml with distilled water. The final pH is 12 ± 0.1 and it is stable for 6 months.]
(l) Wash once in 0.02 M NaOH and three times in water.
(m) Allow to dry in a desiccator before viewing in the TEM.

The method is modified from Springer & Roth (1973) by Poxton & Ip (1981).

Examples of micrographs of thin sections can be seen in Figs 1.1 and 1.12.

Immuno- and lectin-labelling for the TEM

It is possible to define the location of specific antigens or ligands on bacteria in the TEM by use of antibodies or lectins tagged with electron-dense materials. Thin sections are the usual starting material and are prepared in a manner similar to that described above. Full details and examples of the immunogold technique are given in Chapter 6.

Scanning electron microscopy (SEM)

With the advent of the SEM it became possible to view bacteria in their natural surroundings and be able to observe how they associate with other bacteria and with their substratum. For structural investigations of bacteria the SEM is, however, of limited use. Brief details will be given here of the preparation of specimens for the SEM and also some examples of its use. The method is essentially that of Murakami (1974).

(a) Place the specimen in 3% v/v glutaraldehyde in 0.1 M phosphate buffer, pH 7.2, for 1–3 days at room temperature.
(b) Transfer it to a solution containing 2% w/v arginine hydrochloride, 2% w/v glycine, 2% w/v sodium glutamate, 2% w/v sucrose in distilled water (pH 6.2) for 16 h.
(c) Wash the specimen in three 15 minute changes of distilled water and place in 2% w/v aqueous tannic acid (pH 4.0) for 24 h.
(d) Repeat washing, then place in 2% w/v aqueous osmium tetroxide for 6 h.
(e) Wash thoroughly in distilled water, and dehydrate in graded alcohols.
(f) Mount in sample holder.

The following steps require specialized equipment.

(g) Dry in a critical point drier.
(h) Sputter coat with a gold/palladium mixture.
(i) View in SEM.

Examples of scanning electron micrographs are shown in Figure A.1.

Appendix 2
Specialist Suppliers

Albright and Wilson, Whitehaven, Cumbria, UK—**Detergents: Empigen**

Amersham International plc, Lincoln Place, Green End, Aylesbury, Bucks, HP20 2TP, UK—**Radiochemicals, immunochemicals**

Amicon Corporation, 182 Conant Street, Danvers, MA 01923, USA (in UK: Amicon Ltd, Uppermill, Stonehouse, Glos, GL10 2BT)—**Ultrafiltration equipment, concentrators**

Aminco, Silver Springs, Maryland, USA—**French Press**

Bayer AG, D-5090, Leverkusen-Bayerwerk, West Germany—**70% Hydrofluoric acid (HF)**

Bio-Rad, 2200 Wright Avenue, Richmond, California 94804, USA (in UK: Bio-Rad Laboratories Ltd, Caxton Way, Watford Business Park, Watford, Herts, WD1 8RP)—**Immunochemicals, chromatography ('Biogel', 'Aminex') and electrophoresis reagents and equipment**

Biox AB, Box 143, S-175 23, Jarfalla, Sweden (in UK: Life Science Laboratories, Sarum Road, Luton, Beds, LU3 5RA)—**X-Press**

Braun Melsungen International GMBH, Schwartzerbergerweg 23, D-3508, Melsungen, West Germany—**Braun Cell Homogenizer**

Dako Ltd, 22, The Arcade, The Octagon, High Wycombe, Bucks, HP11 2HT, UK—**Immunochemicals**

Flow Laboratories Ltd, Woodcock Hill, Harefield Road, Rickmansworth, Herts WD3 1PQ, UK—**Titertek™ equipment, including Multiskan ELISA plate readers and microtitration systems**

FMC, Marine Colloids Division, Rockland, Maine 04841, USA—**Specialist agaroses and Gelbond™**

ICN Biomedicals Ltd (formerly Miles), Free Press House, Castle Street, High Wycombe, Bucks, HP13 6RN, UK—**Immunochemicals and Biochemicals, Gelbond™**

ISCO, PO Box 5347, Lincoln, NE 68505, USA (in UK: Life Science Laboratories Ltd, Sarum Road, Luton, Beds, LU3 5RA)—**Electrophoretic concentrator**

Janssen Pharmaceutica, Life Sciences Products Division, Turnhoutseweg 30, Beerse, Belgium (in UK: ICN Biomedicals Ltd, see above)—**Immunogold reagents and Aurodye™**

Jencons Ltd, Cherrycourt Way Industrial Estate, Stanbridge Road, Leighton Buzzard, Beds, LU7 8UA, UK—**Glass balotini beads for cell disruption**

London Resin Ltd, PO Box 29, Woking, Surrey, GU21 1AE, UK—**Immunogold reagents and electron microscope embedding resins**

Millipore Corporation, Bedford, MA 01730, USA (in UK: Millipore (UK) Ltd, 11–15 Peterborough Road, Harrow, Middlesex HA1 2YH)—**Pure water systems, filtration and ultrafiltration equipment**

Pharmacia, S-751 82 Uppsala, Sweden (in UK: Pharmacia House, Midsummer Boulevard, Central Milton Keynes, Bucks, MK9 3HP)—**Chromatography and electrophoresis equipment and chemicals ('Sephadex', 'Sepharose', 'Sephacryl')**

Phase Separations Ltd, Deeside Industrial Park, Queensferry, Clwyd CH5 2NU, UK—**Gas chromatography supplies**

Pierce Chemical Co., PO Box 117, Rockford, Illinois 61105, USA (in UK: Pierce and Warriner Ltd, 44 Upper Northgate St, Chester)—**Reactivials™ and materials for I^{125} labelling, HPLC and GC**

Sera-Lab Ltd, Crawley Down, Sussex, RH10 4FF, UK—**Immunochemicals**

Serva Fine Chemicals GMBH, Carl-Benz-Strasse 7, PO Box 105260, D-6900 Heidelberg, West Germany (in UK: Uniscience Ltd, 12–14 St Ann's Crescent, London, SW12 2LS)—**High purity dyes and chemicals for gel electrophoresis**

Schleicher & Schuell GMBH, D-3354 Dassel, West Germany (in UK: Anderman & Co. Ltd, Central Avenue, East Molesey, Surrey, KT8 0QZ)—**Nitrocellulose membranes, dot blotting apparatus**

Sigma Chemical Company, St Louis, MO 63178, USA (in UK: Fancy Road, Poole, Dorset BH17 7TH)—**Biochemicals and immunochemicals**

Wheaton glassware supplied by Aldrich Chemical Co., Old Brickyard, New Road, Gillingham, Dorset, SP8 4JL, UK—**Wheaton MicroV-Vials™**

References

Abdel-Akhar, M., Hamilton, J. K., Montgomery, R. and Smith, F. (1952) A new procedure for the determination of the fine structure of polysaccharides. *J. Am. Chem. Soc.* **74**: 4970–1.

Abe, M., Kimoto, M. and Yoshii, Z. (1983) Structure and chemical characterization of macromolecular arrays in the cell wall of *Bacillus brevis* S1. *FEMS Microbiol. Letts.* **18**: 263–7.

Abraham, S. N., Babu, J. P., Giampapa, C. S., Hasty, D. L., Simpson, W. A. and Beachey, E. H. (1985) Protection against *Escherichia coli*-induced urinary tract infections with hybridoma antibodies directed against type 1 fimbriae or complementary D-mannose receptors. *Infect. Immun.* **48**: 625–8.

Adam, A., Petit, J. F. and Wietzerbin-Falzspan, J. (1969) L'acide *N*-glycolyl-muramique, constituant des parois de *Mycobacterium smegmatis*: identification par spectrometrie de masse. *FEBS Letts.* **4**: 87–92.

Adlam, C., Knights, J. M., Mugridge, A., Lindon, J. C. and Williams, J. M. (1985) Purification, characterization and immunological properties of the serotype-specific capsular polysaccharide of *Pasteurella haemolytica* (serotype T4) organisms. *J. Gen. Microbiol.* **131**: 387–94.

Aksnes, L. and Grov, A. (1974) Studies on the bacteriolytic activity of *Streptomyces albus* culture filtrates. 3: Affinity for chitin. *Acta Path. Micr. Scand.* B **28**: 221–4.

Albright, F. R., White, D. A. and Lennartz, W. J. (1973) Studies on enzymes involved in the catabolism of phospholipids in *Escherichia coli. J. Biol. Chem.* **248**: 3968–77.

Alpenfels, W. F. (1981) A rapid and sensitive method for the determination of monosaccharides as their dansyl hydrazones by high performance liquid chromatography. *Analyt. Biochem.* **114**: 153–7.

Amano, K., Hazama, S., Araki, Y. and Ito, E. (1977) Isolation and characterization of structural components of *Bacillus cereus* AHU 1356 cell walls. *Eur. J. Biochem.* **75**, 513–22.

Ames, G. F., Prody, C. and Kustu, S. (1984) Simple, rapid and quantitative release of periplasmic proteins by chloroform. *J. Bacteriol.* **160**: 1181–3.

Anderson, A. J. and Archibald, A. R. (1975) Poly(glucosylglycerol phosphate) teichoic acid in the walls of *Bacillus stearothermophilus* B65. *Biochem. J.* **151**: 115–20.

Anderson, A. J. and Archibald, A. R. (1981) Duration of teichoic acid incorporation following transient release of phosphate limitation in chemostat culture of *Bacillus subtilis* W23. *FEMS Microbiol. Letts.* **10**: 7–10.

Anderson, A. J., Green, R. S. and Archibald, A. R. (1977) Specific determination of ribitol teichoic acid in whole bacteria and isolated walls of *Bacillus subtilis* W23. *Carbohydrate Res.* **57**: c7–c10.

Anderson, A. J., Green, R. S. and Archibald, A. R. (1978) Wall composition and phage-binding properties of *Bacillus subtilis* W23 grown in chemostat culture in media containing varied concentrations of phosphate. *FEMS Microbiol. Letts.* **4**: 129–32.

Anderson, J. C., Archibald, A. R., Curtis, M. J. and Davey, N. B. (1969) The action of dilute aqueous *N,N*-dimethylhydrazine on bacterial cell walls. *Biochem. J.* **113**: 183–9.

Araki, Y., Nakatani, T., Nakayama, K. and Ito, E. (1972) Occurrence of *N*-unsubstituted glucosamine residues in peptidoglycan of lysozyme-resistant cell walls from *Bacillus cereus*. *J. Biol. Chem.* **247**: 6312–22.

Archibald, A. R. (1976) Cell wall assembly in *Bacillus subtilis*: development of bacteriophage binding properties as a result of the pulsed incorporation of teichoic acid. *J. Bacteriol.* **127**: 956–60.

Archibald, A. R. (1985) Structure and assembly of the cell wall in *Bacillus subtilis*. *Biochem. Soc. Trans.* **13**: 990–2.

Archibald, A. R., Baddiley, J., Heckels, J. E. and Hpetinstall, S. (1971) Further studies on the glycerol teichoic acid of walls of *Staphylococcus lactis*. I3: Location of the phosphodiester groups and their susceptibility to hydrolysis with alkali. *Biochem. J.* **125**: 353–9.

Archibald, A. R., Baddiley, J. and Heptinstall, S. (1973) The alanine ester content and magnesium binding capacity of walls of *Staphylococcus aureus* H grown at different pH values. *Biochim. Biophys. Acta* **291**: 629–34.

Archibald, A. R. and Coapes, H. E. (1971) The interaction of concanavalin A with teichoic acids and bacterial walls. *Biochem. J.* **123**: 665–6.

Archibald, A. R. and Coapes, H. E. (1976) Bacteriophage SP50 as a marker for cell wall growth in *Bacillus subtilis*. *J. Bacteriol.* **125**: 1195–206.

Archibald, A. R., Coapes, H. E. and Stafford, G. H. (1969) The action of dilute alkali on bacterial cell walls. *Biochem. J.* **113**: 899–900.

Aronson, M., Medalia, O., Schori, L., Mirelman, D., Sharon, N. and Ofek, I. (1979) Prevention of colonisation of the urinary tract of mice with *Escherichia coli* by blocking of bacterial adherence with methyl-alpha-D-mannopyranoside. *J. Infect. Dis.* **139**: 329–32.

Avigad, G. (1969) A colorimetric method for the measurement of periodate. *Carbohydr. Res.* **11**: 119–24.

Axelson, N. H., Kroll, J. and Weeke, B. (1973) A manual of quantitative immuno-electrophoresis, methods and applications. *Scand. J. Immunol.* **2**, Suppl. 1.

Baddiley, J., Buchanan, J. G., Handschumacher, R. E. and Prescott, J. F. (1956) Chemical studies in the biosynthesis of purine nucleotides. I: The preparation of *N*-glycylglycosylamines. *J. Chem. Soc.*: 2818–23.

Bates, C. J. and Pasternak, C. A. (1965) Incorporation of labelled aminosugars by *Bacillus subtilis*. *Biochem. J.* **96**: 155–8.

Batley, M., McNichols, P. A. and Redmond, J. W. (1985) Analytical studies of lipopolysaccharide and its derivatives from *Salmonella minnesota* R595. III: Re-appraisal of established methods. *Biochim. Biophys. Acta* **821**: 205–16.

Baumeister, W., Karrenberg, F., Rachel, R., Engel, A., Ten Heggeler, B. and Saxton, W. O. (1982) The major cell envelope protein of *Micrococcus radiodurans* (R1): Structural and chemical characterisation. *Eur. J. Biochem.* **125**: 535–44.

Bax, A., Eagan, W. and Kovac, P. (1984) New n.m.r. techniques for structure determination and resonance assignments of complex carbohydrates. *J. Carbohydr. Chem.* **3**: 593–611.

Bax, A., Freeman, R. and Morris, G. (1981) Correlation of proton chemical shifts by two-dimensional Fourier transform n.m.r. *J. Mag. Reson.* **42**: 164–8.

Beacham, I. R. (1979) Periplasmic enzymes in Gram-negative bacteria. *Int. J. Biochem.* **10**: 877–83.

Bergmeyer, H. U. (1968) *Methods of Enzymatic Analysis*, 3rd edn. Vol. XI, *Antigens and Antibodies*, 2. VCH: Weinheim, West Germany.

Beveridge, T. J. (1979) Surface arrays on the wall of *Sporosarcina ureae. J. Bacteriol.* **139**: 1039–48.

Beveridge, T. J. (1981) Ultrastructure, chemistry and function of the bacterial wall. *Int. Rev. Cytol.* **72**: 229–317.

Beveridge, T. J. and Murray, R. G. E. (1976a) Superficial cell wall layers on *Spirillum* 'Ordal' and their *in vitro* reassembly. *Can. J. Microbiol.* **22**: 567–82.

Beveridge, T. J. and Murray, R. G. E. (1976b) Reassembly *in vitro* of the superficial cell wall components of *Spirillum putridiconchylium. J. Ultrastruct. Res.* **55**: 105–18.

Beveridge, T. J., Stewart, M., Doyle, R. J. and Sprott, G. D. (1985) Unusual stability of *Methanospirillum hungatei* sheath. *J. Bacteriol.* **162**: 728–37.

Blake, M. S. and Gotschlich, E. C. (1982) Purification and partial characterisation of the major outer membrane protein of *Neisseria gonorrhoeae. Infect. Immun.* **36**: 277–83.

Blakeney, A. B., Harris, P. J., Henry, R. J. and Stone, B. A. (1983) A simple and rapid preparation of alditol acetates for monosaccharide analysis. *Carbohyd. Res.* **113**: 281–9.

Bligh, E. G. and Dyer, W. J. (1959) A rapid method of total lipid extraction and purification. *Can. J. Biochem. Physiol.* **37**: 911–17.

Blumenkrantz, N. and Asboe-Hansen, G. (1973) New method for quantitative determination of uronic acids. *Analyt. Biochem.* **54**: 484–9.

Bock, K., Breimer, M. E., Brignole, A., Hansson, G. C., Karlsson, K.-A., Larson, G., Leffler, H., Samuelsson, B. E., Stromberg, N., Svanborg Eden, C. and Thurin, J. (1985) Specificity of binding of a strain of uropathogenic *Escherichia coli* to Gal alpha 1–4 Gal-containing glycosphingolipids. *J. Biol. Chem.* **260**: 8545–51.

Bock, K., Meyer, B. and Vignon, M. (1980) Some consequences of high frequency in n.m.r. spectrometers. *J. Mag. Res.* **38**, 545–51.

Bock, K. and Pedersen, C. (1974) A study of ^{13}C, H coupling constants in hexopyranoses. *J. Chem. Soc. Perkin Trans.* **II**: 293–7.

Bock, K. and Pedersen, C. (1983) Carbon-13 nuclear magnetic resonance spectoscopy of monosaccharides. *Adv. Carbohydr. Chem. Biochem.* **41**: 27–65.

Bock, K. and Pedersen, C. (1985) Determination of one-bond carbon–proton coupling constants through ^{13}C satellites in ^{1}H-n.m.r. spectra. *Carbohydr. Res.* **145**: 135–40.

Bock, K., Pederson, C. and Pedersen, H. (1984) Carbon-13 n.m.r. data for oligosaccharides. *Adv. Carbohydr. Chem. Biochem.* **42**: 193–225.

Bohin, J. P. and Kennedy, E. P. (1984) Mapping of a locus (*mdo A*) that affects the biosynthesis of membrane-derived oligosaccharides in *Escherichia coli. J. Bacteriol.* **157**: 956–7.

Bordet, C. and Perkins, H. R. (1970) Iodinated vancomycin and mucopeptide biosynthesis by cell-free preparations from *Micrococcus lysodeikticus. Biochem. J.* **119**: 877–83.

Bracha, R., Chang, M., Fiedler, F. and Glaser, L. (1978) Biosynthesis of teichoic acids. *Methods in Enzymol.* **50**: 387–402.

Brade, H., Galanos, C. and Luderitz, O. (1983) Differential determination of the 3-deoxy-D-mannooctulosonic acid residues in lipopolyssaccharides of *Salmonella minnesota* rough mutants. *Eur. J. Biochem.* **131**: 195–200.

Bradford, M. M. (1976) A rapid and sensitive method for quantitation of micro-gram quantities of protein utilising the principle of protein-dye binding. *Analyt. Biochem.* **72**: 248–54.

Braun, V. (1973) Molecular organisation of the rigid layer and the cell wall of *Escherichia coli. J. Infect. Diseases* **128**: S9–S16.

Braun, V. (1975) Covalent lipoprotein from the outer membrane of *Escherichia coli. Biochim. Biophys. Acta* **415**: 335–77.

Braun, V. and Bosch, V. (1973) Distribution of murein lipoprotein between the cytoplasmic and outer membranes of *Escherichia coli. FEBS Letts.* **34**: 307–10.

Brennan, P. J., Mayer, H., Aspinall, G. O. and Nam Shin, J. E. (1981) Structures of the glycopeptidolipid antigens from serovars in the *Mycobacterium avium/ Mycobacterium intracellulare/Mycobacterium scrofulaceum* serocomplex. *Eur. J. Biochem.* **115**: 7–15.

Brennan, P. J., Souhrada, M., Ullom, B., McClatchy, K. J. and Goren, M. B. (1978) Identification of atypical mycobacteria by thin-layer chromatography of their surface antigens. *J. Clin. Microbiol.* **8**: 374–9.

Brinton, C. C. (1965) The structure, function, synthesis and genetic control of bacterial pili and a molecular model for DNA and RNA transport in Gram-negative bacteria. *Trans. N.Y. Acad. Sci.* **27**: 1003–54.

Brinton, C. C., Bryan, J., Dillon, J.-A., Guernia, N., Jackobson, L. J., Labik, A., Lee, S., Levine, A., Lim, S., McMichael, J., Polen, S., Rogers, K., To, A. C.-C., and To, S. C.-M. (1978) Uses of pili in gonorrhoea control. In: *Immunobiology of Neiserria gonorrhoeae*, pp. 155–78. Eds. G. F. Brooks, E. C. Gotschlich, K. K. Holmes, W. D. Sawyer and F. E. Young. Washington, D.C.: American Society for Microbiology.

Brinton, C. C., Wood, S. H., Brown, L., Labik, A., Bryan, J. R., Lee, S. W., Polen, S. E., Tramont, E. C., Sadoff, J. and Zollinger, W. (1981) The development of a neisseria pilus vaccine for gonorrhoea and meningococcal meningitis. In: *Seminars in Infectious Diseases* IV, pp. 140–59. Eds. L. Weinstein and B. Fields. New York: Thieme Stratton.

Browder, H. P., Zygmunt, W. A., Young, J. R. and Tavormina, P. A. (1965) Lysostaphin: enzymatic mode of action. *Biochem. Biophys. Res. Commun.* **19**: 383–8.

Brown, A., Hormaeche, C. E., Demarco de Hormaeche, R., Winther, M., Dougan, G., Maskell, D. J. and Stocker, B. A. D. (1987) An attenuated *aro*A *Salmonella typhimurium* vaccine elicits humoral and cellular immunity to cloned β-galactosidase in mice. *J. Infect. Dis.* **155**: 86–91.

Buchanan, T. M. (1977) Surface antigens: pili. In: *The Gonococcus*, pp. 255–72. Ed. R. R. Roberts. New York: John Wiley.

Buchanan, T. M. (1978) Antigen-specific serotyping of *Neisseria gonorrhoeae*. 1: Use of an enzyme-linked immunosorbent assay to quantitate pili as antigens on gonococci. *J. Infect. Dis.* **138**: 319–25.

Buchanan, T. M., Swanson, J., Holmes, K. K., Kraus, S. J. and Gotschlich, E. C. (1973) Quantitative determination of antibody to gonococcal pili. Changes in antibody levels with gonococcal infection. *J. Clin. Investigations* **52**: 2896–909.

Buckmire, F. L. A. and Murray, R. G. E. (1970) Studies on the cell wall of *Spirillum serpens*. 1: Isolation and partial purification of the outermost cell wall layer. *Can. J. Microbiol.* **16**: 1011–22.

Buckmire, F. L. A. and Murray, R. G. E. (1973) Studies on the cell wall of *Spirillum serpens*. 2: Chemical characterisation of the outer structured layer. *Can. J. Microbiol.* **19**: 59–66.

Bundle, D. R., Gerken, M. and Perry, M. B. (1986) Two-dimensional nuclear magnetic resonance at 500 MHz: the structural elucidation of a *Salmonella* serogroup *N* polysaccharide antigen. *Can. J. Chem.* **64**: 255–64.

Bundle, D. R., Jennings, H. J. and Smith, I. C. P. (1973) The carbon-13 nuclear magnetic resonance spectra of 2-acetamino-2-deoxy-D-hexoses and some specifically deuterated, *O*-acetylated and phosphorylated derivatives. *Can. J. Chem.* **51**: 3812–19.

Bundle, D. R., Jennings, H. J. and Smith, I. C. P. (1974) Determination of the structure and conformation of bacterial polysaccharides by carbon-13 n.m.r.: studies on the group-specific antigens of *Neisseria meningitidis* serogroups A and X. *J. Biol. Chem.* **249**: 2275–81.

Bundle, D. R. and Lemieux, R. U. (1976) Determination of anomeric configuration by n.m.r. *Methods in Carbohydr. Chem.* **7**: 79–89.

Burman, L. G. and Park, J. T. (1983) Changes in the composition of *Escherichia coli* murein as it ages during exponential growth. *J. Bacteriol.* **155**: 447–53.

Burman, J. G. and Park, J. T. (1984) Molecular model for elongation of the murein sacculus of *Escherichia coli*. *Proc. Nat. Acad. Sci. USA* **81**: 1844–8.

Burnell, E., van Alphen, L., Verkleij, A., de Kruijff, B. and Lugtenberg, B. (1980) ^{31}P nuclear magnetic resonance and freeze-fracture electron microscopy studies on *Escherichia coli*. III: The outer membrane. *Biochim. Biophys. Acta* **597**: 518–32.

Button, D. K. (1985) Kinetics of nutrient-limited transport and microbial growth. *Microbiol. Revs.* **49**: 270–97.

Button, D. K., Choudry, M. K. and Hemmings, N. L. (1975) *Proc. Soc. Gen. Microbiol.* **2**: 45.

Carof, M., Lebbar, S. and Szabo, L. (1987) Detection of 3-deoxy-2-octulosonic acid in thiobarbiturate-negative endotoxins. *Carbohydrate Res.* **161**: c4–c7.

Chandler, H. M. and Gulaskharam, J. (1974) The protective antigen of a highly immunogenic strain of *Clostridium chauvoei* including an evaluation of its flagella as a protective antigen. *J. Gen. Microbiol.* **84**: 128–34.

Charbit, A., Boulain, J. C., Ryter, A. and Hofnung, M. (1986) Probing the topology of a bacterial membrane protein by genetic insertion of a foreign epitope; expression at the cell surface. *EMBO Journal* **11**: 3029–37.

Chart, H., Buck, M., Stevenson, P. and Griffiths, E. (1986) Iron-regulated outer membrane proteins of *Escherichia coli*. *J. Gen. Microbiol.* **132**: 1373–8.

Chea, S.-C., Hussey, H. and Baddiley, J. (1981) Control of synthesis of wall teichoic acid in phosphate-starved cultures of *Bacillus subtilis* W23. *Eur. J. Biochem.* **118**: 495–500.

Chea, S.-C., Hussey, H., Hancock, I. C. and Baddiley, J. (1982) Control of synthesis of wall teichoic acid during balanced growth of *Bacillus subtilis* W23. *J. Gen. Microbiol.* **128**: 593–9.

Chen, P. S., Toribara, T. Y. and Warner, H. (1956) Microdetermination of phosphorus. *Analyt. Chem.* **28**: 1756–61.

Chiu, T. H., Emdur, L. I. and Platt, D. (1974) Lipoteichoic acids from *Streptococcus sanguis*. *J. Bacteriol.* **118**: 471–80.

Christie, W. W. (1982) *Lipid Analysis*. Pergamon Press, Oxford.

Clarke-Sturman, A. J. and Archibald, A. R. (1982) Cell wall turnover in phosphate and potassium limited chemostat cultures of *Bacillus subtilis* W23. *Arch. Microbiol.* **131**: 375–9.

Clegg, J. C. S. (1982) Glycoprotein detection in nitrocellulose transfers of electrophoretically separated protein mixtures using concanavalin A and peroxidase. *Analyt. Biochem.* **127**: 389–94.

Coley, J., Archibald, A. R. and Baddiley, J. (1977) The presence of N-acetylglucosamine 1-phosphate in the linkage unit that connects teichoic acid to peptidoglycan in *Staphylococcus aureus*. *FEBS Letts.* **80**: 405–7.

Coley, J., Duckworth, M. and Baddiley, J. (1975) Extraction and purification of lipoteichoic acids from Gram-positive bacteria. *Carbohydrate Res.* **40**: 41–52.

Cook, W. R., MacAlister, T. J. and Rothfield, L. I. (1986) Comparmentalisation of the periplasmic space at division sites in Gram-negative bacteria. *J. Bacteriol.* **168**: 1430–8.

Costerton, J. W., Irvin, R. T. and Chen, K. J. (1979) The bacterial glycocalix in nature and disease. *Ann. Rev. Microbiol.* **35**: 299–324.

Cruickshank, R., Dougan, J. P., Marmion, B. P. and Swain, R. H. A. (1975) *Medical Microbiology*, Vol. 2, 12th edn. Churchill Livingstone: Edinburgh.

Dabrowski, J., Hanfland, P., Egge, H. and Dabrowski, U. (1981) Immunochemistry of the Lewis-blood-group system: proton nuclear magnetic resonance study of plasmatic Lewis-blood-group-active glycosphingolipids and related substances. *Arch. Biochem. Biophys.* **210**: 405–11.

Daffe, M., Lanéelle, M. A., Asselineau, C., Levt-Frebault, V. and David, H. (1983) Interet taxonomique des acides gras des mycobacteries: proposition d'une methode d'analyse. *Ann. Microbiol.* **134B**: 241–56.

Darveau, R. P., MacIntyre, S., Buckley, J. T. and Hancock, R. E. W. (1983) Purification and reconstitution in lipid bilayer membranes of an outer membrane, pore-forming protein of *Aeromonas salmonicida*. *J. Bacteriol.* **156**: 1006–11.

Davidson, I. W. (1978) Production of polysaccharide by *Xanthomonas campestris* in continuous culture. *FEMS Microbiol. Letts.* **3**: 347–9.

Davidson, L. A., Draper, P. and Minnikin, D. E. (1982) Studies on the mycolic acids from the walls of *Mycobacterium microti*. *J. Gen. Microbiol.* **128**: 823–8.

Deal, C. D., Tainer, J. A., So, M. and Getzoff, E. D. (1985) Identification of a common structural class for *Neisseria gonorrhoeae* and other bacterial pilins. In: *The Pathogenic Neisseria*, pp. 302–8. Ed. G. K. Schoolnik. Washington, D.C.: American Society for Microbiology.

de Boer, W. R., Kruyssen, F. J. and Wouters, J. T. M. (1981) Cell wall turnover in batch and chemostat cultures of *Bacillus subtilis*. *J. Bacteriol.* **145**: 50–60.

de Boer, W. R., Wouters, J. T. M., Anderson, A. J. and Archibald, A. R. (1978) Further evidence for the structure of the teichoic acids from *Bacillus subtilis* var. niger. *Eur. J. Biochem.* **85**: 433–6.

DeCueninck, B. J., Shockman, G. D. and Swenson, R. M. (1982) Group B, Type III streptococcal cell wall: composition and structural aspects revealed through endo-N-acetylmuramidase-catalysed hydrolysis. *Infect. Immun.* **35**: 572–82.

Dobson, B. C. and Archibald, A. R. (1978) Effect of specific growth limitations on cell wall composition of *Staphylococcus aureus*. *Arch. Microbiol.* **119**: 295–301.

Dobson, G., Minnikin, D. E., Minnikin, S. M., Parlett, J. H., Goodfellow, M., Ridell, M. and Magnusson, M. (1985) Systematic analysis of complex mycobacterial lipids. In: *Chemical Methods in Bacterial Systematics*, pp. 237–65. Eds. M. Goodfellow and D. E. Minnikin. Academic Press, London.

Doddrell, D. M., Pegg, D. T. and Bendall, M. R. (1982) Distortionless enhancement of n.m.r. signals by polarisation transfer. *J. Mag. Reson.* **48**: 323–7.

Domingue, G. J., Roberts, J. A., Laucirica, R., Ratner, M. H., Bell, D. P., Suarez, G. M., Kallenius, G. and Svenson, S. (1985) Pathogenic significance of P-fimbriated *Escherichia coli* in urinary tract infections. *J. Urol.* **133**: 983–8.

Dougan, G., Hormaeche, C. E. and Maskell, D. J. (1987) Live oral salmonella vaccines: potential use of attenuated strains as carriers of heterologous antigens to the immune system. *Parasite Immunol.* **9**: in press.

Doyle, R. J., Streips, U. N., Fan, V. S. C., Brown, W. C., Mobley, H. and Mansfield, J. M. (1977) Cell wall protein in *Bacillus subtilis*. *J. Bacteriol.* **129**: 547–9.

Draper, P., Kandler, O. and Darbre, A. (1987) Peptidoglycan and arabinogalactan of *Mycobacterium leprae*. *J. Gen. Microbiol.* **133**: 1187–94.

Driehuis, F. and Wouters, J. T. M. (1987) Effect of growth rate and cell shape on the peptidoglycan composition of *Escherichia coli*. *J. Bacteriol.* **169**: 97–101.

Drucker, D. B. (1981) *Microbiological Applications of Gas Chromatography*. Cambridge University Press: Cambridge, London and New York.

Dubois, M., Gilles, K. A., Hamilton, J. K. and Rebers, P. A. (1956) Colorimetric method for the determination of sugars and related substances. *Analyt. Chem.* **28**: 350–6.

Duckworth, M., Archibald, A. R. and Baddiley, J. (1972) The location of N-acetylgalactosamine in the walls of *Bacillus subtilis* 168. *Biochem. J.* **130**: 691–6.

Duckworth, M., Jackson, D., Zak, K. and Heckels, J. E. (1983) Structural variations in pili expressed during gonococcal infections. *J. Gen. Microbiol.* **129**: 1593–6.

Duguid, J. P. (1951) The demonstration of bacterial capsules and slime. *J. Path. Bacteriol.* **63**: 673–85.

Duguid, J. P. (1985) Antigens of type-1 fimbriae. In: *Immunology of the Bacterial Cell Envelope*, pp. 301–18. Eds. D. E. S. Stewart-Tull and M. Davies. Chichester: John Wiley.

Duguid, J. P., Clegg, S. and Wilson, M. I. (1979) The fimbrial and non-fimbrial haemagglutinins of *Escherichia coli*. *J. Med. Microbiol.* **12**: 213–27.

Duguid, J. P. and Old, D. C. (1980) Adhesive properties of Enterobacteriaceae. In: *Bacterial Adherence, Receptors and Recognition*, Series B, Vol. 6, pp. 185–216. Ed. E. H. Beachey. Chapman & Hall, London.

Duguid, J. P., Smith, I. W., Dempster, G. and Edmunds, P. N. (1955) Non-flagellar filamentous appendages (fimbriae) and haemagglutinating activity in *Bacterium coli*. *J. Path. Bacteriol.* **70**: 335–48.

Duguid, J. P. and Wilkinson, J. F. (1961) Environmentally induced changes in bacterial morphology. In: *Microbial Reaction to Environment: Symposium 11 of the Society for General Microbiology*, pp. 69–99. Eds. G. G. Meynell and H. Gooder. Cambridge University Press.

Egan, W., Schneerson, R., Werner, K. E. and Zon, G. (1982) Structural studies and chemistry of bacterial capsular polysaccharides: Investigation of phospho-diester-linked capsular polysaccharides isolated from *Haemophilus influenzae* types a, b, c and f. *J. Am. Chem. Soc.* **104**: 2898–910.

Ellwood, D. C. and Tempest, D. W. (1969) Control of teichoic acid and teichuronic acid biosynthesis in chemostat cultures of *Bacillus subtilis* var. niger. *Biochem. J.* **111**: 1–5.

Ellwood, D. C. and Tempest, D. W. (1972) Effects of environment on bacterial wall content and composition. *Adv. Microbial Physiol.* **7**: 83–116.

Emdur, L. I., Saralkar, C., McHugh, J. G. and Chiu, T. H. (1974) Glycerolphosphate-containing cell wall polysaccharides from *Streptococcus sanguis*. *J. Bacteriol.* **120**: 724–7.

Endl, J., Seidl, P. H., Fiedler, F. and Schleifer, K. H. (1984) Determination of cell wall teichoic acid structure of staphylococci by rapid chemical and serological screening methods. *Arch. Microbiol.* **137**: 272–80.

Engvall, E. and Perlmann, P. (1972) Enzyme-linked immunosorbent assay. III: Quantitation of specific antibodies by enzyme-labelled anti-immunoglobulin in antigen coated tubes. *J. Immunol.* **109**: 129–35.

Ensign, J. C. and Wolfe, R. S. (1965) Lysis of bacterial cell walls by an enzyme isolated from a Myxobacter. *J. Bacteriol.* **90**: 395–402.

Eshdat, Y., Silverblatt, F. J. and Sharon, N. (1981) Dissociation and reassembly of *Escherichia coli* type 1 pili. *J. Bacteriol.* **148**: 308–14.

Espersen, F. and Clemmensen, I. (1982) Isolation of a fibronectin-binding protein from *Staphylococcus aureus*. *Infect. Immun.* **37**: 526–31.

Espersen, F., Clemmensen, I. and Barkholt, Y. (1985) Isolation of *Staphylococcus aureus* clumping factor. *Infect. Immun.* **49**: 700–8.

Espinosa, L. R. and Osterland, C. K. (1983) *Circulating Immune Complexes: their Clinical Significance*. Futura Press: New York.

Evenberg, D. and Lugtenberg, B. (1982) Cell surface of the fish pathogenic bacterium *Aeromonas salmonicida*. II: Purification and characterisation of a major envelope protein related to autoagglutination, adhesion and virulence. *Biochim. Biophys. Acta* **648**: 249–54.

Evers, D., Weckesser, J. and Jurgens, U. J. (1986) Cell envelope polymers of the halophilic, phototrophic *Rhodospirillum salexigens*. *Arch. Microbiol.* **145**: 254–8.

Fader, R. C. and Davies, C. P. (1982) *Klebsiella pneumoniae*-induced experimental pyelitis: the effect of piliation on infectivity. *J. Urol.* **128**: 197–201.

Fairweather, N. F., Lyness, V. A., Pickard, D. J., Allen, G. and Thomson, R. O. (1986) Cloning, nucleotide sequencing and expression of tetanus toxin C fragment in *Escherichia coli*. *J. Bacteriol.* **165**: 21–7.

Farrah, S. R. and Unz, R. F. (1976) Isolation of exocellular polymer from *Zoogloea* strains MP6 and 106 and from activated sludge. *Appl. Envir. Microbiol.* **32**: 33–7.

Faye, L. and Chrispeels, M. J. (1985) Characterisation of N-linked oligosaccharides by affinoblotting with concanavalin A-peroxidase and treatment of blots with glycosidases. *Analyt. Biochem.* **149**: 218–24.

Filip, C., Fletcher, G., Wulff, J. L. and Earhart, C. F. (1973) Solubilisation of the cytoplasmic membrane of *Escherichia coli* by the ionic detergent sodium lauryl sarcosinate. *J. Bacteriol.* **115**: 717–22.

Fiedler, F. and Glaser, L. (1974) The synthesis of polyribitol phosphate. I: Purification of polyribitol phosphate polymerase and lipoteichoic acid carrier. *J. Biol. Chem.* **249**: 2684–9.

Firon, N., Ofek, I. and Sharon, N. (1983) Carbohydrate specificity of the surface lectins of *Escherichia coli*, *Klebsiella pneumoniae* and *Salmonella typhimurium*. *Carbohydr. Res.* **120**: 235–49.

Firon, N., Ofek, I. and Sharon, N. (1984) Carbohydrate-binding sites of the mannose-specific fimbrial lectins of enterobacteria. *Infect. Immun.* **43**: 1088–90.

Fischer, H. and Tomasz, A. (1984) Production and release of peptidoglycan and wall teichoic acid polymers in pneumococci treated with β-lactam antibiotics. *J. Bacteriol.* **157**: 507–13.

Fischer, W., Koch, H. U., Rösel, P. and Fiedler, F. (1980) Alanine ester-containing native lipoteichoic acids do not act as lipoteichoic acid carrier. Isolation, structural and functional characterisation. *J. Biol. Chem.* **255**: 4557–62.

Fischer, W., Schmidt, M. A., Jann, B. and Jann, K. (1982) Structure of the *Escherichia coli* K2 capsular antigen. *Biochemistry* **21**: 1279–84.

Fischetti, V. A., Gotschlich, E. C. and Bernheimer, A. W. (1971) Purification and physical properties in Group C streptococcal phage-associated lysin. *J. Exp. Med.*: 1105–17.

Fletcher, J. N., Zak, K., Virji, M. and Heckels, J. E. (1986) Monoclonal antibodies to gonococcal outer membrane protein I: location of a conserved epitope on protein IB. *J. Gen. Microbiol.* **132**: 1611–20.

Forsgren, A. (1969) Protein A from *Staphylococcus aureus*. *Acta. Path. Mic. Scand.* **75**: 481–90.

Fournier, J. M., Vann, W. F. and Karakawa, W. W. (1984) Purification and characterisation of *Staphylococcus aureus* Type 8 capsular polysaccharide. *Infect. Immun.* **45**: 87–93.

Froholm, L. D. and Sletten, K. (1977) Purification and *N*-terminal sequence of a fimbrial protein from *Moraxella nonliquefaciens*. *FEBS Letts.* **73**: 29–32.

Gaastra, W. and de Graaf, F. K. (1982) Host-specific fimbrial adhesins of non-invasive enterotoxigenic *Escherichia coli* strains. *Microbiol. Rev.* **46**: 129–61.

Gabay, J. E., Blake, M., Niles, W. D. and Horwitz, M. A. (1985) Purification of *Legionella pneumophila* major outer membrane protein and demonstration that it is a porin. *J. Bacteriol.* **162**: 85–91.

Galanos, C. and Luderitz, O. (1975) Electrodialysis of lipopolysaccharides and their conversion to uniform salt forms. *Eur. J. Biochem.* **54**: 603–10.

Galanos, C., Luderitz, O. and Westphal, O. (1969) A new method for the extraction of R lipopolysaccharides. *Eur. J. Biochem.* **9**: 245–9.

Garegg, P. J., Lindberg, B. and Kvarnstrom, I. (1979) Preparation and n.m.r. studies or pyruvic acid and related acetals of pyranosides: configuration at the acetal carbon atoms. *Carbohydr. Res.* **77**: 71–8.

Germanier, R. and Furer, E. (1975) Isolation and characterisation of *Gal*E mutant Ty21a of *Salmonella typhi*: a candidate strain for a live oral typhoid vaccine. *J. Infect. Dis.* **131**: 553–9.

Gerwig, G. J., Kammerling, J. P. and Vliegenthart, J. F. G. (1978) Determination of the D- and L-configuration of neutral monosaccharides by high resolution capillary G.L.C. *Carbohydr. Res.* **62**: 349–57.

Ghuysen, J.-M. (1968) Use of bacteriolytic enzymes in determination of wall structure and their role in cell metabolism. *Bact. Revs.* **32**: 425–64.

Ghuysen, J.-M., Dierickx, L., Coyette, J., Leyh-Bouille, M., Guinand, M. and Campbell, J. N. (1969) An improved technique for the preparation of *Streptomyces* peptidases and *N*-acetylmuramyl-L-alanine amidase active on bacterial wall peptidoglycans. *Biochemistry* **8**: 213–22.

Giovannoni, S. J., Godchaux, W., Schabtach, E. and Castenholtz, R. W. (1987) Cell wall and lipid composition of *Isosphera pallida*, a budding eubacterium from hot springs. *J. Bacteriol.* **169**: 2702–7.

Givan, A. L., Glassey, K., Lang, W. K., Anderson, A. J. and Archibald, A. R. (1982) Relation between wall teichoic acid content of *Bacillus subtilis* and efficiency of absorption of bacteriophage SP 50 and φ25. *Arch. Microbiol.* **133**: 318–22.

Glauert, A. M. and Thornley, M. J. (1969) The topography of the bacterial cell wall. *Ann. Rev. Microbiol.* **23**: 159–98.

Glauner, B. and Schwartz, U. (1983) The analysis of murein composition with high pressure liquid chromatography. In: *The Target of Penicillin*, pp. 29–34. Eds. R. Hackenbeck, J.-V. Höltje and H. Labischinski. Berlin: de Gruyter.

Goldman, R. C., White, D., Orskov, F., Orskov, I., Rick, P. D., Bhattacharjee, A. K. and Leive, L. (1982) A surface polysaccharide of *Escherichia coli* 0111 contains *O*-antigen and inhibits agglutination of cells by *O*-antiserum. *J. Bacteriol.* **151**: 1210–21.

Goodell, E. W. and Schwartz, U. (1983) Cleavage and resynthesis of peptide cross bridges in *Escherichia coli* murein. *J. Bacteriol.* **156**: 136–40.

Goren, M. B. and Brennan, P. J. (1979) Mycobacterial lipids: chemistry and biological activities. In: *Tuberculosis*, pp. 63–193. Ed. G. P. Youmans. Philadelphia: Saunders.

Gorin, P. A. J. (1980) ^{13}C-n.m.r. spectroscopy of polysaccharides. *Adv. Carbohydr. Chem. Biochem.* **38**: 13–104.

Gotschlich, E. C. (1984) Meningococcal meningitis. In: *Bacterial Vaccines*, pp. 237–55. Ed. R. Germanier. London: Academic Press.

Gotschlich, E. C., Fraser, B. A., Nishimura, O., Robbins, J. B. and Liu, T. Y. (1981) Lipid on capsular polysaccharides of Gram-negative bacteria. *J. Biol. Chem.* **256**: 8915–21.

Goundry, J., Davison, A. L., Archibald, A. R. and Baddiley, J. (1967) The structure of the cell wall of *Bacillus polymyxa* NCIB 4747. *Biochem. J.* **104**: 1c–2c.

Grov, A., Flandrois, J. P., Fleurette, J. and Oeding, P. (1978) Immunochemical studies on the specific agglutinogens of *Staphylococcus aureus*. *Acta Path. Mic. Scand.* B **86**: 143–7.

Grov, A. and Rude, S. (1967) Immunochemical examination of phenylhydrazine-treated *Staphylococcus aureus* cell walls. *Acta Path. Mic. Scand.* B **71**: 417–21.

Gunner, S. W., Jones, J. K. N. and Perry, M. B. (1961) The gas–liquid partition chromatography of carbohydrate derivatives. 1: The separation of glycitol and glycose acetates. *Can. J. Chem.* **39**: 1892–5.

Guss, B., Uhlen, M., Nilsson, B., Lindberg, M., Sjoquist, J. and Sjodahl, J. (1984) Region X, the cell wall attachment part of staphylococcal protein A. *Eur. J. Biochem.* **143**: 413–20; and correction (1984) *Eur. J. Biochem.* **143**: 685.

Haas, R. and Meyer, T. F. (1986) The repertoire of silent pilus genes in *Neisseria gonorrhoeae*: evidence for gene conversion. *Cell* **44**: 107–15.

Hagberg, L., Enberg, I., Freter, R., Lam, J., Olling, S. and Svanborg Eden, C. (1983) Ascending, unobstructed urinary tract infection in mice caused by pyelonephritogenic *Escherichia coli* of human origin. *Infect. Immun.* **40**: 273–83.

Hagblom, P., Segal, E., Billyard, E. and So, M. (1985) Intragenic recombination leads to pilus antigenic variation in *Neisseria gonorrhoeae*. *Nature* **315**: 156–8.

Hackenbeck, R., Höltje, J.-Y. and Labischinski, H., Eds. (1983) *The Target of Penicillin*. Berlin: de Gruyter.

Hamada, S., Mizuno, J., Kotani, S. and Torii, M. (1980) Distribution of lipoteichoic acids and other amphipathic antigens in oral streptococci. *FEMS Microbiol. Letts.* **8**: 93–6.

Hamada, S., Torii, M., Kotani, S., Masuda, N., Ooshima, T., Yokogawa, K. and Kawata, S. (1978) Lysis of *Streptococcus mutans* cells with mutanolysin, a lytic enzyme prepared from a culture liquor of *Streptomyces globisporus* 1829. *Arch. Oral Biol.* **23**: 543–9.

Hancock, I. C. (1981) The biosynthesis of wall teichoic acid by toluenised cells of *Bacillus subtilis* W23. *Eur. J. Biochem.* **119**: 85–90.

Hancock, I. C. (1983) Activation and inactivation of secondary wall polymers in *Bacillus subtilis* W23. *Arch. Microbiol.* **134**: 222–6.

Hancock, I. C. and Baddiley, J. (1985) Biosynthesis of the bacterial envelope polymers, teichoic acid and teichuronic acid. In: *The Enzymes of Biological Membranes*, Vol. 2, pp. 279–307. Ed. A. N. Martonosi. New York: Plenum Press.

Hancock, I. C. and Williams, K. M. (1986) The outer membrane of *Methylobacterium organophilum*. *J. Gen. Microbiol.* **132**: 599–610.

Hancock, I. C., Wiseman, G. and Baddiley, J. (1976) Biosynthesis of the unit that links teichoic acid to the bacterial wall: inhibition by tunicamycin. *FEBS Letts.* **69**: 75–80.

Hancock, R. E. W. and Carey, A. M. (1980) Protein D1—a glucose-inducible, pore-forming protein from the outer membrane of *Pseudomonas aeruginosa*. *FEMS Microbiol. Letts.* **8**: 105–9.

Hanes, C. S. and Isherwood, F. A. (1949) Separation of the phosphoric esters on the filter paper chromatogram. *Nature* **164**: 1107–12.

Harder, W. and Dijkhuisen, L. (1983) Physiological responses to nutrient limitation. *Ann. Rev. Microbiol.* **37**: 1–24.

Hartmann, R., Höltje, R.-V. and Schwartz, U. (1972) Targets of penicillin action in *Escherichia coli. Nature* **235**: 426–9.

Hase, S. and Matsushima, Y. (1979) The structure of the branching point between acidic polysaccharide and peptidoglycan in *Micrococcus lysodeikticus* cell wall. *J. Biochem.* **81**: 1181–6.

Hash, J. H. (1963) Purification and properties of staphylolytic enzymes from *Chalaropsis* sp. *Arch. Biochem. Biophys.* **102**: 379–85.

Hash, J. H. and Rothlauf, M. V. (1969) The *N,O*-diacetylmuramidase of *Chalaropsis* species. I: Purification and crystallisation. *J. Biol. Chem.* **242**: 5586–90.

Hastie, A. T. and Brinton, C. C. (1979) Isolation, characterization and *in vitro* self assembly of the tetragonally arrayed layer of *Bacillus sphaericus. J. Bacteriol.* **138**: 999–1009.

Hayashi, K. (1975) A rapid determination of sodium dodecyl sulphate with methylene blue. *Analyt. Biochem.* **67**: 503–6.

Heckels, J. E. (1984) Molecular studies on the pathogenesis of gonorrhoea. *J. Med. Microbiol.* **18**: 293–307.

Heckels, J. E., Archibald, A. R. and Baddiley, J. (1975) Studies on the linkage between teichoic acids and peptidoglycan in a bacteriophage-resistant mutant of *Staphylococcus aureus* H. *Biochem. J.* **149**: 637–47.

Heckels, J. E. and Virji, M. (1985) Monoclonal antibodies against gonococcal pili: uses in gonococcal immunochemistry and virulence. In: *Monoclonal Antibodies against Bacteria*, pp. 1–35. Eds. A. J. L. Macario and E. C. de Macario. London: Academic Press.

Herbert, D. (1958) Continuous culture of microorganisms; some theoretical aspects. In: *Continuous Cultivation of Microorganisms: a Symposium*, pp. 45–52. Publishing House of Czeckoslovak Academy of Sciences, Prague.

Herbold, D. R. and Glaser, L. (1975) Interaction of *N*-acetylmuramic acid L-alanine amidase with cell wall polymers. *J. Biol. Chem.* **250**: 7231–8.

Hitchcock, P. J. and Brown, T. M. (1983) Morphological heterogeneity among *Salmonella* lipopolysaccharide chemotypes in silver-stained polyacrylamide gels. *J. Bacteriol.* **154**: 269–77.

Hobot, J. A., Carlemalm, E., Villiger, W. and Kellenberger, E. (1984) Periplasmic gel: new concept resulting from the reinvestigation of bacterial cell envelope ultrastructure by new methods. *J. Bacteriol.* **160**: 143–52.

Hollingshead, S. K., Fischetti, V. A. and Scott, J. R. (1986) Complete nucleotide sequence of Type 6 M-protein of the Group A streptococcus. *J. Biol. Chem.* **261**: 1677–86.

Holt, S. C. and Canale-Parola, E. (1967) Fine structure of *Sarcina maxima* and *Sarcina ventriculi. J. Bacteriol.* **93**: 399–410.

Höltje, R.-V., Mirelman, D., Sharon, D. and Schwartz, U. (1975) Novel type of murein transglycosylase in *Escherichia coli. J. Bacteriol.* **124**: 1067–76.

Höltje, R.-V. and Schwartz, U. (1985) Biosynthesis and growth of the murein sacculus. In: *Molecular Cytology of Escherichia coli*, pp. 77–119. Ed. N. Nanninga. London: Academic Press.

Honda, S. (1984) High-performance liquid chromatography of mono- and oligo-saccharides. *Analyt. Biochem.* **140**: 1–47.

Honda, T. and Finkelstein, R. A. (1979) Selection and characterization from a novel

Vibrio cholerae mutant lacking the A (ADP-ribosylating) portion of cholera enterotoxin. *Proc. Nat. Acad. Sci. USA* **76**: 2052–6.

Hosieth, S. K. and Stocker, B. A. D. (1981) Aromatic-dependent *Salmonella typhimurium* are non-virulent and effective as live oral vaccines. *Nature* **291**: 238–40.

Howard, L. and Tipper, D. J. (1973) A polypeptide bacteriophage receptor: modified cell wall protein subunits in bacteriophage-resistant mutants of *Bacillus sphaericus* strain P-1. *J. Bacteriol.* **113**: 1491–504.

Hoyle, B. D. and Beveridge, T. J. (1984) Metal binding by the peptidoglycan sacculus of *Escherichia coli* K12. *Can. J. Microbiol.* **40**: 204–11.

Hughes, A. H., Hancock, I. C. and Baddiley, J. (1973) The function of teichoic acids in cation control in bacterial membranes. *Biochem. J.* **132**: 83–90.

Hughes, R. C. (1970a) The cell wall of *Bacillus licheniformis* NCTC 6346. *Biochem. J.* **117**: 431–9.

Hughes, R. C. (1970b) Autolysis of isolated cell walls of *Bacillus licheniformis* NCTC 6346 and *Bacillus subtilis* Marburg Strain 168. *Biochem. J.* **119**: 849–60.

Hughes, R. C. (1971) Autolysis of *Bacillus cereus* cell walls and isolation of structural components. *Biochem. J.* **121**: 791–802.

Hughes, R. C., Pavlik, J. G., Rogers, H. J. and Tanner, P. J. (1968) Organisation of polymers in the cell walls of some Bacilli. *Nature* **219**: 642–4.

Hughes, R. C. and Tanner, P. J. (1968) The action of dilute alkali on some bacterial cell walls. *Biochem. Biophys. Res. Commun.* **33**: 22–8.

Hughes, R. C., Thurman, P. F. and Stokes, E. (1975) *Zeit. Immunol.* **149**: 126–35.

Hull, S., Clegg, S., Svanborg Eden, C. and Hull, R. (1985) Multiple forms of genes in pyelonephritogenic *Escherichia coli* encoding adhesins binding globoseries glycolipid receptors. *Infect. Immun.* **47**: 80–3.

Hultgren, S. J., Porter, T. N., Schaeffer, A. J. and Duncan, J. L. (1985) Role of pili and effects of phase variation on lower urinary tract infections produced by *Escherichia coli*. *Infect. Immun.* **50**: 370–7.

Hunter, S. W., Murphy, R. C., Clay, K., Goren, M. B. and Brennan, P. J. (1983) Trehalose-containing lipooligosaccharides. *J. Biol. Chem.* **258**: 10481–7.

Hurrell, J. G. R. (1982) *Monoclonal Hybridoma Antibodies: Techniques and Applications*. Boca Raton, Florida: CRC Press.

Hussey, H., Brooks, D. and Baddiley, J. (1969) Direction of chain extension during the biosynthesis of teichoic acids in bacterial cell walls. *Nature* **221**: 665–6.

Hussey, H., Sueda, S., Chea, S.-C. and Baddiley, J. (1978) Control of teichoic acid synthesis in *Bacillus licheniformis* ATCC 9945. *Eur. J. Biochem.* **82**: 169–74.

Hyslop, P. A. and York, D. A. (1980) The use of 1,6-diphenylhexatriene to detect lipids on thin-layer chromatograms. *Analyt. Biochem.* **101**: 75–7.

Inouye, M., Shaw, J. and Shen, C. (1972) The assembly of a structural lipoprotein in the envelope of *Escherichia coli*. *J. Biol. Chem.* **247**: 8154–9.

Irschik, H. and Reichenbach, H. (1978) Intracellular location of flexirubins in *Flexibacter elegans* (Cytophagales). *Biochem. Biophys. Acta* **510**: 1–10.

Ishidate, K., Creeger, E. S., Zrike, J., Deb, S., Glauner, B., MacAlister, T. J. and Rothfield, L. I. (1986) Isolation of differentiated membrane domains from *Escherichia coli* and *Salmonella typhimurium*, including a fraction containing attachment sites between the inner and outer membranes and the murein skeleton of the cell envelope. *J. Biol. Chem.* **261**: 428–43.

Ivatt, R. J. and Gilvarg, C. (1979) The primary structure of the teichuronic acid of *Bacillus megaterium*. *J. Biol. Chem.* **254**: 2759–65.

Iwahi, T., Abe, Y., Nakao, M., Imada, A. and Tsuchiya, K. (1983) Role of type 1 fimbriae in the pathogenesis of ascending urinary tract infection induced by *Escherichia coli* in mice. *Infect. Immun.* **39**: 1307–15.

Iwasaki, H., Shimada, A. and Ito, E. (1986) Comparative studies of lipoteichoic acids from several *Bacillus* strains. *J. Bacteriol.* **167**: 508–16.

Izaki, K. and Strominger, J. L. (1968) Biosynthesis of the peptidoglycan of bacterial cell walls. XIV: Purification and properties of two D-alanine carboxypeptidases from *Escherichia coli*. *J. Biol. Chem.* **243**: 3193–201.

Izhar, M., Nuchamowitz, Y. and Mirelman, D. (1982) Adherence of *Shigella flexneri* to guinea pig intestinal cells is mediated by a mucosal adhesin. *Infect. Immun.* **35**: 1110–18.

Jacques, N. A., Hardy, L., Campbell, L. K., Knox, K. W., Evans, J. D. and Wicken, A. J. (1979) Effect of carbohydrate source and growth conditions on the production of lipoteichoic acid by *Streptococcus mutans* Ingbritt. *Infect. Immun.* **26**: 1079–87.

Jann, K. (1985) Isolation and characterisation of capsular polysaccharides (K antigens) from *Escherichia coli*. In: *The Virulence of Escherichia coli*, pp. 375–9. Ed. M. Sussman. London: Academic Press.

Jansson, P.-E., Kenne, L., Liedgren, H., Lindberg, B. and Lonngren, J. (1976) A practical guide to the methylation analysis of carbohydrates. *Chem. Commun.* **1976**, no. 8, Stockholm.

Jansson, P.-E., Kenne, L. and Widalm, G. (1986) Casper—a program for computer-assisted n.m.r. spectrum evaluation of poly- and oligosaccharides. Abstracts XIII, International Carbohydrate Symposium, Ithaca, N.Y., p. 185.

Jayappa, H. G., Goodnow, R. A. and Geary, S. J. (1985) Role of *Escherichia coli* type 1 pilus in colonization of porcine ileum and its protective nature as a vaccine antigen in controlling colibacillosis. *Infect. Immun.* **48**: 350–4.

Jennings, H. J., Rosell, K.-G. and Kenny, C. P. (1979) Structural elucidation of the capsular polysaccharide antigen of *Neisseria meningitidis* serogroup Z using ^{13}C-nuclear magnetic resonance. *Can. J. Chem.* **57**: 2902–7.

Jennings, H. J. and Smith, I. C. P. (1978) Polysaccharide structures using carbon-13 nuclear magnetic resonance. *Methods in Enzymol.* **50C**: 39–50.

Kallenius, G., Mollby, R., Svenson, S. B., Winberg, J. and Hultberg, H. (1980) Identification of a carbohydrate receptor recognised by uropathogenic *Escherichia coli*. *Infection* **8**, Suppl. 3: 288–93.

Kandler, O. and Hippe, H. (1977) Lack of peptidoglycan in the cell walls of *Methanosarcina barkeri*. *Arch. Microbiol.* **113**: 57–60.

Kandler, O. and König, H. (1978) Chemical composition of the peptidoglycan-free cell walls of methanogenic bacteria. *Arch. Microbiol.* **118**: 141–52.

Kandler, O. and König, H. (1985) Cell envelopes of archaebacteria. In: *The Bacteria*, Vol. VIII, *Archaebacteria*, pp. 413–57. Eds. C. R. Woese and R. S. Wolfe. London: Academic Press.

Kaper, J. B., Lockman, H., Baldini, M. M. and Levine, M. M. (1984) Recombinant non-enterotoxigenic *Vibrio cholerae* strains are attenuated cholera vaccine candidates. *Nature* **308**: 655–8.

Kasper, D. L., Weintraub, A., Lindberg, A. A. and Lonngren, J. (1983) Capsular polysaccharides and lipopolysaccharides from two *Bacteroides fragilis* strains: chemical and immunochemical characterisation. *J. Bacteriol.* **153**: 991–7.

Kates, M. (1986) *Techniques of Lipidology*. Amsterdam: Elsevier.

Kato, K. and Strominger, J. L. (1968) Structure of the cell wall of *Staphylococcus aureus*. IX: Mechanism of hydrolysis by the L-11 enzyme. *Biochemistry* **7**: 2754–62.

Kato, K., Umemoto, T., Sagawa, H., and Kotani, S. (1979) *Curr. Microbiol.* **3**: 147–51.

Kaya, S., Araki, Y. and Ito, E. (1985) Characterisation of a novel linkage unit between ribitol teichoic acid and peptidoglycan in *Listeria monocytogenes* cell walls. *Eur. J. Biochem.* **146**: 517–22.

Kaya, S., Yokoyama, K., Araki, Y. and Ito, E. (1984) *N*-acetylmannosamine (1–4)*N*-acetylglucosamine, a linkage unit between glycerol teichoic acid and peptidoglycan in cell walls of several *Bacillus* strains. *J. Bacteriol.* **158**: 990–6.

Keck, W. and Schwartz, U. (1979) *Escherichia coli* murein-DD-endopeptidase insensitive to β-lactam antibiotics. *J. Bacteriol.* **139**: 770–4.

Keith, B. R., Maurer, L., Spears, P. A. and Orndorff, P. E. (1986) Receptor-binding function of type 1 pili effects bladder colonisation by a clinical isolate of *Escherichia coli. Infect. Immun.* **53**: 693–6.

Kessler, R. E. and Shockman, G. D. (1979) Precursor-product relationship of intracellular and extracellular lipoteichoic acids of *Streptococcus faecium. J. Bacteriol.* **137**: 869–77.

Kiener, A., König, H., Winter, J. and Leisinger, T. (1987) Purification and use of *Methanobacterium wolfei* pseudomurein endopeptidase for lysis of *Methanobacterium thermoautotrophicum. J. Bacteriol.* **169**: 1010–16.

Kipps, T. J. and Hertzenberg, L. A. (1986) Schemata for production of monoclonal antibody-producing hybridomas. In: *Handbook of Experimental Immunology*, 4th edn, Vol. 4, Chap. 108. Ed. D. M. Weir. Oxford: Basil Blackwell.

Kist, M. L. and Murray, R. G. E. (1984) Components of the regular surface array of *Aquaspirillum serpens* MW5 and their assembly *in vitro. J. Bacteriol.* **157**: 599–606.

Klemm, P. (1985) Fimbrial adhesins in *Escherichia coli. Rev. Infect. Dis.* **7**: 321–40.

Klemm, P., Ørskov, I. and Ørskov, F. (1982) F7 and type 1-like fimbriae from three *Escherichia coli* strains isolated from urinary tract infections: protein chemical and immunological aspects. *Infect. Immun.* **36**: 462–8.

Klipstein, F. A., Engert, R. F. and Houghton, R. A. (1983) Protection in rabbits immunised with a vaccine of *Escherichia coli* heat stable toxin cross-linked to the B subunit of heat-labile toxin. *Infect. Immun.* **40**: 888–93.

Knox, K. W. and Wicken, A. J. (1985) Environmentally induced changes in the surfaces of oral streptococci and lactobacilli. In: *Molecular Basis of Oral Microbial Adhesion*, pp. 212–19. Eds. S. E. Merganhagen and B. Rosan. Washington D.C.: American Society for Microbiology.

Koch, A. L. and Doyle, R. J. (1985) Inside-to-outside growth and turnover of the wall of Gram-positive rods. *J. Theor. Biol.* **117**: 137–57.

Kojima, N., Araki, Y. and Ito, E. (1983) Structure of the linkage region between ribitol teichoic acid and peptidoglycan in cell walls of *Staphylococcus aureus* H. *J. Biol. Chem.* **258**: 9043–5.

Kojima, N., Iida, J., Araki, Y. and Ito, E. (1985a) Structural studies on the linkage unit between poly(*N*-acetylglucosamine 1-phosphate) and peptidoglycan in cell walls of *Bacillus pumilus* AHU 1650. *Eur. J. Biochem.* **149**: 331–6.

Kojima, N., Araki, Y. and Ito, E. (1985b) Structure of the linkage region between teichoic acid and peptidoglycan in *Lactobacillus plantarum. Eur. J. Biochem.* **148**: 29–34.

Kojima, N., Uchikawa, K., Araki, Y. and Ito, E. (1985c) A common linkage saccharide unit between teichoic acid and peptidoglycan in cell walls of *Bacillus coagulans. J. Biochem.* **97**: 1085–92.

König, H. (1985) Influence of aminoacids on growth and cell wall composition of methanobacteriales. *J. Gen. Microbiol.* **131**: 3271–5.

König, H. and Kandler, O. (1979) The aminoacid sequence of the peptide moiety of the pseudomurein from *Methanobacterium thermoautotrophicum*. *Arch. Microbiol.* **121**: 271–5.

König, H., Kralik, R. and Kandler, O. (1982) Structure and modifications of pseudomurein in *Methanobacteriales*. *Zbl. Bakt. Hyg.*, 1 Abt. Orig. C3, 1982, 179–91.

König, H., Schlesner, H. and Hirsch, P. (1984) Cell wall studies on budding bacteria of the *Planctomyces/Pasteuria* group and on a *Prosthecomicrobium* species. *Arch. Microbiol.* **138**: 200–5.

König, H., Semmler, R., Lerp, C. and Winter, J. (1985) Evidence for the occurrence of autolytic enzymes in *Methanobacterium wolfei*. *Arch. Microbiol.* **141**: 177–80.

Korhonen, T. K., Vaisanen-Rehn, V., Rehn, M., Pere, A., Parkkinen, J. and Finne, J. (1984) *Escherichia coli* fimbriae recognising sialyl galactosides. *J. Bacteriol.* **159**: 762–6.

Korhonen, T. K., Valtonen, M. V., Parkinnen, J., Vaisanen-Rehn, V., Finne, J., Ørskov, I., Svenson, S. B. and Makela, P. H. (1985) Serotypes, haemolysin production and receptor recognition of *Escherichia coli* strains associated with neonatal sepsis and meningitis. *Infect. Immun.* **48**: 486–91.

Koval, S. F. and Murray, R. G. E. (1983) Solubilisation of the surface protein of *Aquaspirillum serpens* by chaotropic agents. *Can. J. Microbiol.* **29**, 146–50.

Koval, S. F. and Murray, R. G. E. (1984) The isolation of surface array proteins from bacteria. *Can. J. Biochem. Cell Biol.* **62**: 1181–9.

Kropinski, A. M. B., Lewis, V. and Berry, D. (1987) Effect of growth temperature on the lipids, outer membrane proteins and lipopolysaccharides of *Pseudomonas aeruginosa* PAO. *J. Bacteriol.* **169**: 1960–6.

Kruyssen, F. J., de Boer, W. R. and Wouters, J. T. M. (1980) Effects of carbon source and growth rate on cell wall composition of *Bacillus subtilis* subsp. *niger*. *J. Bacteriol.* **144**: 238–46.

Kuo, J. S., Doelling, V. W., Graveline, J. F. and McCoy, D. W. (1985) Evidence for covalent attachment of phospholipid to the capsular polysaccharide of *Haemophilus influenzae* type b. *J. Bacteriol.* **163**: 769–73.

Kupcu, Z., Marz, L., Messner, P. and Sleytr, U. B. (1984) Evidence for the glycoprotein nature of the crystalline cell wall surface layer of *Bacillus stearothermophilus* strain NRS2004/3a. *FEBS Letts.* **173**: 185–90.

Kuriyama, S. M. and Silverblatt, F. J. (1986) Effect of Tamm–Horsfall urinary glycoprotein on phagocytosis and killing of type 1 fimbriated *Escherichia coli*. *Infect. Immun.* **51**: 193–8.

Labigne-Roussel, A. F., Lark, D., Schoolnik, G. and Falkow, S. (1984) Cloning and expression of an afimbrial adhesin (AFA-1) responsible for P blood group-independent, mannose-resistant hemagglutination from a pyelonephritic *Escherichia coli* strain. *Infect. Immun.* **46**: 251–9.

Lacey, B. W. (1961) Non-genetic variation of surface antigens in *Bordetella* and other microorganisms. In: *Microbial Reaction to Environment* (Symposium 11 of the Society for General Microbiology), pp. 343–90. Ed. G. G. Meynell and H. Gooder. Cambridge University Press.

Laemmli, U. K. (1970) Cleavage of structural proteins during the assembly of the head protein of bacteriophage T4. *Nature* **277**: 680–5.

Lambden, P. R. (1982) Biochemical comparison of pili from variants of *Neisseria gonorrhoeae* P9. *J. Gen. Microbiol.* **128**: 2105–11.

Lambden, P. R. and Heckels, J. E. (1979) Outer membrane protein composition and colonial morphology of *Neisseria gonorrhoeae* strain P9. *FEMS Microbiol. Letts.* **5**: 263–5.

Lambden, P. R., Heckels, J. E., McBride, H. and Watt, P. J. (1981) The identification and isolation of novel pilus types produced by variants of *Neisseria gonorrhoeae* P9 following selection *in vivo*. *FEMS Microbiol. Letts.* **10**: 339–41.

Lambden, P. R., Robertson, J. N. and Watt, P. J. (1980) Biological properties of two distinct pilus types produced by isogenic variants of *Neisseria gonorrhoeae* P9. *J. Bacteriol.* **141**: 393–6.

Lambert, P. A., Hancock, I. C. and Baddiley, J. (1977) Occurrence and function of membrane teichoic acids. *Biochem. Biophys. Acta* **472**: 1–12.

Lane-Smith, R. and Gilkerson, E. (1979) Quantitation of glycosaminoglycan hexosamine using 3-methyl-2-benzothiazolone hydrazone hydrochloride. *Analyt. Biochem.* **98**: 478–80.

Lang, W. K. and Archibald, A. R. (1982) Length of teichoic acid chains incorporated into walls of *Bacillus subtilis* grown under conditions of differing phosphate supply. *FEMS Microbiol. Letts.* **13**: 93–7.

Lang, W. K. and Archibald, A. R. (1983) Relation between wall teichoic acid content and concanavalin A binding in *Bacillus subtilis* 168. *FEMS Microbiol. Letts.* **20**: 163–6.

Lang. W. K., Glassey, K. and Archibald, A. R. (1982) Influence of phosphate supply on teichoic acid and teichuronic acid content of *Bacillus subtilis* cell walls. *J. Bacteriol.* **151**: 367–75.

Lapchine, L. (1976) Ultrastructure de la paroie de *Pseudomonas acidovorans*. *J. Microscopie Biol. Cell.* **25**: 67–72.

Lerner, R. A. (1982) Tapping the immunological repertoire to produce antibodies of predetermined specificity. *Nature* **299**: 592–6.

Leutgeb, W. and Weidel, W. (1963) Uber ein in Coli-Zellwand praparaten zurück-gehaltenes Glykogen. *Z. Naturforsch.* **18b**: 1060–2.

Levine, M. M., Black, R. E., Brinton, C. C., Clements, M. L., Fusco, P., Hughes, T. P., O'Donnell, S., Robins-Browne, R., Wood, S. and Young, C. R. (1982) Reactogenicity, immunogenicity and efficacy studies of *Escherichia coli* type 1 somatic pili parenteral vaccine in man. *Scand. J. Infect. Dis.* **33**: 83–95.

Levine, M. M., Nataro, J. P., Karch, H., Baldinin, M., Kaper, J. B., Black, R. E., Clements, M. L. and O'Brien, A. D. (1985) The diarrheal response of humans to some classic serotypes of enteropathogenic *Escherichia coli* is dependent on a plasmid encoding an enteroadhesiveness factor. *J. Infect. Dis.* **152**: 550–9.

Liesack, W., König, H., Schlesner, H. and Hirsch, P. (1986) Chemical composition of the peptidoglycan-free cell envelopes of budding bacteria of the *Pirella/Planctomyces* group. *Arch. Microbiol.* **145**: 361–6.

Lifely, M. R., Tarelli, E. and Baddiley, J. (1980) The teichuronic acid from the walls of *Bacillus licheniformis* ATCC 9945. *Biochem. J.* **191**: 305–8.

Liu, T. Y. and Gotschlich, E. C. (1967) Muramic acid phosphate as a component of the mucopeptide of Gram-positive bacteria. *J. Biol. Chem.* **242**: 471–6.

Locht, C. and Keith, J. M. (1986) Cloning and sequencing of the pertussis toxin operon. *Science* **232**: 1258–63.

Logan, S. M. and Trust, T. J. (1983) Molecular identification of surface protein antigens of *Campylobacter jejuni*. *Infect. Immun.* **42**: 675–82.

Lomberg, H., Cedergren, B., Leffler, H., Nilsson, B., Carstrom, A.-S. and Svanborg Eden, C. (1986) Influence of blood group on the availability of receptors for attachment of uropathogenic *Escherichia coli*. *Infect. Immun.* **51**: 919–26.

Lotan, R., Sharon, N. and Mirelman, C. (1975) Interaction of wheat germ agglutinin with bacterial cells and cell wall polymers. *Eur. J. Biochem.* **55**: 257–62.

Lugtenberg, B. and van Alphen, L. (1983) Molecular architecture and functioning of the outer membrane of *Escherichia coli* and other Gram-negative bacteria. *Biochem. Biophys. Acta* **737**: 51–115.

Lugtenberg, E. J. J., van Schijndel-van Dam, A. and van Bellegem, H. M. (1971) *In vivo* and *in vitro* action of new antibiotics interfering with the utilization of *N*-acetylglucosamine-*N*-acetylmuramyl pentapeptide. *J. Bacteriol.* **108**: 20–29.

MacArthur, A. E. and Archibald, A. R. (1984) Effect of culture pH on the D-alanine ester content of lipoteichoic acid in *Staphylococcus aureus*. *J. Bacteriol.* **160**: 792–3.

MacIntyre, S., Lucken, R. and Owen, P. (1986) Smooth lipopolysaccharide is the major protective antigen for mice in the surface extract from IATC serotype 6 contributing to the polyvalent *Pseudomonas aeruginosa* vaccine PEV. *Infect. Immun.* **52**: 76–84.

Mannel, D. and Mayer, H. (1978) Isolation and chemical characterization of the enterobacterial common antigen. *Eur. J. Biochem.* **86**: 361–70.

Mantle, M. and Allen A. (1978) A colorimetric assay for glycoproteins based on the periodic acid/Schiff stain. *Biochem. Soc. Trans.* **6**: 607–9.

Marquis, R. E. (1968) Salt-induced contraction of bacterial cell walls. *J. Bacteriol.* **95**: 775–81.

Marquis, R. E. (1973) Immersion refractometry of isolated bacterial cell walls. *J. Bacteriol.* **116**: 1273–9.

Martin, M. L., Delpeuch, J.-J. and Martin, G. J. (1980) *Practical N.M.R. Spectroscopy*. London: Heyden.

Masuda, K. and Kawata, T. (1979) Ultrastructure and partial characterization of a regular array in the cell wall of *Lactobacillus brevis*. *Microbiol. Immunol.* **23**: 941–53.

Maurelli, A. T., Baudry, B., d'Hauteville, H., Hale, T. L. and Sansonetti, P. J. (1985) Cloning of plasmid DNA sequences involved in invasion of HeLa cells by *Shigella flexneri*. *Infect. Immun.* **49**: 164–71.

Maurer, L. and Orndorff, P. E. (1985) A new locus, *pil*E, required for the binding of type 1 piliated *Escherichia coli* to erythrocytes. *FEMS Microbiol. Lett.* **30**: 59–66.

McArthur, H. A. I. and Reynolds. P. E. (1980) Purification and properties of the D-alanyl-D-alanine carboxypeptidase of *Bacillus coagulans* 9365. *Biochim. Biophys. Acta* **612**: 107–18.

McNary, J. E., Carnahan, J. and Brinton, C. C. (1968) Ultrastructure of the cell envelope of a *Bacillus* species. *Bact. Proc.* **65**: 65.

Mekalanos, J. J., Swartz, O., Pearson, G. D. N., Hartford, N., Groyne, F. and de Wilde, M. (1983) Cholera toxin genes: nucleotide sequence, deletion analysis and vaccine development. *Nature* **306**: 551–6.

Mescher, M. F. and Strominger, J. L. (1976a) Purification and characterization of a prokaryotic glycoprotein from the cell envelope of *Halobacterium salinarium*. *J. Biol. Chem.* **251**: 2005–14.

Mescher, M. F. and Strominger, J. L. (1976b) Structural (shape-determining) role of the cell surface glycoprotein of *Halobacterium salinarium*. *Proc. Nat. Acad. Sci. USA* **73**: 2687–91.

Meyer, H.-P., Kappeli, O. and Fiechter, A. (1985) Growth control in microbial cultures. *Ann. Rev. Microbiol.* **39**: 299–319.

Mindich, L. (1970) Membrane synthesis in *Bacillus subtilis*. I: Isolation and properties of strains bearing mutations in glycerol metabolism. *J. Mol. Biol.* **49**: 415–32.

Minnikin, D. E. (1982) Lipids: complex lipids, their chemistry, biosynthesis and roles. In: *The Biology of the Mycobacteria*, pp. 95–184. Eds. C. Ratledge & J. L. Stanford. London: Academic Press.

Minnikin, D. E., Abdulrahimzadeh, H. and Baddiley, J. (1972) Variation of polar lipid composition of *Bacillus subtilis* (Marburg) with different growth conditions. *FEBS Letts.* **27**: 16–18.

Minnikin, D. E., Dobson, G. and Draper, P. (1985b) The free lipids of *Mycobacterium leprae* harvested from experimentally infected nine-banded armadillos. *J. Gen. Microbiol.* **131**: 2007–11.

Minnikin, D. E., Hutchinson, I. G., Caldicott, A. B. and Goodfellow, M. (1980) Thin-layer chromatography of methanolysates of mycolic acid-containing bacteria. *J. Chromatog.* **188**: 221–33.

Minnikin, D. E., Minnikin, S. M., O'Donnell, A. G. and Goodfellow, M. (1984a) Extraction of mycobacterial mycolic acids and other long-chain compounds by an alkaline methanolysis procedure. *J. Microbiol. Methods* **2**: 243–9.

Minnikin, D. E., Minnikin, S. M., Parlett, J. H. and Goodfellow, M. (1985a) Mycolic acid patterns of some rapidly-growing species of *Mycobacterium*. *Zbl. Bakt. Hyg.* A. **259**: 446–60.

Minnikin, D. E. and O'Donnell, A. G. (1984) Actinomycete envelope lipid and peptidoglycan composition. In: *The Biology of Actinomycetes*, pp. 337–88. Eds. M. Goodfellow, M. Mordarski and S. T. Williams. London: Academic Press.

Minnikin, D. E., O'Donnell, A. G., Goodfellow, M., Alderson, G., Athalye, M., Schaal, A. and Parlett, J. H. (1984b) An integrated procedure for the extraction of bacterial isoprenoid quinones and polar lipids. *J. Microbiol. Methods* **2**: 233–41.

Minnikin, D. E. and Polgar, N. (1966) Stereochemical studies on the mycolic acids. *Chem. Commun.*, 648–9.

Mizuno, T., Chou, M.-Y. and Inouye, M. (1983) A comparative study on the genes for three porins of the *Escherichia coli* outer membrane. *J. Biol. Chem.* **258**: 6932–40.

Mobley, H. L., Doyle, R. J. and Joliffe, L. K. (1983) Cell wall polypeptide complexes in *Bacillus subtilis*. *Carbohydrate Res.* **116**: 113–25.

Monod, J. (1950) La technique de culture continue: theorie et applications. *Ann. Inst. Pasteur, Paris* **79**: 390–410.

Moorhouse, R., Winter, W. T., Arnott, S. and Bayer, M. E. (1977) Conformation and molecular organisation in fibers of the capsular polysaccharide from *Escherichia coli* M41 mutant. *J. Mol. Biol.* **109**: 373–91.

Morrissey, P. and Dougan, G. (1985) Molecular studies on virulence factors associated with enterotoxigenic *Escherichia coli* isolated from domestic animals: applications to vaccine development. *Veterinary Microbiol.* **10**: 241–51.

Movitz, I. (1976) Formation of extracellular protein A by *Staphylococcus aureus*. *Eur. J. Biochem.* **68**: 291–9.

Murakami, T. (1974) A revised tannin–osmium method for non-coated scanning electron microscope specimens. *Arch. Histol. Japan* **36**: 189–95.

Murray, R. G. E. (1984) The genus *Lampropedia*. In: *Bergey's Manual of Systematic Bacteriology*, Vol. 1, pp. 402–6. Eds. N. R. Krieg and J. G. Holt. Baltimore and London: Williams and Wilkins.

Murray, P. A., Levine, M. J., Reddy, M. S., Tabak, L. A. and Bergey, E. J. (1986) Preparation of a sialic acid-binding protein from *Streptococcus mitis* KS32AR. *Infect. Immun.* **53**: 359–65.

Nakae, T. (1976) Outer membrane of *Salmonella*: isolation of a protein complex that produces transmembrane channels. *J. Biol. Chem.* **251**: 2176–8.

Nasir-ud-Din, Lhermitte, M., Lamblin, G. and Jeanloz, R. W. (1985) The phosphate diester linkage of the peptidoglycan polysaccharide moieties of *Micrococcus lysodeikticus* cell wall. *J. Biol. Chem.* **260**: 9981–7.

Nataro, J. P., Scaletsky, I. C. A., Kaper, J. B., Levine, M. M. and Trabulsi, L. R.

(1983) Plasmid-mediated factors conferring diffuse and localized adherence of enteropathogenic *Escherichia coli*. *Infect. Immun.* **48**: 378–83.

Naumann, D., Barnickel, G., Bradaczek, H., Labischinski, H. and Giesbrecht, P. (1982) Infrared spectroscopy, a tool for probing bacterial peptidoglycan. *Eur. J. Biochem.* **125**, 505–15.

Naumova, I. B., Zaretskaya, M. Sh., Dmitrieva, N. F. and Streshinskaya, G. M. (1978) Structural features of teichoic acids of certain *Streptomyces* species. In: *Nocardia and Streptomyces* (Proceedings of the International Symposium on Nocardia and Streptomyces), pp. 261–8. Eds. M. Mordarski and J. Jelijaszewicz. Stuttgart: Gustav Fischer.

Ndule, A. N. and Flandrois, J. P. (1983) Immunochemical studies of *Staphylococcus aureus* Oeding–Haukenes antigen a_5: a phosphorus-containing polysaccharide. *J. Gen. Microbiol.* **129**: 3603–10.

Nermut, M. V. and Murray, R. G. E. (1967) Ultrastructure of the cell wall of *Bacillus polymyxa*. *J. Bacteriol.* **93**: 1949–65.

Nesser, J.-R., Koellreuter, B. and Wuersch, P. (1986) Oligomannoside-type glycopeptides inhibiting adhesion of *Escherichia coli* strains mediated by type 1 pili: preparation of potent inhibitors from plant glycoproteins. *Infect. Immun.* **52**: 428–36.

Nikaido, H. and Yaara, M. (1985) The molecular basis of bacterial outer membrane permeability. *Microbiol. Revs.* **49**: 1–32.

Normark, S., Lindberg, F., Lund, B., Baga, M., Ekbak, G., Goransson, M., Morner, S., Norgren, M., Marklund, B. and Ulin, B. (1986) In: *Protein–Carbohydrate Interactions in Biological Systems*, pp. 3–12. Ed. D. L. Lark. London: Academic Press.

Nossal, N. G. and Heppel, L. A. (1966) The release of enzymes by osmotic shock from *Escherichia coli* in exponential phase. *J. Biol. Chem.* **241**: 3055–62.

Oakley, B. R., Kirsch, D. R. and Morris, R. N. (1980) A simplified ultrasensitive silver stain for detecting proteins in polyacrylamide gels. *Analyt. Biochem.* **105**: 361–3.

Ofek, I., Simpson, W. A. and Beachey, E. H. (1982) Formation of molecular complexes between a structurally defined M-protein and acylated or deacylated lipoteichoic acid of *Streptococcus pyogenes*. *J. Bacteriol.* **149**: 426–33.

Oguchi, M. and Oguchi, M. S. (1979) Tetraborate concentration on Morgan–Elson reaction and an improved method for hexosamine determination. *Analyt. Biochem.* **98**: 433–7.

O'Hanley, P., Low, D., Romero, I., Lark, D., Vosti, K., Flakow, S. and Schoolnik, G. (1985) Gal–gal binding and hemolysin phenotypes and genotypes associated with uropathogenic *Escherichia coli*. *N. Engl. J. Med.* **313**: 414–20.

Okumura, K. and Habu, S. (1986) Growing hybridoma and producing monoclonal antibodies *in vivo*. In: *Handbook of Experimental Immunology*, Vol. 4, Chapter 111. Ed. D. M. Weir. Oxford: Blackwell.

Olafson, R. W., McCarthy, P. J., Bhatti, A. R., Dooley, J. S. G., Heckels, J. E. and Trust, T. J. (1985) Structural and antigenic analysis of meningococcal piliation. *Infect. Immun.* **48**: 336–42.

Old, D. C. (1972) Inhibition of the interaction between fimbrial haemagglutinins and erythrocytes by D-mannose and other carbohydrates. *J. Gen. Microbiol.* **71**: 149–57.

Old, D. C. (1985) Haemagglutination methods in the study of *Escherichia coli*. In: *The Virulence of Escherichia coli*, pp. 287–313. Ed. M. Sussman. London: Academic Press.

Old, D. C., Roy, A. I. and Tavendale, A. (1986) Differences in adhesiveness among type 1 fimbriate strains of Enterobacteriaceae revealed by an *in vitro* HEp2 cell adhesion model. *J. Appl. Bacteriol.* **61**: in press.

Ombaka, E. A., Cozens, R. M. and Brown, M. R. W. (1983) Influence of nutrient limitation of growth on stability and production of virulence factors of mucoid and non-mucoid strains of *Psuedomonas aeruginosa. Revs. Infect. Diseases* **5**: S880–8.

Orndorff, P. E. and Dworkin, M. (1980) Separation and properties of the cyto-plasmic and outer membranes of vegetative cells of *Myxococcus xanthus. J. Bacteriol.* **141**: 914–27.

Ørskov, I., Birch-Anderson, A., Duguid, J. P., Stenderup, J. and Ørskov, F. (1985) An adhesive protein capsule of *Escherichia coli. Infect. Immun.* **47**: 191–200.

Ørskov, I. and Ørskov, F. (1985) *Escherichia coli* in extraintestinal infections. *J. Hyg.* **95**: 551–75.

Ørskov, I., Ørskov, F., Jann, B. and Jann, K. (1977) Serology, genetics and chemistry of O and K antigens of *Escherichia coli. Bacteriol. Rev.* **41**: 667–710.

Ørskov, I., Ferencz, A. and Ørskov, F. (1980) Tamm–Horsfall protein or uromucoid is the normal urinary slime that traps type 1 fimbriated *Escherichia coli. Lancet* **1**: 887.

Ouchterlony, O. (1948) *In vitro* method for testing the toxin-producing capacity of diphtheria bacteria. *Acta Path. Mic. Scand.* **25**: 186–91.

Palva, E. T. (1979) Protein interactions in the outer membrane of *Escherichia coli. Eur. J. Biochem.* **93**: 495–503.

Parkkinen, J., Rogers, G. N., Korhonen, T., Dahr, W. and Finne, J. (1986) Identification of the O-linked sialyl oligosaccharides of glycophorin A as the erythrocyte receptors for S-fimbriated *Escherichia coli. Infect. Immun.* **54**: 37–42.

Parr, T. R., Poole, K., Crockford, G. W. K. and Hancock, R. E. W. (1986) Lipopolysaccharide-free *Escherichia coli* OmpF and *Pseudomonas aeruginosa* pro-tein P porins are functionally active in lipid bilayer membranes. *J. Bacteriol.* **165**: 523–6.

Parratt, D., McKenzie, H., Nielson, K. H. and Cobb, S. J. (1982) *Radioimmunoassay of Antibody and its Clinical Applications*. Chichester: John Wiley.

Pavlik, J. G. and Rogers, H. J. (1973) Selective extraction of polymers from cell walls of Gram-positive bacteria. *Biochem. J.* **131**: 619–21.

Pazur, J. H. (1982) β-D-glucose 1-phosphate. A structural unit and an immunologi-cal determinant of a glycan from streptococcal cell walls. *J. Biol. Chem.* **257**: 589–91.

Pere, A., Vaisanen-Rehn, V., Rehn, M., Tenhunen, J. and Korhonen, T. K. (1986) Analysis of P-fimbriae on *Escherichia coli* 02, 04, and 06 strains by immuno-precipitation. *Infect. Immun.* **51**: 618–25.

Perera, V. Y., Penn, C. W. and Smith, H. (1982) Variant pili of autoagglutinating *Neisseria gonorrhoeae. FEMS Microbiol. Letts.* **13**: 313–16.

Perkins, H. R. (1967) The use of photolysis of dinitrophenyl peptides in structural studies on the cell wall mucopeptide of *Corynebacterium poinsettiae. Biochem. J.* **102**: 29c–32c.

Perlin, A. S. and Casu, B. (1982) Spectroscopy methods. In: *The Polysaccharides*, Vol. 1, pp. 133–93. Ed. G. O. Aspinall. London: Academic Press.

Perry, M. B., Bundle, D. R., MacClean, L., Perry, J. A. and Griffith, D. W. (1986) The structure of the antigenic lipopolysaccharide O-chains produced by *Sal-monella urbana* and *Salmonella godesberg. Carbohydr. Res.* **156**: 107–22.

Peterson, A. A. and McGroarty, E. J. (1985) High molecular weight components in lipopolysaccharides of *Salmonella typhimurium*, S. *minnesota* and *Escherichia coli*. *J. Bacteriol.* **162**: 738–45.

Petit, J. F., Adam, A., Wietzerbin-Falszpan, J. and Lederer, E. (1969) Chemical structure of the cell wall of *Mycobacterium smegmatis*. I: Isolation and partial characterization of the peptidoglycan. *Biochem. Biophys. Res. Common.* **35**: 478–85.

Phipps, B. M., Trust, T. J., Ishiguro, E. E. and Kay, W. W. (1983) Purification and characterisation of the cell surface virulent A protein from *Aeromonas salmonicida*. *Biochemistry* **22**: 2934–9.

Pinney, A. M. and Widdowson, J. P. (1977) Characteristics of the extracellular M-proteins of Group A streptococci. *J. Med. Microbiol.* **10**: 415–29.

Pommier, M. T. and Michel, G. (1981) Structure of 2',3'-di-o-acyl-α-D-gluco-pyranosyl-(1–2)-D-glyceric acid, a new glycolipid from *Nocardia caviae*. *Eur. J. Biochem.* **118**: 329–333.

Pooley, H. M. (1976) Turnover and spreading of old wall during surface growth of *Bacillus subtilis*. *J. Bacteriol.* **125**: 1127–38.

Poxton, I. R. and Brown, R. (1986) Immunochemistry of the surface carbohydrate antigens of *Bacteroides fragilis* and definition of a common antigen. *J. Gen. Microbiol.* **132**: 2475–81.

Poxton, I. R. and Ip, M. K.-Y. (1981) The cell surface antigens of *Bacteroides fragilis*. *J. Gen. Microbiol.* **126**: 103–9.

Poxton, I. R., Tarelli, E. and Baddiley, J. (1978) The structure of the C-polysaccharide from the walls of *Streptococcus pneumoniae*. *Biochem. J.* **175**: 1033–42.

Poxton, I. R., Aronnson, B., Mollby, R., Nord, C. E. and Collee, J. G. (1984) Immunochemical fingerprinting of *Clostridium difficile* strains from an outbreak of antibiotic-associated colitis and diarrhoea. *J. Med. Microbiol.* **17**: 317–24.

Poxton, I. R., Bell, G. T. and Barclay, G. R. (1985) The association on SDS-polyacrylamide gels of lipopolysaccharide and outer membrane proteins of *Pseudomonas aeruginosa* as revealed by monoclonal antibodies and Western blotting. *FEMS Microbiol. Letts.* **27**: 247–51.

Primosigh, J., Pelzer, H., Maass, D. and Weidel, W. (1961) Chemical characterization of muropeptides released from the *Escherichia coli* B cell wall by enzymatic action. *Biochim. Biophys. Acta* **46**: 68–80.

Punsalang, A. P. and Sawyer, W. D. (1973) Role of pili in the virulence of *Neisseria gonorrhoeae*. *Infect. Immun.* **8**: 255–63.

Pyle, S. W. and Schill, W. B. (1985) Rapid serological analysis of bacterial LPS by electrotransfer to nitrocellulose. *J. Immunol. Meth.* **85**: 371–82.

Qureshi, N., Takayama, K. and Ribi, E. (1982) Purification and structural determination of non-toxic lipid A obtained from the lipopolysaccharide of *Salmonella typhimurium*. *J. Biol. Chem.* **257**: 11808–15.

Ragan, I. and Cherry, R. J. (1986) *Techniques for Analysis of Membrane Proteins*. London: Chapman and Hall.

Razin, S. (1973) The physiology of mycoplasmas. *Adv. Microbial Physiol.* **10**: 1–80.

Razin, S. (1978) The mycoplasmas. *Microbiol. Revs.* **42**: 414–70.

Read, S. M. and Northcote, D. H. (1981) Minimisation of variation in the response to different proteins of the Coomassie blue G dye—binding assay for protein. *Analyt. Biochem.* **116**: 53–64.

Reis, K. J., Ayoub, E. M. and Boyle, M. D. P. (1984) Streptococcal Fc receptors. 1: Isolation and partial characterization of the receptor from a group C streptococcus. *J. Immunol.* **132**: 3091–7.

Reis, K. J., Ayoub, E. M. and Boyle, M. D. P. (1985) A rapid method for the isolation and characterisation of a homogeneous population of streptococcal Fc receptors. *J. Microbiol. Meth.* **4**: 45–58.

Reusch, V. M. (1982) Isolation and analysis of sacculi from *Streptococcus sanguis. J. Bacteriol.* **151**: 1543–55.

Ridell, M., Minnikin, D. E., Parlett, J. H. and Mattsby-Baltzer, I. (1986) Detection of mycobacterial lipid antigens by a combination of thin-layer chromatography and immunostaining. *Letts. Appl. Microbiol.* **2**: 89–92.

Rippka, R., Deruelles, J., Waterbury, J. B., Herdman, M. and Stanier, R. Y. (1979) Generic assignments, strain histories and properties of pure cultures of cyanobacteria. *J. Gen. Microbiol.* **111**: 1–61.

Robbin, J. B., Schneerson, R. and Pittman, M. (1984) *Haemophilus influenzae* type B infection. In: *Bacterial Vaccines*, pp. 289–316. Ed. R. Germanier. London: Academic Press.

Robertson, J. N., Vincent, P. and Ward, M. E. (1977) The preparation and properties of gonococcal pili. *J. Gen. Microbiol.* **102**: 169–77.

Robson, R. L. and Baddiley, J. (1976) Role of teichuronic acid in *Bacillus licheniformis*: defective autolysis due to deficiency of teichuronic acid in a novobiocin-resistant mutant. *J. Bacteriol.* **129**: 1051–8.

Rogers, H. J. (1979). In: *Microbial Polysaccharides and Polysaccharases*, pp. 237–68. Eds. R. C. W. Berkeley, G. W. Gooday and D. C. Ellwood. London: Academic Press.

Rogers, H. J. and Forsberg, C. W. (1971) Role of autolysins in the killing of bacteria by some bactericidal antibiotics. *J. Bacteriol.* **108**: 1235–43.

Rogers, H. J., Perkins, H. R. and Ward, J. B. (1980) *Microbial Cell Walls and Membranes*. London: Chapman Hall.

Rosenbusch, J. P. (1974) Characterisation of the major envelope protein from *Escherichia coli. J. Biol. Chem.* **249**: 8019–29.

Rothbard, J. B., Fernandez, R. and Schoolnik, G. K. (1984) Strain-specific and common epitopes of gonococcal pili. *J. Exp. Med.* **160**: 208–21.

Russell, R. R. B. (1979) Wall-associated protein antigens of *Streptococcus mutans. J. Gen. Microbiol.* **114**: 109–15.

Russell, R. R. B., Peach, S. L., Colman, G. and Cohen, B. (1983) Antibody responses to antigens of *Streptococcus mutans* in monkeys (*Macaca fascicularis*) immunized against dental caries. *J. Gen. Microbiol.* **129**: 865–75.

Russell, R. R. B. and Smith, K. (1986) Effect of subculturing on *Streptococcus mutans* antigens. *FEMS Microbiol. Letts.* **35**: 319–23.

Saastry, P. A., Pearlstone, L. B., Smillie, L. B. and Paranchych, W. (1983) Aminoacid sequence of pilin isolated from *Pseudomonas aeruginosa* PAK. *FEBS Letts.* **151**: 253–6.

Salit, I. E. and Gotschlich, E. C. (1977) Hemagglutination of purified *Escherichia coli* Type 1 pili. *J. Exp. Med.* **146**: 1182–94.

Sawardeker, J. S., Sloneker, J. H. and Jeanes, A. (1965) Quantitative determination of monosaccharides as their alditol acetates by gas liquid chromatography. *Analyt. Chem.* **37**: 1602–4.

Saxen, H., Nurminen, M., Kuusi, N., Svenson, S. B. and Makela, P. H. (1986) Evidence for the importance of *O* antigen specific antibodies in mouse protective *Salmonella* outer membrane protein (porin) antisera. *Microbiol, Path.* **1**: 433–41.

Scaletsky, I. C. A., Silva, M. L. M. and Trabulsi, L. R. (1984) Distinctive patterns of adherence of enteropathogenic *Escherichia coli* to HeLa cells. *Infect. Immun.* **45**: 534–6.

Scherrer, R. and Gerhardt, P. (1971) Molecular sieving by the *Bacillus megaterium* cell wall and protoplast. *J. Bacteriol.* **107**: 718–35.

Schindler, M., Mirelman, D. and Schwartz, U. (1976) Quantitative determination of N-acetylglucosamine residues at the non-reducing ends of peptidoglycan chains by the enzymatic attachment of [^{14}C]-D-galactose. *Eur. J. Biochem.* **71**: 131–4.

Schleifer, K. H., Hammes, W. P. and Kandler, O. (1976) Effects of exogenous and endogenous factors on the primary structures of bacterial peptidoglycan. *Adv. Microbial. Physiol.* **13**: 245–88.

Schleifer, K. H. and Kandler, O. (1972) Peptidoglycan types of bacterial cell walls and their taxonomic implications. *Bact. Rev.* **36**: 407–77.

Schleifer, K. H. and Stackebrandt, E. (1983) Molecular systematics of prokaryotes. *Ann. Rev. Microbiol.* **37**: 143–87.

Schmidt, G., Jann, B. and Jann, K. (1969) Immunochemistry of R lipopolysaccharides in *Escherichia coli*. *Eur. J. Biochem.* **10**: 501–10.

Schmidt, M. A. and Jann, K. (1982) Phospholipid substitution of capsular (K) polysaccharide antigens from *Escherichia coli* causing extraintestinal infections. *FEMS Microbiol. Letts.* **14**: 69–74.

Schnaitman, C. A. (1970) Examination of the protein composition of the cell envelope of *Escherichia coli* by polyacrylamide gel electrophoresis. *J. Bacteriol.* **104**: 882–9.

Schnaitman, C. A. (1971) Solubilisation of the cytoplasmic membrane of *Escherichia coli* by Triton X-100. *J. Bacteriol.* **108**: 545–52.

Scholler, M., Klein, J. P., Sommer, P. and Frank, R. (1983) Protoplast and cytoplasmic membrane preparations from *Streptococcus sanguis* and *Streptococcus mutans*. *J. Gen. Microbiol.* **129**: 3271–9.

Schulman, H. and Kennedy, E. P. (1979) Localisation of membrane-derived oligosaccharides in the outer envelope of *Escherichia coli* and their occurrence in other Gram-negative bacteria. *J. Bacteriol.* **137**: 686–8.

Schwalbe, R. S., Sparling, P. F. and Cannon, J. G. (1985) Variation of *Neisseria gonorrhoeae* protein II among isolates from an outbreak caused by a single gonococcal strain. *Infect. Immun.* **49**: 250–2.

Scott, B. B. and Barclay, G. R. (1987) Endotoxin-polymyxin complexes in an improved enzyme-linked immunosorbent assay for IgG antibodies in blood-donor sera to Gram-negative endotoxin core glycolipids. *Vox Sanguinis* **52**: 272–80.

Segal, E. Hagblom, P., Seifert, H. S. and So, M. (1986) Antigenic variation of gonococcal pilus involves assembly of separated silent gene segments. *Proc. Nat. Acad. Sci. USA* **83**: 2177–81.

Shaw, D. (1984) *Fourier Transform N.M.R. Spectroscopy*, 2nd edn. Amsterdam: Elsevier.

Sherman, P. M., Houston, W. L. and Boedeker, E. C. (1985) Functional heterogeneity of intestinal *Escherichia coli* strains expressing type 1 somatic pili (fimbriae): assessment of bacterial adherence to intestinal membranes and surface hydrophobicity. *Infect. Immun.* **49**: 797–804.

Shockman, G. D. and Barrett, J. F. (1983) Structure, function and assembly of cell walls of Gram-positive bacteria. *Ann. Rev. Microbiol.* **37**: 501–27.

Siegel, J. L., Hurst, S. F., Liberman, E. S., Coleman, S. E. and Bleiweis, A. S. (1981) Mutanolysin-induced spheroplasts of *Streptococcus mutans* are true protoplasts. *Infect. Immun.* **31**: 808–15.

Silvestri, L. J., Hurst, R. E., Simpson, L. and Settine, J. M. (1982) Analysis of sulphate in complex carbohydrates. *Analyt. Biochem.* **123**: 303–9.

Sjoquist, J., Melhoun, B. and Hjelm, H. (1972) Protein A isolated from *Staphylococcus aureus* after digestion with lysostaphin. *Eur. J. Biochem.* **29**: 572–8.

Skoblilova, N. K., Agre, N. S. and Naumova, I. B. (1982) Structure of the teichoic acids of the cell walls of *Streptomyces roseoflavus* var. *roseofungini* and its *Nocardia*-like variant. *Microbiology* (USSR) **50**: 772–6.

Slabyj, B. H. and Panos, C. (1973) Teichoic acid of a stabilized L-form of *Streptococcus pyogenes*. *J. Bacteriol.* **114**: 934–42.

Sleytr, U. B. (1976) Self-assembly of the hexagonally and tetragonally arranged subunits of bacterial surface layers and their reattachment to cell walls. *J. Ultrastruct. Res.* **55**: 360–77.

Sleytr, U. B. and Glauert, A. M. (1976) Ultrastructure of the cell walls of two closely related clostridia that possess different regular arrays of surface subunits. *J. Bacteriol.* **126**: 869–82.

Sleytr, U. B. and Messner, P. (1983) Crystalline surface layers on bacteria. *Ann. Rev. Microbiol.* **37**: 311–39.

Sleytr, U. B., Messner, P., Sara, M. and Pum, D. (1986a) Crystalline cell envelopes on archaebacteria. *System. Appl. Microbiol.* **7**: 310–13.

Sleytr, U. B., Messner, P., Sara, M. and Pum, D. (1986b) Structural and chemical characterisation of S-layers of selected strains of *Bacillus stearothermophilus* and *Desulfotomaculum nigrificans*. *Arch. Microbiol.* **146**: 19–24.

Sleytr, U. B. and Plohberger, R. (1980) The dynamic process of assembly of two-dimensional arrays of macromolecules on bacterial cell walls. In: *Electron Microscopy at Molecular Dimensions*, pp. 36–47. Eds. W. Baumeister and W. Vogell. Berlin: Springer Verlag.

Sleytr, U. B. and Thorne, K. J. I. (1976) Chemical characterisation of the regularly arranged surface layers of *Clostridium thermosaccharolyticum* and *Clostridium thermohydrosulfuricum*. *J. Bacteriol.* **127**: 377–83.

Smit, J., Grano, D. A., Glaeser, R. M. and Agabian, N. (1981) Periodic surface array in *Caulobacter crescentus*: fine structure and chemical analysis. *J. Bacteriol.* **146**: 1135–50.

Smit, J., Kamio, Y. and Nikaido, H. (1975) Outer membrane of *Salmonella typhimurium*: chemical analysis and freeze-fracture studies with lipopolysaccharide mutants. *J. Bacteriol.* **129**: 942–58.

Smyth, C. J. (1984) Serologically distinct fimbriae on enterotoxigenic *Escherichia coli* of serotype O6:K15:H16 or H⁻. *FEMS Microbiol. Lett.* **21**: 51–7.

Smyth, C. J. (1986) Fimbrial variation in *Escherichia coli*. In: *Antigenic Variation in the Course of Infectious Diseases*, pp. 95–125. Eds. C. W. Penn and T. H. Birkbeck. Oxford: IRL Press.

Smyth, C. J., Siegel, J., Salton, M. R. and Owen, P. (1978) Immunochemical analysis of inner and outer membranes of *Escherichia coli* by crossed immunoelectrophoresis. *J. Bacteriol.* **133**: 306–19.

Sparling, P. F., Cannon, J. G. and So, M. (1986) Phase and antigenic variation of pili and outer membrane protein II of *Neisseria gonorrhoeae*. *J. Infect. Dis.* **153**: 196–201.

Springer, E. L. and Roth, I. L. (1973) The ultrastructure of the capsules of *Diplococcus pneumoniae* and *Klebsiella pneumoniae* stained with ruthenium red. *J. Gen. Microbiol.* **74**: 21–31.

Stackebrandt, E., Ludwig, W., Schubert, W., Klink, F., Schlesner, H., Roggentin, T. and Hirsch, P. (1984) Molecular genetic evidence for early evolutionary origin of budding peptidoglycan-less eubacteria. *Nature* **307**: 735–7.

Stark, N. J., Levy, G. N., Rohr, T. E. and Anderson, J. S. (1977) Initial reactions of teichuronic acid synthesis in *Micrococcus lysodeikticus*. *J. Biol. Chem.* **252**: 3466–72.

Steber, J. and Schleifer, K. H. (1975) *Halococcus morrhuae*: a sulphated heteropoly-saccharide as the structural component of the bacterial cell wall. *Arch. Microbiol.* **105**: 173–7.

Stephens, D. S., Whitney, A. M., Rothbard, J. and Schoolnik, G. K. (1985) Pili of *Neisseria meningitidis*. Analysis of structure, and investigation of structural and antigenic relationships to gonococcal pili. *J. Exp. Med.* **161**: 1539–53.

Stock, J. B., Rauch, B. and Roseman, S. (1977) Pepriplasmic space in *Salmonella typhimurium* and *Escherichia coli*. *J. Biol. Chem.* **252**: 7850–61.

Sumper, M. (1987) Halobacterial glycoprotein biosynthesis. *Biochim. Biophys. Acta* **906**: 69–79.

Sutherland, I. W. (1982) Biosynthesis of microbial exopolysaccharides. *Adv. Microbial Physiol.* **23**: 79–150.

Sutherland, I. W. (1985) Biosynthesis and composition of Gram-negative bacterial extracellular and wall polysaccharides. *Ann. Rev. Microbiol.* **39**: 243–70.

Sutherland, I. W. and Kennedy, A. F. D. (1986) Comparison of bacterial lipopoly-saccharide by high-performance liquid chromatography. *Appl. Env. Microbiol.* **52**: 948–50.

Svendsen, P. J. (1973) Cited by Weeke, B. in *Scand. J. Immunol.* **2** (Suppl. 1), Chap. 1.

Swanson, J. (1978) Studies on gonococcus infection. XIV: Cell wall protein differences among color/opacity colony variants of *Neisseria gonorrhoeae*. *Infect. Immun.* **21**: 292–302.

Swanson, J. and Berrera, O. (1983) Gonococcal pilus subunit size heterogeneity correlates with transitions in colony piliation phenotype, not with changes in colony opacity. *J. Exp. Med.* **158**: 1459–72.

Takemoto, H., Hase, S. and Ikenaka, T. (1985) Microquantitative analysis of neutral and aminosugars as fluorescent pyridylamino derivatives by high performance liquid chromatography. *Analyt. Biochem.* **145**: 245–50.

Taku, A. and Fan, D. P. (1979) Dissociation and reconstitution of membranes synthesising the peptidoglycan of *Bacillus megaterium*. *J. Biol. Chem.* **254**: 3991–9.

Tamura, T., Imae, Y. and Strominger, J. L. (1976) Purification to homogeneity and properties of two D-alanine carboxypeptidase I from *Escherichia coli*. *J. Biol. Chem.* **251**: 414–23.

Tarelli, E. and Coley, J. (1979) ^{13}C-NMR spectra of some ribitol teichoic acids. *Carbohydr. Res.* **75**: 31–7.

Tavendale, A. and Old, D. C. (1985) Haemagglutinins and adhesion of *Escherichia coli* to HEp2 epithelial cells. *J. Med. Microbiol.* **20**: 345–53.

Taylor, R. L. and Conrad, H. E. (1972) Stoichiometric depolymerisation of polyuronides and glycosaminoglycans to monosaccharides following reduction of their carbodiimide-activated carboxyl groups. *Biochemistry* **11**: 1383–8.

Tempest, D. W., Dicks, J. W. and Ellwood, D. C. (1968) Influence of growth condition on the concentration of potassium in *Bacillus subtilis* var. niger and its possible relationship to cellular nucleic acid, teichoic acid and teichuronic acid. *Biochem. J.* **106**: 237–43.

Thompson, B. G. and Murray, R. G. E. (1981) Isolation and characterisation of the plasma membrane and the outer membrane of *Deinococcus radiodurans* strain Stark. *Can. J. Microbiol.* **27**: 729–34.

Thorne, K. J. I., Thornley, M. J., Naisbitt, P. and Glauert, A. M. (1975) The nature of the attachment of a regularly arranged surface protein to the outer membrane of an *Acinetobacter* sp. *Biochim. Biophys. Acta* **389**: 97–116.

Thornley, M. J., Thorne, K. J. I. and Glauert, A. M. (1974) Detachment and chemical characterisation of the regularly arranged subunits from the surface of *Acinetobacter*. *J. Bacteriol.* **118**: 654–62.

Tille, D., Chartwal, G. S. and Blobel, H. (1986) Release of Fc receptors after streptococcal lysis induced by a lytic enzyme from *Streptomyces globisporus*. *Med. Microbiol. Immunol.* **175**: 35–41.

Tinsley, C. R. and Heckels, J. E. (1986) Variation in the expression of pili and outer membrane protein by *Neisseria meningitidis* during the course of meningococcal infection. *J. Gen. Microbiol.* **132**: 2483–90.

Tipper, D. J. and Gauthier, M. (1972) In: *Spores* V, pp. 3–12. Ed. H. O. Halvorson. Washington, D.C.: American Society for Microbiology.

Tomasz, A. and Westphal, M. (1971) Abnormal autolytic enzyme in a pneumococcus with altered teichoic acid composition. *Proc. Nat. Acad. Sci. USA* **68**: 2627–30.

Torbet, J. and Norton, M. Y. (1982) The structure of the cell wall of *Staphylococcus aureus* studied with neutron scattering and magnetic birefringence. *FEBS Letts.* **147**: 201–6.

Torres, C.-R. and Hart, G. W. (1984) Topography and polypeptide distribution of terminal N-acetylglucosamine residues on the surfaces of intact lymphocytes. *J. Biol. Chem.* **159**: 3308–17.

Towbin, H., Staehelin, T. and Gordon, J. (1979) Electrophoretic transfer of proteins from polyacrylamide gels to nitrocellulose sheets: procedure and some applications. *Proc. Nat. Acad. Sci. USA* **76**: 4350–4.

Townsend, R. and Plaskitt, K. A. (1985) Immunogold localization of p55-fibril protein and p25-spiralin in *Spiraplasma* cells. *J. Gen. Microbiol.* **131**: 983–92.

Trevelyan, W. E., Procter, D. P. and Harrison, J. S. (1950) Detection of sugars on paper chromatograms. *Nature* **166**: 444–5.

Tsai, C.-M. and Frasch, C. E. (1982) A sensitive silver stain for detecting lipopolysaccharide in polyacrylamide gels. *Analyt. Biochem.* **119**: 115–19.

Tsien, H. C., Shockman, G. D. and Higgins, M. L. (1978) Structural arrangement of polymers within the wall of *Streptococcus faecalis*. *J. Bacteriol.* **133**: 372–86.

Tsuboi, A., Tsukagoshi, N. and Udaka, S. (1982) Reassembly *in vitro* of hexagonal arrays in a protein-producing bacterium, *Bacillus brevis* 47. *J. Bacteriol.* **151**: 1485–97.

Tuomanen, E., Cozens, R., Tosch, W., Zak, O. and Tomasz, A. (1986) The rate of killing of *Escherichia coli* by beta-lactam antibiotics is strictly proportional to the rate of bacterial growth. *J. Gen. Microbiol.* **132**: 1297–304.

Uhlen, M., Guss, B., Nilsson, B., Gatenbeck, S., Philipson, L. and Lindberg, M. (1984) Complete sequence of the staphylococcus gene encoding protein A: a gene evolved through multiple duplications. *J. Biol. Chem.* **259**: 1695–702.

Umeda, A., Ueki, V. and Amako, K. (1987) Structure of *Staphylococcus aureus* cell wall determined by the freeze-substitution method. *J. Bacteriol.* **169**: 2482–7.

Vachon, V., Lyew, D. J. and Coulton, J. W. (1985) Transmembrane permeability channels across the outer membrane of *Haemophilus influenzae* Type B. *J. Bacteriol.* **162**: 918–24.

Vaisanen, V., Korhonen, T. K., Jokinen, M., Gahmberg, C. G. and Enholm, C. (1982) Blood group M specific haemagglutination in pyelonephritogenic *Escherichia coli*. *Lancet* **1**: 1192.

Vaisanen-Rhen, V., Elo, J., Vaisanen, E., Siitonen, A., Ørskov, F., Svenson, S. B., Makela, P. H. and Korhonen, T. (1984) P-fimbriated clones among uropathogenic *Escherichia coli* strains. *Infect. Immun.* **43**: 149–55.

Vaisanen-Rhen, V., Korhonen, T. K. and Finne, J. (1983) Novel cell-binding activity specific for N-acetyl-D-glucosamine in an *Escherichia coli* strain. *FEBS Lett.* **159**: 233–6.

Van Alphen, L., Havekes, L. and Lugtenberg, B. (1977) Major outer membrane protein d of *Escherichia coli* K12. Purification and *in vitro* activity of bacteriophage k3 and f-pilus-mediated conjugation. *FEBS Letts.* **75**: 285–90.

van de Rijn, I. and Fischetti, V. A. (1981) Immunochemical analysis of intact M-protein secreted from cell wall-less streptococci. *Infect. Immun.* **32**: 86–91.

van Die, I., Zuidweg, E., Hoekstra, W. and Bergmans, H. (1986) The role of fimbriae in uropathogenic *Escherichia coli* as carriers of the adhesin involved in mannose-resistant hemagglutination. *Microbial Pathogenesis* **1**: 51–6.

Virji, M., Everson, J. S. and Lambden, P. R. (1982) Effect of anti-pilus antisera on virulence of variants of *Neisseria gonorrhoeae* for cultured epithelial cells. *J. Gen. Microbiol.* **128**: 1095–100.

Virji, M. and Heckels, J. E. (1984) The role of common and type-specific pilus antigenic domains in adhesion and virulence of gonococci for human epithelial cells. *J. Gen. Microbiol.* **130**: 1089–95.

Virji, M., Heckels, J. E. and Watt, P. J. (1983) Monoclonal antibodies to gonococcal pili: studies on antigenic determinants on pili from variants of strain P9. *J. Gen. Microbiol.* **129**: 1965–73.

Voller, A., Bartlett, A. and Bidwell, D. (1981) *Immunoassays for the Eighties.* Lancaster: MTP.

Voller, A., Bidwell, D. and Bartlett, A. (1976) *Manual of Clinical Immunology*, Chap. 69, pp. 506–12. Eds. N. R. Rose and H. Friedman. Washington, D.C.: American Society for Microbiology.

Wang, W. T., LeDonne, N. C., Ackerman, B. and Sweeley, C. C. (1984) Structural characterization of oligosaccharides by HPLC, fast-atom bombardment mass spectrometry and exoglycoside digestion. *Analyt. Biochem.* **141**: 366–81.

Ward, J. B. (1973) The chain length of the glycans in bacterial cell walls. *Biochem. J.* **133**: 395–8.

Ward, J. B. and Curtis, C. A. M. (1982) The biosynthesis and linkage of teichuronic acid to peptidoglycan in *Bacillus licheniformis*. *Eur. J. Biochem.* **122**: 125–32.

Ward, J. B. and Perkins, H. R. (1973) The direction of glycan synthesis in a bacterial peptidoglycan. *Biochem. J.* **135**: 721–8.

Warth, A. D. and Strominger, J. L. (1972) Structure of the peptidoglycan from spores of *Bacillus subtilis*. *Biochemistry* **11**: 1383–8.

Watanabe, H. and Nakamura, A. (1986) Identification of *Shigella sonnei* form 1 plasmid genes necessary for cell invasion and their conservation among *Shigella* species and enteroinvasive *Escherichia coli*. *Infect. Immun.* **53**: 352–8.

Waxman, D. J., Yu, W. and Strominger, J. L. (1980) Linear, uncrosslinked peptidoglycan secreted by penicillin-treated *Bacillus subtilis*. *J. Biol. Chem.* **255**: 11577–87.

Weckesser, J., Zalman, L. S. and Nikaido, H. (1984) Porin from *Rhodopseudomonas sphaeroides*. *J. Bacteriol.* **159**: 199–205.

Weerkamp, A. H. and Jacob, T. (1982) Cell wall-associated protein antigens of *Streptococcus salivarius*: purification, properties and function in adherence. *Infect. Immun.* **38**: 233–42.

Wehrli, F. W. and Wirthlin, T. (1976) *Interpretation of Carbon-13 N.M.R. Spectra.* London: Heyden.

Weir, D. M. (1978) *Handbook of Experimental Immunology*, 3rd edn, Vol. 3: *Application of immunological methods*. Oxford: Blackwell.

Weiss, R. L. (1974) Subunit cell wall of *Sulfolobus acidocaldarius*. *J. Bacteriol.* **118**: 275–84.

Westphal, O. and Luderitz, O. (1954) Chemische erforschung von lipopolysacchariden gramnegativer bacterien. *Angew. Chem.* **66**: 407–17.

Wheat, J., Kohler, R. B., Garten, M. and White, A. (1984) Commercially available (ENDO-STAPH) assay for teichoic acid antibodies. *Arch. Int. Med.* **144**: 261–4.

White, P. J. and Gilvarg, C. (1977) A teichuronic acid containing rhamnose from cell walls of *Bacillus megaterium. Biochemistry* **16**: 2428–35.

Wicken, A. J. and Knox, K. W. (1975) Characterization of Group N streptococcus lipoteichoic acid. *Infect. Immun.* **11**: 973–81.

Wicken, A. J., Evans, J. D. and Knox, K. W. (1986) Critical micelle concentrations of lipoteichoic acids. *J. Bacteriol.* **166**: 72–7.

Wickus, G. G. and Strominger, J. L. (1972) Penicillin-sensitive transpeptidation during peptidoglycan biosynthesis in cell-free preparations from *Bacillus megaterium. J. Biol. Chem.* **247**: 5297–306.

Wieland, F., Dompert, W., Bernhardt, G. and Sumper, M. (1980) Halobacterial glycoprotein saccharides contain covalently linked sulphate. *FEBS Letts.* **120**: 110–14.

Wieland, F., Lechner, J. and Sumper, M. (1982) The cell wall glycoprotein of Halobacteria: structural, functional and biosynthetic aspects. *Zbl. Bakt. Hyg.* I. *Abt. Orig.* **C3**: 161–70.

Wientjes, F. B., Pas, E., Taschner, P. E. M. and Woldringh, C. L. (1985) Kinetics of uptake and incorporation of meso-diaminopimelic acid in different *Escherichia coli* strains. *J. Bacteriol.* **164**: 331–7.

Williams, P., Brown, M. R. W. and Lambert, P. A. (1984) Effect of iron deprivation on the production of siderophores and outer membrane proteins in *Klebsiella aerogenes. J. Gen. Microbiol.* **130**: 2357–65.

Williams, P. H., Knutton, S., Brown, M. G. M., Candy, D. C. A. and McNeish, A. S. (1984) Characterization of non-fimbrial mannose resistant protein haemagglutinins of two *Escherichia coli* strains isolated from infants with enteritis. *Infect. Immun.* **44**: 592–8.

Winther, M. and Dougan, G. (1984) Recombinant DNA techniques for vaccine production. In: *Biotechnology and Genetic Engineering Reviews*, Vol. 2, pp. 1–39. Ed. G. E. Russell. Incept.

Yano, I. (1985) Analysis of bacterial metabolites and components by computerized GC/MS system—from shorter chain acids to very long-chain compounds up to C_{80}. In: *Rapid Methods and Automation in Microbiology and Immunology*, pp. 239–47. Ed. K.-O. Habermehl.

Yoneyama, H. and Nakae, T. (1986) A small diffusion pore in the outer membrane of *Pseudomonas aeruginosa. Eur. J. Biochem.* **157**: 33–8.

Zak, K., Diaz, J.-L., Jackson, D. and Heckels, J. E. (1984) Antigenic variation during infection with *Neisseria gonorrhoeae*: dilution of antibodies to surface proteins in sera of patients with gonorrhoea. *J. Infect. Dis.* **149**: 166–74.

Zaman, Z. and Verwilghen, R. L. (1979) Quantitation of proteins solubilized in SDS–mercaptoethanol–Tris electrophoresis buffer. *Analyt. Biochem.* **100**: 64–9.

Index

Accessory wall polymers, 179–190
4-Acetamido-4,6-dideoxymannose,
  NMR, 155
N-Acetylaminouronic acids,
  determination, 271–272
N-Acetylation, of hexosamines, 139,
  271
N-Acetylgalactosamine, NMR, 154
N-Acetylglucosamine, determination,
  271
N-Acetylhexosamine, determination,
  270–272
N-Acetylhexosamine phosphate,
  determination, 271–272
N-Acetylhexosaminidase, 82–84
N-Acetylmannosamine, 13
N-Acetylmuramyl-L-alanine amidase,
  82–84
N-Acetyltalosaminuronic acid, 22
Acid extraction, of accessory polymers,
  75, 77, 79–81
Acid hydrolysis, 59, 75, 182
  of peptidoglycan, 171
  of polysaccharides, 138–139, 182
'AcrylAide' (FMC) for SDS-PAGE, 281
Adhesins, in bacterial envelope, 228
  in cell adhesion, 227
  of *Escherichia coli*, 244
  in outer membrane, 229, 236–237
Adhesion, in chemostats, 41
  in pathogenesis, 29, 227–240, 246
Adhesion zones, between inner and
  outer membrane, 1
Affinity chromatography, 84
  of IgG, 194
  by immunoabsorption, 221–225
  using lectins, 86–88
  of lipopolysaccharide, 178–179
  of teichoic acid, 86–87

AL-1 peptidase, 83
Alanine, *see* alanyl esters
Alaninol, chromatography, 172
Alanyl esters, determination, 186–187
  in lipoteichoic acid, 88
  pH sensitivity, 35, 59, 63
  in teichoic acid, 8, 63
Alginate, 123
Alkaline extraction, of accessory
  polymers, 81–82
  of mycolic acids, 127–128
Alkaline phosphatase, 183, 188, 210
Alumina, for cell disruption, 56
Amicon ultrafiltration membranes, 89
Amidase, autolytic, 82–84
D-Amino acid, determination, 186–187
Amino acids, effects on peptidoglycan
  composition, 34
  as M pilus receptors, 235
Amino sugar, determination, 270–272
Aminoarabinose, 18
p-Aminobenzoic ethyl ester, 142
Aminouronic acids, 27, 269, 271
Amylase, for cell wall preparation, 62
1,6-Anhydromuramic acid, 3, 164–165
Anhydroribitol, 180, 184, 186
Anomeric configuration, by NMR, 152
Antibodies, biotin-labelled, 214
  to cell wall proteins, 106
  to commensal bacteria, 259
  fluorescent labelling, 216
  IgA, 246
  IgM, 249, 254
Antibody–antigen complexes,
  dissociation, 225
Antibody–gold conjugates, 220
Antifoam, 62
Antigen, heterologous in *Salmonella*
  solid-phase, for ELISA, 262